A STUDENT'S COMPANION

J. RICHARD CHRISTMAN
U.S. Coast Guard Academy

to accompany

FUNDAMENTALS OF
PHYSICS

SIXTH EDITION

DAVID HALLIDAY
University of Pittsburgh

ROBERT RESNICK
Rensselaer Polytechnic Institute

JEARL WALKER
Cleveland State University

D0160929

JOHN WILEY & SONS, INC.
New York • Chichester • Weinheim • Brisbane • Singapore • Toronto

COVER PHOTO   © Tsuyoshi Nishiinoue/Orion Press

To order books or for customer service call 1-800-CALL-WILEY (225-5945).

Copyright © 2001 John Wiley & Sons, Inc. All rights reserved.

No part of this publication may be reproduced, stored in a retrieval system or transmitted
in any form or by any means, electronic, mechanical, photocopying, recording, scanning
or otherwise, except as permitted under Sections 107 or 108 of the 1976 United States
Copyright Act, without either the prior written permission of the Publisher, or
authorization through payment of the appropriate per-copy fee to the Copyright
Clearance Center, 222 Rosewood Drive, Danvers, MA 01923, (978) 750-8400, fax
(978) 750-4470. Requests to the Publisher for permission should be addressed to the
Permissions Department, John Wiley & Sons, Inc., 605 Third Avenue, New York, NY
10158-0012, (212) 850-6011, fax (212) 850-6008, E-Mail: PERMREQ@WILEY.COM.

ISBN 0-471-36042-2

Printed in the United States of America

10 9 8 7 6 5 4 3 2

Printed and bound by Courier Stoughton, Inc.

TO THE STUDENT:
HOW TO USE A STUDENT'S COMPANION
TO FUNDAMENTALS OF PHYSICS

A Student's Companion to Fundamentals of Physics is designed to be used closely with the text *FUNDAMENTALS OF PHYSICS* by Halliday, Resnick, and Walker. There are 4 overview chapters, corresponding to major sections of the text: mechanics, thermodynamics, electricity and magnetism (including optics), and modern physics. Modern physics chapters are included in the extended version of the text only. Read the appropriate overview when you start to study a section and refer back to it as your study proceeds. Read it again when you finish the section. The overviews are designed to help you understand how the important topics fit together and how the text is organized.

Other chapters in the Companion correspond to chapters in the text and should be read along with the text. Most of the Companion chapters are divided into 3 sections: Basic Concepts, Problem Solving, and Notes. Some chapters also contain a Mathematical Skills section.

BASIC CONCEPTS.   This section deals with two important types of information you should obtain from your reading of the text. The first consists of definitions of physical concepts used in the chapter; the second consists of the laws of physics (relationships between the concepts).

A firm understanding of the basic concepts is important for understanding how nature behaves, for working problems, for doing well on exams, and for understanding following chapters. Rather than just list the definitions and laws, this section asks you to do the important part of the work—write the key phrases by filling in blanks. To derive the greatest benefit, don't copy information from the text. Rather, read the text first, then try to fill in the blanks with your own words, without reference to the text. Thinking about what to write and writing it well should help you retain the important information. Comparing what you have written with what the text says will help you pinpoint any misconceptions you might have. If you have trouble expressing an idea you probably don't understand it very well. Go back and study the appropriate section of the text.

Try to write your responses carefully. The section, as completed by you, will serve later as a review. Before working a problem assignment and while studying for an exam, read over the completed section. If there are parts you don't understand you might want to write them more carefully. The better the job you do, the better this section will serve you when you review.

The concepts of physics are best learned in small doses. Try to obtain a firm understanding of each concept before moving on to the next. The Basic Concepts section will help you.

PROBLEM SOLVING. You cannot claim to understand a definition or law of physics unless you can apply it in a variety of different situations. The purpose of the problems is to provide you with various situations so you can test your understanding. There are three main parts to solving a physics problem. The first and probably the most difficult for many students is identifying the physical concepts or physical laws involved in the problem. Once the concepts and laws are identified, you are ready for the second part, writing the equations associated with them. The third part of problem solving involves carrying out mathematical manipulations to obtain an answer.

This section of the Companion concentrates on helping you identify the concepts and laws involved in problem of a chapter and gives you some general problem solving techniques. The various types of problems that can be associated with the concepts of the chapter are discussed and classified. As you read a Problem Solving section be sure you also study the sample problems that are referenced. As you work problems, you will learn more details about the concepts and laws involved. If necessary, go back to the Basic Concepts section and revise your responses there.

Detailed hints for solving many of the problems in each chapter are given on the website that accompanies the Companion. The website also contains quizzes you can use to test your reading and animated illustrations. Your password and the address of the site are given on the card that is bound with your copy of the Companion.

MATHEMATICAL SKILLS. Here you will find a list of the mathematical skills required to solve the problems of the chapter. You will have learned most of these in algebra, trigonometry, geometry, and calculus classes. Review the list to see if you need to brush up on any of the skills. If you do, consult your math texts for complete discussions and for practice problems. Your goal should be to become facile enough with the required mathematical techniques that the math does not hinder your ability to solve problems.

NOTES. Part or all of a page is left blank for you to record any additional notes you think will be beneficial when you review. You might write some detail of a definition or law that is not covered in the Basic Concepts section and that you have trouble remembering. You might also record any details of problem solving that give you special trouble. Try to write your notes so you can understand them later when you review.

EXAM REVIEWS. The last few pages of the Companion can be used to keep a list of the important concepts and laws you need to know for each exam. Here you reduce the important ideas to brief sentences so each page becomes an outline that can be read to remind you of the details you learned while reading the text, filling in the BASIC CONCEPTS sections, and solving problems. You might mark the ideas you had trouble with. Start your review for the final exam by reading the EXAM pages.

The author wholeheartedly thanks Joan Kalkut, who contributed in many ways to this project. He also owes a debt of gratitude to the many good people of Wiley who helped with this and earlier editions of the Student's Companion, especially Cliff Mills, Stuart Johnson, Catherine Donovan, and Thomas Hempstead. He is also grateful to Karen Christman, who proofread an earlier edition of the manuscript, and is extremely thankful for the support and encouragement of his wife, Mary Ellen.

TABLE OF CONTENTS

OVERVIEW I
MECHANICS

Wherever we look, from the submicroscopic world of fundamental particles to the grand scale of galaxies, we see objects in motion, influencing the motions of other objects.

Think of a gardener pushing a wheelbarrow. Each of the objects involved (Earth's surface, the gardener, the wheelbarrow, and the air) influences the motion of the others. If the gardener stops pushing, the wheelbarrow might coast for a while but it eventually stops, chiefly because of friction and air resistance. To keep himself and the wheelbarrow going, the gardener must push against the ground with his feet. Both the ground and wheelbarrow push on him. The ground also pushes on the wheels and makes them turn.

From a microscopic viewpoint, each of these objects consists of a myriad of particles, mainly electrons, protons, and neutrons, in continual motion and continually exerting an influence over the other particles.

On a grander scale Earth makes its yearly trip around the Sun because the Sun affects its motion. The Sun itself travels around the center of our galaxy, the Milky Way, under the influence of other stars. The presence of other galaxies makes the motion of our galaxy different from what it otherwise would be.

Mechanics is the study of motion. The goal is to understand exactly what aspects of the motion of one object are changed by the presence of other objects and exactly what properties objects must have in order to influence the motion of each other. The fun-

damental problem of mechanics is: given the relevant properties of a group of interacting objects, what are their motions?

Mechanics divides neatly into two parts: kinematics, a study of the description of motion, and dynamics, a study of the causes of motion.

First you will learn to describe the motion of an object and to use the ideas of displacement, velocity, and acceleration to predict changes in the position of an object. You will find that if you know the position and velocity of an object at some initial time and the acceleration of the object at all times you can predict its position and velocity at all times. In Chapter 2 you concentrate on motion in one dimension so you can master the important concepts without the geometric complications of more complex motions. In Chapter 4 the concepts are extended to describe motion in two and three dimensions.

Displacement, velocity, and acceleration in more than one dimension have direction as well as magnitude. They behave like mathematical quantities called vectors. Vectors are so important for understanding physics that the whole of Chapter 3 is devoted to explaining their properties and describing how they are manipulated mathematically.

Newton's laws of motion, introduced in Chapter 5, are at the heart of classical mechanics. Here you will learn that the interaction between two objects can be described in terms of a force and that the net force on an object accelerates it. That is, the net force changes the velocity of the object. The fundamental problem splits into two parts: (1) given the properties of the objects, find the forces they exert on each other and (2) given the net force on an object, find its motion.

You will be introduced to a few simply described forces: the gravitational force of Earth on an object near its surface and the force exerted by a spring on an object in contact with one end, for example. The gravitational force is described in detail in Chapter 14 and later on you will learn about other forces: the electrical force in Chapter 22 and the magnetic force in Chapter 29.

You will also learn to calculate the accelerations of objects in a wide variety of situations. The study of Newton's laws of motion is continued in Chapter 6, where you will concentrate on frictional forces and the force required to make an object move on a circular path.

One important quantity that characterizes a system of interacting objects is its energy. Energy may take several forms: kinetic energy is introduced in Chapter 7, potential energy and thermal energy in Chapter 8, and kinetic energy of rotation in Chapter 11. As objects move and exert forces on each other the values of their individual energies may change and the form of the total energy may change, from kinetic to potential, for example. You will learn that the mechanism for these changes is the work done by the forces of interaction. You will learn how to compute the work done by a force and also the changes in the energies of the objects.

Energy is important because, under certain conditions, the total energy of a collection (or system) of objects does not change. As the objects move the energy may change form and may be transferred from one object to another but if certain conditions hold the total remains the same. This is one of the great conservation principles of mechanics and in Chapter 8 you will learn when it applies and when it does not. You will also learn to use it to answer questions about the motions of objects.

When the total energy of a system does change the change can be accounted for by the work done on objects in the system by objects outside the system. We may think of energy flowing into or out of the system as objects outside interact with objects inside.

Another quantity that behaves in much the same way is the momentum of a system. In Chapter 9 momentum is defined and the conditions for which it is conserved are discussed. In Chapter 10 the principle of momentum conservation is applied to collisions between objects.

Chapters 11 and 12 are devoted to rotational motion. Here Newton's laws are applied to wheels spinning on fixed axes and rolling on surfaces, for example. The plan is like the one followed for linear motion: first kinematics, then dynamics. In addition, you will learn about the kinetic energy of rotation and how it changes when work is done on the rotating object.

The most important concept introduced here is angular momentum, another quantity that is conserved in certain situations. You should learn to identify those situations and to use the principle of angular momentum conservation to help with your understanding of rotational motion.

In Chapter 13 Newton's laws for linear and rotational motion are used to discuss the special case of an object at rest. Here you will learn to calculate the forces that must act on an object to hold it at rest. You will also learn about deformations of objects produced by forces acting on them. These topics are enormously important for engineering applications. Bridges, buildings, and automobiles, for example, must be designed to withstand the loads to which they are subjected during use.

In Chapter 14 you will study the force of gravity, one of the fundamental forces. This discussion provides an excellent example of the dependence of a force on properties of the interacting bodies. The principles of dynamics are then applied to the motions of objects moving under the influence of gravity: planets, satellites, and spacecraft. You will bring to bear many of the concepts you learned earlier, most notably conservation of energy and angular momentum. Electrical and gravitational forces are mathematically quite similar. Much of what you study here will be put to use when you study Chapter 22 and later chapters.

Chapter 15 deals with fluids, both at rest and in motion. Density and pressure are defined, then the principles of mechanics are used to understand the variation of pressure with depth in a body of water and with altitude in Earth's atmosphere, as well as the variation of pressure along a pipe in which a fluid is flowing. You will also be able understand, for example, why some objects float while others sink when they are placed in a fluid and why the fluid speed increases when the nozzle opening of a hose is decreased.

Motions that repeat, called oscillations, are discussed in Chapter 16 using Newton's laws and the conservation of energy principle. Oscillations are prevalent in nature and among man-made artifacts. The swaying of trees, buildings, and bridges, the motion of a clock pendulum, and the bouncing of a car as it rides over a pothole are all examples. In addition, what you learn here will be of great use when you study waves.

Wave motion, in which a disturbance created at one place is propagated to another, is the basis for transmitting information (the form of the disturbance) and energy. Sound, light, radio signals, x rays, and microwaves are all examples of waves. Fundamental concepts developed in Chapter 17 are applied to sound waves in Chapter 18 and later on, in Chapter 34, the ideas are used to discuss electromagnetic waves, such as light.

Chapter 1
MEASUREMENT

I. BASIC CONCEPTS

Physics is an experimental science and relies strongly on accurate measurements of physical quantities. All measurements are comparisons, either direct or indirect, with standards. This means that for every quantity you must not only have a qualitative understanding of what the quantity represents but also an understanding of how it is measured. A length measurement is a familiar example. You should know that the length of an object represents its extent in space and also that length might be measured by comparison with a meter stick, say, whose length is accurately known in terms of the SI standard for the meter. Make a point of understanding both aspects of each new quantity as it is introduced.

Systems of units and standards.   A system of units consists of a unit for each physical quantity, organized so that all can be derived from a small number of independent base units. Standards are associated with base units but not with derived units. Ideally a standard should have the following properties: _____

_____

The three International System <u>base</u> <u>units</u> used in mechanics are:

time: _____ (abbreviation: ___)
length: _____ (abbreviation: ___)
mass: _____ (abbreviation: ___)

The <u>standards</u> for these units are:

length: _____

mass: _____

time: _____

Notice that the SI unit for length is defined in terms of the speed of light and the time standard. The speed of light is by definition exactly $c =$ _____ m/s. Assume you have an instrument that accurately measures any time interval. Briefly explain how you can, in principle, calibrate a meter stick in terms of SI standards: _____

_____

At the atomic level the second, non-SI, standard for mass is _____

_____ .

The atomic mass unit is related to the kilogram by $1\,u =$ _____ kg.

Some examples of quantities with derived units are speed (SI unit: _____), acceleration (SI unit: _____), and force (SI unit: _____).

To appreciate the magnitudes of quantities discussed in this course, you should have an intuitive feeling for the size of a meter, kilogram, and second. Search Tables 1–3, 1–4, and 1–5 for familiar objects and try to visualize them as you remember their sizes. Use the tables or seek elsewhere for quantities that are about 1 m long, 1 kg in mass, or 1 s in duration. List them here and remember them as examples:

objects about 1 m long: _____

objects with about 1 kg of mass: _____

time intervals of about 1 s: _____

SI prefixes. SI prefixes are used extensively throughout this course. The following are used the most. For each of them write the associated power of ten and the symbol used as a prefix.

PREFIX	POWER OF TEN	SYMBOL
kilo:	10^{-}	_____
mega:	10^{-}	_____
centi:	10^{-}	_____
milli:	10^{-}	_____
micro:	10^{-}	_____
nano:	10^{-}	_____
pico:	10^{-}	_____

Memorize them. When evaluating an algebraic expression, substitute the value using the appropriate power of ten. That is, for example, if a length is given as 25 μm, substitute 25×10^{-6} m. One catch: the SI unit for mass is the <u>kilogram</u>. Thus, a mass of 25 kg is substituted directly, while a mass of 25 g is substituted as 25×10^{-3} kg.

Unit conversion. Carefully study Section 1–3 to see how a quantity given in one system of units is converted to another. Turn to Appendix D and become familiar with the conversion tables there. A good habit to cultivate is to say the words associated with a conversion. Suppose you want to convert 50 ft to meters. The length table in the appendix tells you that 1 ft is equivalent to 0.3048 m. Say "If 1 ft is equivalent to 0.3048 m, then 50 ft must be equivalent to $(50\,\text{ft}) \times (0.3048\,\text{m/ft}) = 15\,\text{m}$".

For practice verify the following:

2.90 in = 73.7 mm	4.50 ft = 1.37 m	2.10 mi = 3.38 km
36.0 mi/h = 57.9 km/hr	36.0 mi/h = 16.1 m/s	45.0 ft/s = 13.7 m/s
32.0 ft/s^2 = 9.75 m/s^2	100 lb = 445 N	5.10 slugs = 74.4 kg

In many countries automobile speed limits are given in kilometers per hour and the conversion to meters per second is useful for solving some of the problems in this text. Verify that 1 km/h = $(1/3.6)$ m/s = 0.2778 m/s.

II. PROBLEM SOLVING

Most of the problems at the end of this chapter are exercises in unit conversion, powers of 10 arithmetic, and SI prefixes. In addition, some deal with calculations of areas and volumes. See the Mathematical Skills section below for a discussion of powers of ten arithmetic and some useful equations from geometry.

III. MATHEMATICAL SKILLS

Powers of 10 arithmetic. Powers of ten arithmetic is handled automatically by your calculator. Nevertheless you should have some facility with the process. It will help you check on the result your calculator displays and thus see if you keyed the numbers in correctly. In many cases you can estimate an answer by approximating the input numbers to the nearest power of ten and carrying out the calculation in your head.

When you multiply two numbers expressed as powers of ten, multiply the numbers in front of the tens, then multiply the tens themselves. This last operation is carried out by adding the powers. Thus, $(1.6 \times 10^3) \times (2.2 \times 10^2) = (1.6 \times 2.2) \times (10^3 \times 10^2) = 3.5 \times 10^5$ and $(1.6 \times 10^3) \times (2.2 \times 10^{-2}) = (1.6 \times 2.2) \times (10^3 \times 10^{-2}) = 3.5 \times 10 = 35$.

When you divide two numbers, divide the numbers in front of the tens, then divide the tens. The last operation is carried out by subtracting the power in the denominator from the power in the numerator. Thus, $(1.6 \times 10^3)/(2.2 \times 10^2) = (1.6/2.2) \times (10^3/10^2) = 0.73 \times 10 = 7.3$ and $(1.6 \times 10^3)/(2.2 \times 10^{-2}) = (1.6/2.2) \times (10^3/10^{-2}) = 0.73 \times 10^5 = 7.3 \times 10^4$.

When you add or subtract two numbers, first convert them so the powers of ten are the same, then add or subtract the numbers in front of the tens and multiply the result by 10 to the common power. Thus, $1.6 \times 10^3 + 2.2 \times 10^2 = 1.6 \times 10^3 + 0.22 \times 10^3 = 1.8 \times 10^3$.

This means you must know how to write the same number with different powers of ten. Remember that multiplication by 10 is equivalent to moving the decimal point one digit to the right and division by 10 is equivalent to moving the decimal point one digit to the left. Thus, $1.6 \times 10^3 = 16 \times 10^2 = 0.16 \times 10^4$. In the first case we multiplied 1.6 by 10 and divided 10^3 by 10. In the second we divided 1.6 by 10 and multiplied 10^3 by 10.

You should be able to verify the following:

$$512 \times 10^2 = 5.12 \times 10^4$$
$$0.00512 = 5.12 \times 10^{-3}$$
$$(3.4 \times 10^2) \times (2.0 \times 10^4) = 6.8 \times 10^6$$
$$(3.4 \times 10^2)/(2.0 \times 10^4) = 1.7 \times 10^{-2}$$
$$(3.4 \times 10^4) + (2.0 \times 10^3) = (3.4 \times 10^4) + (0.20 \times 10^4) = 3.6 \times 10^4$$

Significant digits. Always express your answers to problems using the proper number of significant digits. Some students unthinkingly copy all 8 or 10 digits displayed by their calculators, thus demonstrating a lack of understanding. A calculated value cannot be more precise than the data that went into the calculation. Here is what you must remember about significant digits:

a. Leading zeros are not counted as significant. Thus, 0.00034 has two significant digits.

b. Following zeros after the decimal point count. Thus, 0.000340 has three significant digits.

c. Following zeros before the decimal point may or may not be significant. Thus, 500 might contain one, two, or three significant digits. Use powers of ten notation to avoid ambiguities: 5.0×10^2, for example, unambiguously contains two significant digits.

d. When two numbers are added or subtracted, the number of significant digits in the result is obtained by locating the position (relative to the decimal point) of the least significant digit in the numbers that are added or subtracted. The least significant digit of the result is at the same position.

e. When two numbers are multiplied or divided, the number of significant digits in the result is the same as the least number of significant digits in the numbers that are multiplied or divided.

Geometry. You should be familiar with some geometric concepts.

a. The circumference of a rectangle is given by $2(a + b)$, where a and b are the lengths of its sides.

b. The area of a rectangle is given by ab.

c. The area of a triangle is given by $\frac{1}{2}\ell h$, where ℓ is the length of one side and h is the length of the perpendicular line from that side to the vertex opposite that side (the altitude).

d. The volume of a rectangular solid is given by abc, where a, b, and c are the lengths of its sides. The volume of a cube is given by a^3, where a is the length of one of its sides.

e. The circumference of a circle is given by $2\pi r$, where r is its radius. The value of π is about 3.14159.

f. The area of a circle is πr^2.

g. The surface area of a sphere is given by $4\pi r^2$.

h. The volume of a sphere is given by $\frac{4}{3}\pi r^3$.

i. The area of the curved surface of a right circular cylinder is the product of the circumference of an end and the cylinder length: $2\pi r \ell$. The ends are circles and each has an area of πr^2.

j. The volume of a right circular cylinder is the product of the area of an end and the length: $\pi r^2 \ell$.

Carefully note that all circumferences have units of length, all areas have units of length squared, and all volumes have units of length cubed.

IV. NOTES

Chapter 2
MOTION ALONG A STRAIGHT LINE

I. BASIC CONCEPTS

This chapter introduces you to some of the concepts used to describe motion; most important are those of position, displacement, velocity, and acceleration. Pay particular attention to their definitions and to the relationships between them.

Definitions. In this section of the text objects are treated as particles. Tell in your own words what a particle is: _____

_____
Can a particle rotate? Can a particle have parts that move relative to each other? _____ An object can be treated as a particle if _____

_____ .

If an extended object can be treated as a particle we may pick one point on the object and follow its motion. The position of a crate, for example, means the position of the point on the crate we have chosen to follow, perhaps one of its corners or its center.

The motion of a particle in one dimension can be described by giving its coordinate x as a function of time t. You must carefully distinguish between an <u>instant</u> of time and an <u>interval</u> of time. The symbol t represents an instant and has no extension. Thus, t might be *exactly* 12 min, 2.43 s after noon on a certain day. At any other time, no matter how close, t has a different value. On the other hand, an interval extends from some initial time to some final time: *two* instants of time are required to describe it. Note that a value of time may be positive or negative, depending on whether the instant is after or before the instant designated $t = 0$.

Similarly, a value of the coordinate x specifies a <u>point</u> on the x axis. It has no extension in space. On the other hand, two values are required to specify a <u>displacement</u>. A coordinate may be positive or negative, depending on where the point lies relative to the origin.

Suppose a particle has coordinate x_1 at time t_1 and coordinate x_2 at a later time t_2. Then, its displacement Δx over the interval from t_1 to t_2 is given by

$$\Delta x =$$

The magnitude of the displacement during a time interval may be quite different from the distance traveled during the interval. Suppose a particle starts at time $t_1 = 0$ with coordinate $x_1 = 5.0$ m, arrives at $x_2 = 15.0$ m at time $t_2 = 2.0$ s, then turns around and arrives at $x_3 = 10.0$ m at time $t_3 = 3.0$ s. Note that it travels first in the positive x direction, then in the negative x direction. From t_1 to t_2 its displacement is $(15.0\,\text{m} - 5.0\,\text{m}) = 10.0$ m; from t_2 to t_3 its displacement is $(10.0\,\text{m} - 15.0\,\text{m}) = -5.0$ m; and from t_1 to t_3 its displacement is $(10.0\,\text{m} - 5.0\,\text{m}) = 5.0$ m. In contrast, the total distance traveled from t_1 to t_3 is at least 15 m.

If a particle goes from x_1 at time t_1 to x_2 at time t_2, its <u>average velocity</u> v_{avg} in the interval from t_1 to t_2 is given by

$$v_{avg} =$$

Write down in words the steps you would take to find the average velocity in the interval from t_1 to t_2 if you are given the function $x(t)$ as an algebraic expression: _____

_____

_____

On a graph of x versus t the average velocity over the interval from t_1 to t_2 is related to a certain line you might draw. Describe the line and tell what property gives the average velocity:

_____

_____

You must distinguish average velocity and <u>average speed</u>. The average speed over a time interval Δt is defined by

$$s_{avg} = \frac{d}{\Delta t},$$

where d is _____.

Describe the limiting process used to obtain the <u>instantaneous velocity</u> at time t by applying it to a series of average velocities: _____

_____

_____

If the function $x(t)$ is known as an algebraic expression, the instantaneous velocity at any time t_1 is found by _____

_____ and evaluating the result for $t =$ _____. On a graph of x versus t, the instantaneous velocity at any time t_1 is related to a line you might draw. Describe the line and tell what property gives the instantaneous velocity: _____

_____

_____

The term "velocity" means instantaneous velocity. The modifier "instantaneous" is implied.

For each of the functions $x(t)$ shown graphically below, tell if the average velocity is positive or negative in the interval from t_1 to t_2. Also give the sign of the instantaneous velocity at t_1 and t_2 or state that it is zero if it is.

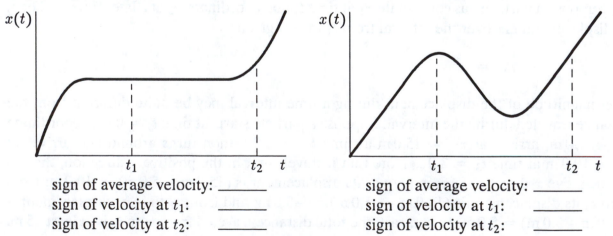

sign of average velocity: _____
sign of velocity at t_1: _____
sign of velocity at t_2: _____

sign of average velocity: _____
sign of velocity at t_1: _____
sign of velocity at t_2: _____

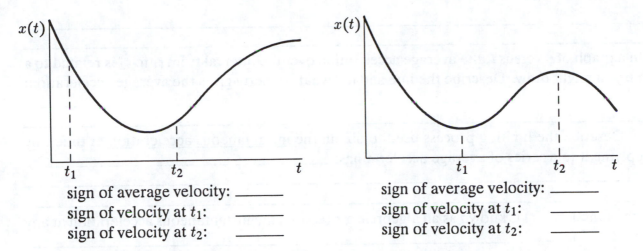

sign of average velocity: _____
sign of velocity at t_1: _____
sign of velocity at t_2: _____

sign of average velocity: _____
sign of velocity at t_1: _____
sign of velocity at t_2: _____

On each of the first three diagrams indicate a time t_3 such that the average velocity from t_1 to t_3 is zero.

On the axes below sketch graphs of the velocity $v(t)$ for the motions represented by the first two of the x versus t graphs above. Be sure you get the sign of v correct. Also be sure your graphs indicate roughly where the magnitude of v is large and where it is small or zero.

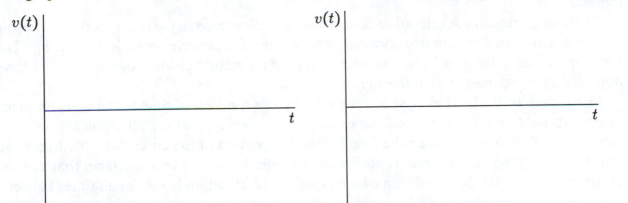

The sign of the velocity indicates the direction of travel. Describe the direction of particle motion if the velocity is positive and if the velocity is negative. Warning! Do not assume the x axis is positive to the right.

positive velocity: _____

_____

negative velocity: _____

_____

Define <u>speed</u>: _____

Carefully note that speed and average speed may be quite different.

If the velocity of a particle changes from v_1 at time t_1 to v_2 at a later time t_2 then its <u>average acceleration</u> a_{avg} over the interval from t_1 to t_2 is given by

$$a_{avg} = $$

Write down in words the steps you would take to find the average acceleration in the interval from t_1 to t_2, given the function $x(t)$ as an algebraic expression: _____

On a graph of v versus t, the average acceleration over the interval from t_1 to t_2 is related to a line you might draw. Describe the line and tell what property gives the average acceleration: _____

_____

Describe the limiting process used to obtain the <u>instantaneous acceleration</u> at time t by applying it to a series of average accelerations: _____

_____

If the function $x(t)$ is known as an algebraic expression, the instantaneous acceleration at any time t_1 is found by _____

_____ and evaluating the result for $t = $ _____. On a graph of v versus t, the instantaneous acceleration at any time t_1 is related to a line you might draw. Describe the line and tell what property gives the instantaneous acceleration: _____

_____

The term "acceleration" means instantaneous acceleration. The modifier "instantaneous" is implied.

Note that a positive acceleration does not necessarily mean the particle speed is increasing and a negative acceleration does not necessarily mean the particle speed is decreasing. The speed increases if the velocity and acceleration have the same sign and decreases if they have opposite signs, no matter what the signs are.

The graph to the left below shows the velocity as a function of time for an object moving along a straight line. On the t axis mark two times (t_1 and t_2, say) such that the acceleration is negative at all times between them and label that portion of the curve "$a < 0$". Mark two times such that the acceleration is positive at all times between them and label that portion of the curve "$a > 0$". Mark two times such that the acceleration is zero at all times between them and label that portion of the curve "$a = 0$". There is one time for which the acceleration is zero only for that instant and is not zero for neighboring times. Label it. On the axes to the right below sketch the acceleration as a function of time.

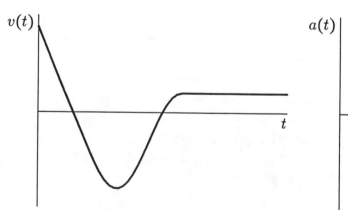

At the instant a particle is momentarily at rest its acceleration is not necessarily zero. To remind yourself that $v = 0$ does not imply $a = 0$, write down a function $x(t)$ for which $v = 0$

but $a \neq 0$ at $t = 0$:

$$x(t) =$$

A particle is initially moving in the positive x direction. Describe in words what its motion might be if at some instant it has zero velocity but non-zero acceleration: _____

_____

For comparison, describe in words what its motion might be if at some instant it has zero velocity and thereafter it has zero acceleration: _____

_____

Motion with constant acceleration. If a particle is moving along the x axis with constant acceleration a, its coordinate is given as a function of time by

$$x(t) =$$

and its velocity is given as a function of time by

$$v(t) =$$

The coordinate and velocity of the particle at time $t = 0$ should appear in your equations. Give the symbol used for each: coordinate at $t = 0$: _____; velocity at $t = 0$: _____.

Eq. 2–16 of the text is also extremely useful. Write it here:

$$v^2 =$$

It can be obtained from the equations for $x(t)$ and $v(t)$ you wrote above by _____

_____ .

You should memorize these three equations. Since the second is the derivative with respect to time of the first and so can be derived quickly, you probably need to memorize only the first and third. Before using these equations always ask if the problem specifies or implies that the acceleration is constant. They are not valid if the acceleration varies with time.

On the coordinate axes below draw possible graphs of $x(t)$ for a particle moving along the x axis with constant acceleration. Take the initial position to be $x = 0$ in all cases. For the first curve suppose the particle starts with a positive velocity and has a positive acceleration; for the second suppose it starts with a negative velocity and has a positive acceleration; for the third suppose it starts with a positive velocity and has a negative acceleration; and for the fourth suppose it starts with a negative velocity and has a negative acceleration. Label each graph with the signs of the initial velocity and the acceleration. Also label points where the particle momentarily stops to start moving in the opposite direction.

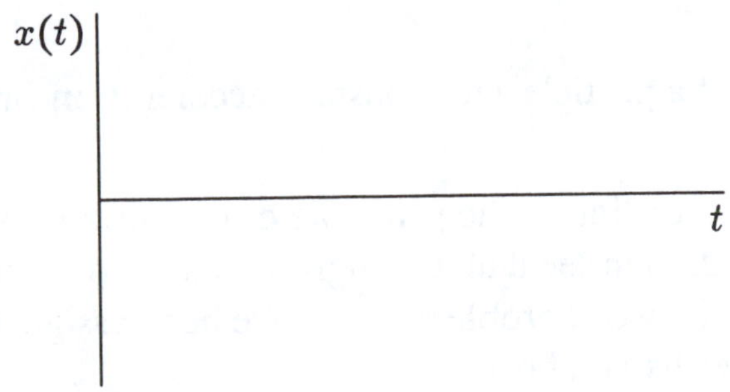

$x(t)$ |

                              t

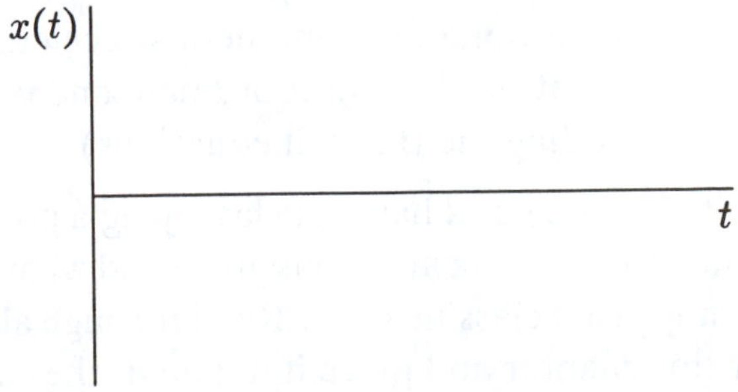

$x(t)$ |

                              t

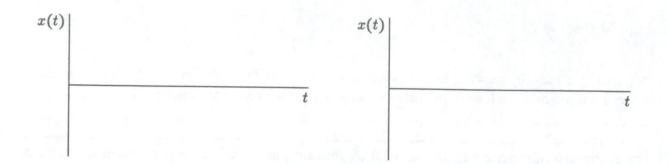

Free fall. Specialize the constant-acceleration equations for the case of an object in free fall. Take the y axis to be positive in the upward direction, away from Earth, and suppose the particle moves along that axis. Then, the y coordinate and the velocity of the particle, as functions of time, are given by

$$y(t) =$$

$$v(t) =$$

Here $g =$ _____ m/s^2 is the magnitude of the free-fall acceleration near the surface of Earth.

Remember that, in the absence of air resistance, *all* objects in free fall have the same acceleration, regardless of their masses. Also remember that their acceleration is the same throughout their motion from the time they are released or thrown to the time they hit something. Their acceleration is g downward while they are going up, when they are at their highest points, and while they are going down.

These equations are for constant acceleration only. To emphasize this point, describe a situation in which an object is thrown upward or downward near the surface of Earth and these equations are *not* valid: _____

_____

_____

II. PROBLEM SOLVING

Nearly all the problems at the end of this chapter can be solved using one or more of the following:

- a. definition of average velocity
- b. definition of instantaneous velocity
- c. definition of average acceleration
- d. definition of instantaneous acceleration
- e. equations for the coordinate and velocity of a particle with constant acceleration (including the free fall equations)

Cultivate the good habit of classifying a problem according to the principle or definition that it illustrates. Look at what is given and what is asked, then see if all the ingredients are present for any given classification. Read through all the homework problems you have been assigned for this chapter and place it in one of the categories listed above.

All constant-acceleration problems can be solved using only the equations $x(t) = x_0 + v_0 t + \frac{1}{2}at^2$ and $v(t) = v_0 + at$. Six quantities appear in these equations: $x(t)$, $v(t)$, x_0, v_0, a, and t. In most cases all but two are given and you are asked to solve for one or both of the others. Mathematically a typical constant-acceleration problem involves identifying the known and unknown quantities, then simultaneously solving the two kinematic equations for the unknowns. Finding solutions to simultaneous algebraic equations is discussed in the MATHEMATICAL SKILLS section for this chapter.

Other kinematics equations are derived in the text by eliminating one or another of the kinematic quantities from the expressions for $x(t)$ and $v(t)$. See Table 2–1 of the text. You may use them or solve the two fundamental equations simultaneously.

Constant-acceleration problems that deal with a single object describe two events, each of which has a time, a coordinate, and a velocity associated with it. The acceleration is the additional quantity that enters the problem. It is often helpful in solving a kinematics problem to fill in a table such as:

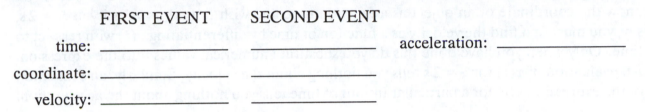

FIRST EVENT   SECOND EVENT

time: _____   _____   acceleration: _____

coordinate: _____   _____

velocity: _____   _____

For any problem write numerical values next to given quantities and question marks next to unknown quantities. When you are finished, you should have no more than two question marks. If you do you missed some information.

You should be aware that you may assign the value 0 to the time when the particle is at *any* point along its trajectory: when the particle starts out, when it reaches a particular point, when it has a particular velocity, or any other point. A negative value for t simply means a time before the instant you chose as $t = 0$. In the kinematic equations x_0 is the coordinate of the object and v_0 is its velocity at time $t = 0$, not necessarily when the particle starts out.

You usually have the option of selecting x_0 to be zero; this selection simply places the origin of the coordinate system at the position of the object when $t = 0$. If you do this you need deal with only the other 5 kinematic quantities.

Some problems deal with two objects. You must now write down two sets of kinematic equations, one for each object: $x_1(t) = x_{01} + v_{01}t + \frac{1}{2}a_1 t^2$ and $v_1(t) = v_{01} + a_1 t$ for object 1 and $x_2(t) = x_{02} + v_{02}t + \frac{1}{2}a_2 t^2$ and $v_2(t) = v_{02} + a_2 t$ for object 2. At $t = 0$ object 1 is at x_{01} and has velocity v_{01} while object 2 is at x_{02} and has velocity v_{02}. These equations can be solved for 4 unknown quantities.

Free-fall problems are exactly the same as other constant-acceleration problems for which the acceleration is given. The notation is different because we choose the y axis to be vertical, so the object moves along that axis rather than the x axis. You should realize that this is an superficial difference. The acceleration is always known (g, downward) and is usually not given explicitly in the problem statement.

III. MATHEMATICAL SKILLS

The following is a listing of mathematical knowledge you will need to understand and solve problems in this chapter. Refer to a mathematics text if you are rusty on any of it.

Functions.   You should understand the notation used for a function. The symbol $f(t)$, for example, indicates that the value of f depends on the value of t. The coordinate of a moving object, for example, is different at different times so it is a function of time. Sometimes the functional dependence is given by means of an equation, such as $x(t) = v_0 t + \frac{1}{2}at^2$. Sometimes it is given by means of a graph or a table of values.

If you know the dependence of x on t you can select a value for t and find the corresponding value for x. Since you can choose the value for t, t is called the <u>independent</u> <u>variable</u>. Since the value for x is determined through the functional relationship by the value of t, x is called the <u>dependent</u> <u>variable</u>.

You must carefully distinguish between a function and its value for a particular value of the independent variable. If $x(t) = 5 - 7t$, for example, then $x(2) = 5 - 7 \times 2 = -9$. If you know the coordinate of an object as a function time and wish to find its velocity at $t = 2\,\text{s}$, say, you must first find the velocity as a function of time by differentiating $x(t)$ with respect to time. Only when you have done this do you substitute numerical values into the expression. An evaluation of $x(t)$ for $t = 2\,\text{s}$ tells you nothing about the velocity. Similarly, an evaluation of the expression $v(t)$ for a particular instant of time tells you nothing about the acceleration.

Derivatives.   You should understand the meaning of a derivative as a limit of a ratio. This is useful for a solid understanding of instantaneous velocity and acceleration. The formula $v(t) = dx/dt$ gives the velocity at any instant of time. To find the velocity at a given instant, you must substitute a value for t. If the object is accelerating, its velocity just before or just after the given instant is different from the velocity at the instant.

You should be able to write down derivatives of polynomials. For example, $d(A + Bt + Ct^2 + Dt^3)/dt = B + 2Ct + 3Dt^2$ if A, B, C, and D are constants. Your instructor may also require you to know how to evaluate the derivative with respect to t of other functions such as e^{At}, $\sin(At)$, and $\cos(At)$. The derivatives are Ae^{At}, $A\cos(At)$, and $-A\sin(At)$ respectively.

You should know the product and quotient rules for differentiation. If $f(t)$ and $g(t)$ are two functions of t, then

$$\frac{d}{dt}[f(t)g(t)] = \frac{df(t)}{dt}g(t) + f(t)\frac{dg(t)}{dt}$$

and

$$\frac{d}{dt}\left[\frac{f(t)}{g(t)}\right] = \frac{1}{g(t)}\frac{df(t)}{dt} - \frac{f(t)}{g^2(t)}\frac{dg(t)}{dt}.$$

You should also know the chain rule. Suppose $u(t)$ is a function of t and $f(u)$ is a function of u. Then, f depends on t through u and

$$\frac{df}{dt} = \frac{df}{du}\frac{du}{dt}.$$

The chain rule was used above to find the derivatives of e^{At}, $\sin(At)$, and $\cos(At)$. In each case u is taken to be At.

You should be able to interpret the slope of a line tangent to a curve as a derivative. The instantaneous velocity is the slope of the coordinate as a function of time, the instantaneous acceleration is the slope of the velocity as a function of time.

Simultaneous equations.

You should be able to solve two simple simultaneous equations for two unknowns. For example, given any four of the algebraic symbols in $x = x_0 + v_0 t + \frac{1}{2}at^2$ and $v = v_0 + at$, you should be able to solve for the other two.

One way is to solve one of the equations algebraically for one of the unknowns, thereby obtaining an expression for the chosen unknown in terms of the second unknown. Substitute the expression into the second equation, replacing the first unknown wherever it occurs. You now have a single equation with only one unknown. Solve in the usual way, then go back to the expression you obtained from the first equation and evaluate it for the first unknown.

With a little practice you will learn some of the shortcuts. If the problem asks for only one of the two unknowns, eliminate the one that is *not* requested. If one equation contains only one of the unknowns, solve it immediately and use the result in the second equation to obtain the value of the second unknown. If one equation is linear in the unknown you wish to eliminate but the other equation is quadratic, use the linear equation to eliminate the unknown from the quadratic equation, rather than vice versa.

Quadratic equations.

You should be able to solve algebraic equations that are quadratic in the unknown. If $At^2 + Bt + C = 0$, then

$$t = \frac{-B \pm \sqrt{B^2 - 4AC}}{2A}.$$

When the quantity under the radical sign does not vanish, there are two solutions. Always examine both to see what physical significance they have, then decide which is required to answer the particular problem you are working. If the quantity under the radical sign is negative, the solutions are complex numbers and probably have no physical significance for problems in this course. Check to be sure you have not made a mistake.

IV. NOTES

Chapter 3
VECTORS

I. BASIC CONCEPTS

You will deal with vector quantities throughout the course. In this chapter you will learn about their properties and how they can be manipulated mathematically. A solid understanding of this material will pay handsome dividends later.

To understand this chapter, you should have a firm grasp of the concept of displacement, defined in Chapter 2 for one-dimensional motion. If necessary, review that chapter.

Definitions.   What properties distinguish a vector from a scalar? _____
_____
_____

List below some examples of physical quantities that are scalars and some that are vectors:

SCALARS   VECTORS

_____   _____
_____   _____
_____   _____

A vector is represented graphically by an arrow in the direction of the vector, with length proportional to the magnitude of the vector (according to some scale). As an algebraic symbol in the text, a vector is always written in italic with an arrow over the symbol ($\vec{a}$). The magnitude of $\vec{a}$ is written a, in italic without an arrow, or as $|\vec{a}|$. Be sure you follow this convention. It helps you distinguish vectors from scalars and components of vectors. It helps you communicate properly with your instructors and exam graders. Do *not* write $a + b$ when you mean $\vec{a} + \vec{b}$, for example. They have entirely different meanings!

Displacement vectors are used as examples of vectors in this chapter. Tell in words what a displacement vector is: _____
_____

Note that a displacement vector tells us nothing about the path of the object, only the relationship between the initial and final positions. When you need an example to illustrate addition or subtraction of vectors, think of displacement vectors.

Graphical vector addition and subtraction.   You will need to know the physical significance of the sum and difference of two vectors as well as how to carry out vector addition and subtraction using both graphical and analytical techniques.

When two displacement vectors, one from point A to point B and the other from point B to point C, are added, the result is the displacement vector from point _____ to point _____ . Except in special circumstances, the magnitude of the resultant vector is *not* the sum of the magnitudes of the vectors entering the sum and the direction of the resultant vector is *not* in the direction of any of the vectors entering the sum.

Suppose the incomplete diagram on the right is meant to demonstrate the addition of two vectors $\vec{a}$ and $\vec{b}$. Place arrows on two sides of the triangle and label them $\vec{a}$ and $\vec{b}$. Place an arrow on the third side and label it $\vec{a} + \vec{b}$. Be sure the arrows are placed correctly so the triangle represents vector addition.

Now, describe in words the steps you must take to add two vectors graphically. Be sure you mention how the vectors must be placed relative to each other and how the resultant vector is drawn: _____

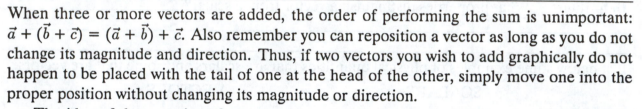

_____

_____

_____

When three or more vectors are added, the order of performing the sum is unimportant: $\vec{a} + (\vec{b} + \vec{c}) = (\vec{a} + \vec{b}) + \vec{c}$. Also remember you can reposition a vector as long as you do not change its magnitude and direction. Thus, if two vectors you wish to add graphically do not happen to be placed with the tail of one at the head of the other, simply move one into the proper position without changing its magnitude or direction.

The idea of the negative of a vector is used to define vector subtraction. How are the magnitude and direction of the negative of a vector related to the magnitude and direction of the original vector?

magnitude: _____

direction: _____

If $\vec{c} = \vec{a} - \vec{b}$, then $\vec{c}$ is found by adding the negative of $\vec{b}$ to $\vec{a}$: $\vec{c} = \vec{a} + (-\vec{b})$. This defines vector subtraction. Write down the steps to subtract two vectors graphically (be sure to include a description of how they are positioned relative to each other): _____

_____

_____

Notice that vector subtraction is defined so that if $\vec{a} + \vec{b} = 0$, then $\vec{a} = -\vec{b}$ and if $\vec{c} = \vec{a} + \vec{b}$, then $\vec{a} = \vec{c} - \vec{b}$. Just subtract $\vec{b}$ from both sides of each equation. Vector subtraction is clearly useful for solving vector equations.

In the space on the right draw two vectors such that the magnitude of their sum equals the sum of their magnitudes.

In the space on the right draw two vectors such that the magnitude of their sum equals the difference of their magnitudes.

In the space on the right draw two vectors such that the magnitude of their difference equals the sum of their magnitudes.

In the space to the right draw two vectors such that the magnitude of their difference equals the difference of their magnitudes.

Analytic vector addition and subtraction. To carry out vector addition and subtraction analytically, you will need to know how to find the components of a vector. For each of the vectors shown below, illustrate the x and y components by marking their lengths along the axes. Write expressions that give the components in terms of the magnitude and angle shown. Evaluate the expressions. Notice that the components of a vector can be positive or negative.

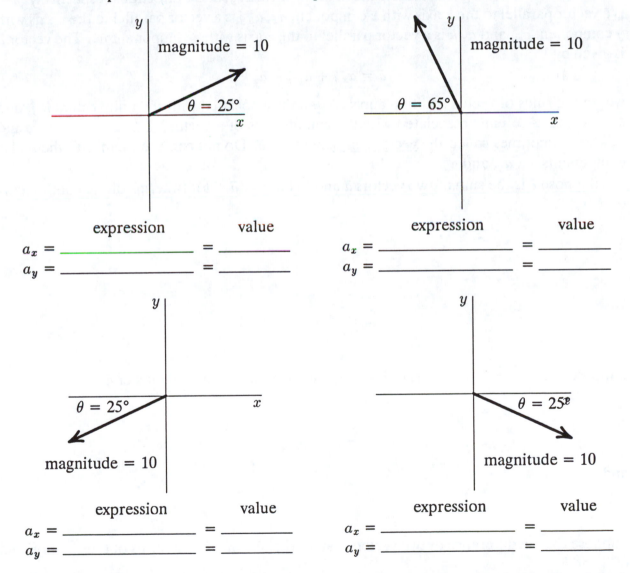

	expression		value
$a_x =$	_____	=	_____
$a_y =$	_____	=	_____

	expression		value
$a_x =$	_____	=	_____
$a_y =$	_____	=	_____

	expression		value
$a_x =$	_____	=	_____
$a_y =$	_____	=	_____

	expression		value
$a_x =$	_____	=	_____
$a_y =$	_____	=	_____

Describe in words the steps you can take to find the components of a vector, given its magnitude and direction and given a coordinate system: _____

You must also be able to find the magnitude and orientation of a vector when you are given its components. Suppose a vector $\vec{a}$ lies in the xy plane and its components a_x and a_y are given. In terms of the components the magnitude of $\vec{a}$ is given by

$$a =$$

and the angle $\vec{a}$ makes with the x axis is given by

$$\theta =$$

The unit vectors $\hat{\imath}$, $\hat{\jmath}$, and $\hat{k}$ are used when a vector is written in terms of its components. These vectors have magnitude 1 and are in the positive x, y, and z directions respectively. $a_x\,\hat{\imath}$ is a vector parallel to the x axis with x component a_x, $a_y\,\hat{\jmath}$ is a vector parallel to the y axis with y component a_y, and $a_z\,\hat{k}$ is a vector parallel to the z axis with z component a_z. The vector $\vec{a}$ is given by

$$\vec{a} = a_x\,\hat{\imath} + a_y\,\hat{\jmath} + a_z\,\hat{k},$$

where the rules of vector addition apply. Are units associated with the unit vectors $\hat{\imath}$, $\hat{\jmath}$, and $\hat{k}$? _____ Are units associated with the components of a vector? _____ $a_x\,\hat{\imath}$, $a_y\,\hat{\jmath}$, and $a_z\,\hat{k}$ are sometimes called the <u>vector</u> <u>components</u> of $\vec{a}$. Do not confuse them with the scalar components a_x, a_y, and a_z.

Suppose $\vec{c}$ is the sum of two vectors $\vec{a}$ and $\vec{b}$ (i.e. $\vec{c} = \vec{a} + \vec{b}$). In terms of the components of $\vec{a}$ and $\vec{b}$:

$$c_x =$$

$$c_y =$$

and

$$c_z =$$

Suppose $\vec{c}$ is the negative of $\vec{a}$ (i.e. $\vec{c} = -\vec{a}$). In terms of the components of $\vec{a}$:

$$c_x =$$

$$c_y =$$

and

$$c_z =$$

Suppose $\vec{c}$ is the difference of two vectors $\vec{a}$ and $\vec{b}$ (i.e. $\vec{c} = \vec{a} - \vec{b}$). In terms of the components of $\vec{a}$ and $\vec{b}$:

$$c_x =$$

$$c_y =$$

and

$$c_z =$$

Many pocket calculators calculate the magnitude and direction of a two-dimensional vector from its components with a single keystroke (after the components have been entered). They also calculate the components from the magnitude and direction in the same manner. Some pocket calculators perform vector addition and subtraction with similar ease. If you have such a calculator you should learn how to use it to perform these calculations.

Multiplication involving vectors. Vectors can be multiplied by scalars. Let $\vec{a}$ be a vector and s a scalar. Then $s\vec{a}$ is a vector. If s is positive, its direction is _____ and its magnitude is _____. If s is negative, the direction of $s\mathbf{a}$ is _____ and its magnitude is _____. If $\vec{c} = s\vec{a}$, then in terms of components

$$c_x =$$

$$c_y =$$

and

$$c_z =$$

Division of a vector by a scalar is just multiplication by the reciprocal of the scalar.

The <u>scalar</u> <u>product</u> (or dot product) of two vectors is defined in terms of the magnitudes of the two vectors and the angle ϕ between them when they are drawn with their tails at the same point: $\vec{a} \cdot \vec{b} = ab \cos \phi$. For each of two cases shown below write an expression for the scalar product of $\vec{a}$ and $\vec{b}$ in terms of the given quantities. To evaluate the expressions, take $a = 10$ and $b = 5$.

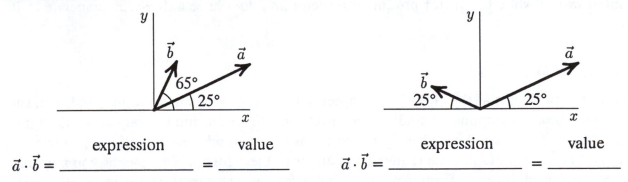

	expression	value		expression	value
$\vec{a} \cdot \vec{b} =$	_____	= _____	$\vec{a} \cdot \vec{b} =$	_____	= _____

The scalar product can be interpreted in terms of the component of one vector along the direction of the other. For the scalar product $\vec{a} \cdot \vec{b}$, write that interpretation in words:

_____

_____

In terms of components the scalar product is given by

$$\vec{a} \cdot \vec{b} =$$

Notice that this expression gives $\vec{a} \cdot \vec{a} = a_x^2 + a_y^2 = a^2$ for a vector in the xy plane.

Let ϕ be the angle between $\vec{a}$ and $\vec{b}$ when they are drawn with their tails at the same point. The sign of the scalar product $\vec{a} \cdot \vec{b}$ is positive if ϕ is in the range from _____ to _____ and is negative if ϕ is in the range from _____ to _____ . Also $\vec{a} \cdot \vec{b} = 0$ if ϕ is _____ .

Write the equation that gives the magnitude of the <u>vector product</u> $\vec{a} \times \vec{b}$ in terms of the magnitudes of $\vec{a}$ and $\vec{b}$:

$$|\vec{a} \times \vec{b}| =$$

Carefully describe how to determine the angle that occurs in this expression: _____
_____
_____

For two vectors with given magnitudes the magnitude of $\vec{a} \times \vec{b}$ is the greatest when the angle between them is _____ and is zero when the angle between them is _____ or _____ .

Describe the right hand rule used to find the direction of $\vec{a} \times \vec{b}$: _____
_____
_____

In terms of the cartesian components of $\vec{a}$ and $\vec{b}$,

$$\vec{a} \times \vec{b} =$$

Remember that the direction of the vector product depends on the order in which the vectors appear: $\vec{a} \times \vec{b}$ and $\vec{b} \times \vec{a}$ are in opposite directions.

The magnitude of the vector product can also be interpreted in terms of the component of one vector along a certain direction perpendicular to the other vector. The direction is in the plane defined by the two vectors in the product. For the vector product $\vec{a} \times \vec{b}$ write the interpretation in words: _____
_____

Remember that the scalar product of two vectors is a scalar and has no direction associated with it while the vector product is a vector and does have a direction associated with it.

II. PROBLEM SOLVING

All of the problems at the end of the chapter deal with vector manipulations: addition, subtraction, finding components, finding magnitude and direction, and the various kinds of multiplication. The vectors are given either in terms of magnitude and direction or in terms of components; answers may be requested in either of these forms. This means you may need to convert from the given form to a form suitable for the vector operation, then convert again to obtain the form required for the answer.

If a is the magnitude of a vector $\vec{a}$ in the xy plane and θ is the angle that the vector makes with the positive x axis, then the components of $\vec{a}$ are $a_x = a \cos \theta$ and $a_y = a \sin \theta$. If you know the components, you can find the magnitude and the angle with the positive x axis. The magnitude is given by

$$a = \sqrt{a_x^2 + a_y^2}$$

and the angle is given by

$$\theta = \arctan(a_y/a_x).$$

When you have found values for a and θ, check to be sure $a \cos \theta$ has the proper sign for the x component and $a \sin \theta$ has the proper sign for the y component. If they have the wrong signs, add 180° to the value you used for the angle.

The basic prescription for vector addition is: if $\vec{c} = \vec{a} + \vec{b}$, then $c_x = a_x + b_x$, $c_y = a_y + b_y$, and $c_z = a_z + b_z$. The basic prescription for vector subtraction is: if $\vec{c} = \vec{a} - \vec{b}$, then $c_x = a_x - b_x$, $c_y = a_y - b_y$, and $c_z = a_z - b_z$.

You must also know how to carry out the multiplication of a vector by a scalar, the scalar product of two vectors, and the vector product of two vectors. Know the results of multiplication involving vectors in terms of magnitudes and directions and in terms of components.

III. MATHEMATICAL SKILLS

Trigonometry is a large portion of the mathematics used in this chapter. Here is a listing of the important elements you should know very well.

The Pythagorean theorem   The square of the hypotenuse of a right triangle equals the sum of squares of the other two sides. In the diagram $C^2 = A^2 + B^2$. The theorem is true only if the triangle contains a right angle (90°). The theorem is used, for example, to calculate the magnitude of a vector in the xy plane given its x and y components.

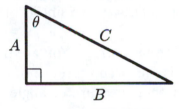

Trigonometric functions.   For the triangle shown, $A = C \cos \theta$ and $B = C \sin \theta$. Remember these relations in the following form: the length of the side adjacent to an interior angle is the product of the hypotenuse and the cosine of the angle and the length of the opposite side is the product of the hypotenuse and the sine of the angle. The relations follow directly from the definition of the sine and cosine and are used to find the components of a vector, given the magnitude and the angle it makes with an axis.

Also know that for the triangle above $\tan \theta = B/A$. In a right triangle the tangent of an interior angle is the length of the opposite side divided by the adjacent side. This relationship, in the form $\theta = \arctan(a_y/a_x)$, is used to find the angle a vector makes with a coordinate axis.

WARNING! For any values of a_x and a_y the equation $\theta = \arctan(a_y/a_x)$ has two solutions for θ. If you use a calculator to evaluate θ, it will give the solution closest to 0. The other solution is the one given by the calculator plus or minus 180°. When solving for θ, first make a sketch of the vector pointing in roughly the right direction, so its components have the correct signs, as given, then check the answer displayed by your calculator. If necessary, add 180° to the calculator answer or subtract 180° from it to make the answer agree with the sketch. Alternatively, once θ has been found, calculate $a \cos \theta$ and $a \sin \theta$ to be sure they give the original components and not their negatives.

Memorize these special values of the trigonometric functions:

$$\cos(0) = 1 \qquad \cos(90°) = 0 \qquad \cos(180°) = -1 \qquad \cos(270°) = 0$$

$$\sin(0) = 0 \qquad \sin(90°) = 1 \qquad \sin(180°) = 0 \qquad \sin(270°) = -1$$
$$\tan(0) = 0 \qquad \tan(90°) = \pm\infty \qquad \tan(180°) = 0 \qquad \tan(270°) = \pm\infty$$

Take special care that you don't get them confused.

Trigonometric identities. You should also know the following trigonometric identities:

$$\sin(-A) = -\sin A$$
$$\cos(-A) = \cos A$$
$$\sin(A + B) = \sin(A)\cos(B) + \cos(A)\cos(B)$$
$$\cos(A + B) = \cos(A)\cos(B) - \sin(A)\sin(B)$$

The identities given above are useful for evaluating $\sin(180° - \phi)$, $\sin(90° - \phi)$, $\sin(90° + \phi)$, $\sin(180° + \phi)$, and the corresponding cosines, for example. Using these relations you should be able to show that the components of the vector in the diagram are $a_x = a\cos(90° + \phi) = -a\sin\phi$ and $a_y = a\sin(90° + \phi) = a\cos\phi$.

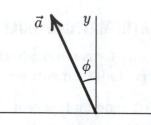

Radian measure. You are probably familiar with the degree as a measure of angle: an arc that is the circumference of a circle divided by 360 subtends an angle of 1 degree. A radian is another measure of angle. The angle subtended by an arc, in radians, is the arc length divided by the radius of the circle. Since the circumference of a circle with radius r is given by $2\pi r$, 360° is equivalent to $2\pi r/r = 2\pi$ rad. You should also know that 180° is equivalent to π rad, 90° is equivalent to $\pi/2$ rad, and 45° is equivalent to $\pi/4$ rad. 1 rad is equivalent to $180/\pi \approx 57.30°$.

Since an angle in radians is one length (the arc length) divided by another (the radius) the radian is actually unitless. However, instead of writing simply a number, without a unit, for an angle, the label "rad" is often appended to remind you that the angle is given in radians.

IV. NOTES

Chapter 4
MOTION IN TWO
AND THREE DIMENSIONS

I. BASIC CONCEPTS

The ideas of position, velocity, and acceleration that were introduced earlier in connection with one-dimensional motion are now extended. You should pay close attention to the definitions and relationships discussed in this chapter. There are three main topics: projectile motion, circular motion, and relative motion. First, however, some general concepts are discussed.

Before studying this chapter review the concepts of displacement, average and instantaneous velocity, and average and instantaneous acceleration, given for one-dimensional motion in Chapter 2. Also review finding the components of a vector from its magnitude and direction and finding the magnitude and direction from the components, as discussed in Chapter 3.

Definitions. Consider a particle moving in two or three dimensions. The fundamental concept used to describe its motion is its position vector. The tail of this vector is always at ____ _____ and at any instant the head is at _____. The cartesian components of the position vector are the coordinates of the particle. As the particle moves its position vector changes and so is a function of time.

A displacement vector is used to describe a change in a position vector. If the particle has position vector $\vec{r}_1$ at time t_1 and position vector $\vec{r}_2$ at a later time t_2, then the displacement vector for this interval is

$$\Delta \vec{r} =$$

If the particle has coordinates x_1, y_1, z_1 at time t_1 and coordinates x_2, y_2, z_2 at time t_2, then the components of the displacement vector are given by

$$(\Delta \vec{r})_x = \Delta x = \qquad\qquad (\Delta \vec{r})_y = \Delta y = \qquad\qquad (\Delta \vec{r})_z = \Delta z =$$

In writing these equations be sure to get the order of the subscripts right. A displacement vector is a position vector at a *later* time minus a position vector at an *earlier* time.

In terms of the displacement vector $\Delta \vec{r}$ the average velocity of the particle in the interval from t_1 to t_2 is

$$\vec{v}_{\text{avg}} =$$

The average velocity has components that are given by

$$v_{\text{avg } x} = \qquad\qquad v_{\text{avg } y} = \qquad\qquad v_{\text{avg } z} =$$

To use the definition to calculate the average velocity over the time interval from t_1 to t_2, you must know _____ for the beginning and end of the interval. Just as for one–dimensional motion, it is important to realize that the components of the position vector are coordinates and represent points on a coordinate axis. They are not necessarily related in any way to the distance traveled by the particle.

The <u>instantaneous velocity</u> $\vec{v}$ at any time t is the limit of the average velocity over a time interval that includes t, as the duration of the interval becomes vanishingly small. In terms of the position vector, it is given by the derivative

$$\vec{v} =$$

In terms of the particle coordinates its components are

$$v_x = \qquad\qquad v_y = \qquad\qquad v_z =$$

You should be aware that the instantaneous velocity, unlike the average velocity, is associated with a single instant of time. At any other instant, no matter how close, the instantaneous velocity might be different. The term "instantaneous" is usually implied: "velocity" means "instantaneous velocity".

To use the definition to calculate the instantaneous velocity, you must know the position vector as a function of time. This is identical to knowing the coordinates as functions of time. The information may be given in algebraic form or as a graph. You should remember that the instantaneous velocity vector at any time is tangent to the path at the position of the particle at that time. If you are asked for the direction the particle is traveling at a certain time, you automatically calculate the components of its _____ for that time. <u>Speed</u> is the magnitude of the instantaneous velocity and, if the velocity components are given, can be calculated using

$$v =$$

In terms of the velocity $\vec{v}_1$ at time t_1 and the velocity $\vec{v}_2$ at a later time t_2 the <u>average acceleration</u> over the interval from t_1 to t_2 is given by

$$\vec{a}_{\text{avg}} =$$

In terms of velocity components the components of the average acceleration are

$$a_{\text{avg } x} = \qquad\qquad a_{\text{avg } y} = \qquad\qquad a_{\text{avg } z} =$$

To use the definition to calculate the average acceleration over the interval from t_1 to t_2, you must know _____.

The <u>instantaneous acceleration</u> $\vec{a}$ at any time t is the limit of the average acceleration over an interval that includes t, as the duration of the interval becomes vanishingly small. In terms of the velocity vector, it is given by the derivative

$$\vec{a} =$$

In terms of the velocity components its components are

$$a_x = \qquad\qquad a_y = \qquad\qquad a_z =$$

The terms "instantaneous acceleration" and "acceleration" mean the same thing. To use the definition to calculate the acceleration, you must know the velocity vector as a function of time.

A non-zero velocity indicates that the _____ vector of the particle is changing with time. A non-zero acceleration indicates that the _____ vector of the particle is changing with time. Remember that these changes may be changes in magnitude, in direction, or both. Describe a possible motion in which the magnitude of the position vector does not change but the velocity does not vanish: _____

_____

Describe a possible motion in which the speed does not change but the acceleration does not vanish: _____

_____

Projectile motion.   Projectile motion with negligible air resistance is an important example of constant acceleration kinematics in two dimensions. The acceleration (magnitude and direction) of a particle in projectile motion with negligible air resistance is _____

_____. Remember that this is constant as long as the projectile is in flight; it has the same magnitude and direction at the top of the trajectory as it does when the projectile is launched. It changes, of course, when the projectile hits something.

Consider a projectile near the surface of Earth, moving with negligible air resistance. Assume it moves in the xy plane, that the y axis is vertical with the positive direction upward, and that the x axis is horizontal in the plane of the motion. Take the initial coordinates to be x_0 and y_0 and components of the initial velocity to be v_{0x} and v_{0y}. Then the coordinates and velocity components at time t are given by

$$x(t) = \qquad\qquad\qquad\qquad v_x(t) =$$

$$y(t) = \qquad\qquad\qquad\qquad v_y(t) =$$

Check to be sure these equations are consistent with the magnitude and direction of the acceleration. According to the projectile motion equations the motion of a projectile may be considered to be a combination of two independent motions that take place simultaneously: as far as the y components of its position and velocity vectors are concerned, it is in free fall; as far as the x components are concerned, it is moving with constant velocity.

Another important equation results when $v_y = v_{0y} - gt$ is used to eliminate t from $y = y_0 + v_{0y}t - \frac{1}{2}gt^2$. It is

$$v_y^2 - v_{0y}^2 = -2g(y - y_0).$$

This relates the y coordinate and the y component of velocity.

Sometimes the initial speed v_0 and the launch angle θ_0 are known, rather than the x and y components of the initial velocity. Describe the launch angle: _____

_____

If the initial speed v_0 and launch angle θ_0 are given, components of the initial velocity can be found using $v_{0x} = $ _____ and $v_{0y} = $ _____ .

The trajectory of a projectile is shown below. Suppose $t = 0$ when the projectile is launched. On the graph label the initial coordinates x_0 and y_0. Draw an arrow to show the initial velocity and label it $\vec{v}_0$. Label the launch angle. Draw an arrow to show the velocity of the projectile just before it hits the target and label it $\vec{v}_f$. Draw both the velocity and acceleration vectors at the point marked with a dot and at the highest point on the trajectory; label them $\vec{v}$ and $\vec{a}$, as appropriate.

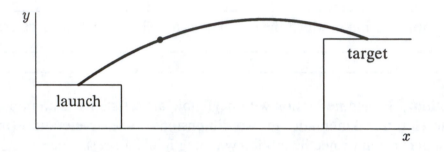

There are two special conditions you should remember. At the highest point of its trajectory the velocity of a projectile is horizontal and $v_y = $ _____. To find the time when the projectile is at its highest point, you solve _____ for t. When a projectile returns to the original launch height, $y = $ _____. To find the time when the projectile returns to the launch height, you solve _____ for t.

The <u>range</u> of a projectile is _____

_____ .

Uniform circular motion.   Uniform circular motion, in which a particle moves around a circle with constant speed, is the second important example of motion in a plane. Remember that the velocity vector is always tangent to the path and therefore continually changes direction. This means the acceleration is *not* zero.

If the radius of the circle is r and the speed is v, the acceleration of the particle has magnitude

$$a =$$

and always points from particle toward _____ . This means the direction of the acceleration continually changes as the particle moves around the circle. The term "centripetal acceleration" is applied to this acceleration to indicate it is directed toward the center of the orbit. You should not forget it is the rate of change of velocity, as are all accelerations.

Suppose a particle travels counterclockwise with constant speed around the circular path shown to the right. At each of the points A, B, and C draw a vector that gives the direction of its velocity and another that gives the direction of its acceleration. Label the vectors $\vec{v}$ and $\vec{a}$, as appropriate.

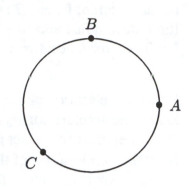

You should be aware that the magnitude of the acceleration is proportional to the *square* of the speed. If, for example, the speed is doubled without changing the orbit radius, the acceleration is multiplied by a factor of 4.

Sometimes the period of the motion is given. The period is the time for the particle to _____. If r is the radius of the orbit and T is the period of the motion, then the speed of the particle is given by

$$v =$$

Thus the period might be given instead of the speed or instead of the radius. In the first case, you can use the equation you just wrote to calculate the speed. In the second case, you can use it to calculate the radius.

Relative motion. This topic deals with a comparison of the values obtained when the position, velocity, or acceleration of a particle is measured using two coordinate systems (or reference frames) that are moving relative to each other. Given the position, velocity, and acceleration of the particle in one frame and the relative motion of the frames, you should be able to calculate the position, velocity, and acceleration in the other frame.

According to the diagram on the right, at any instant of time the position of the particle P relative to the origin of coordinate frame A is given by

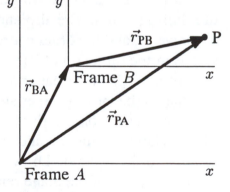

$$\vec{r}_{PA} =$$

in terms of its position $\vec{r}_{PB}$ relative to the origin of frame B and the position vector $\vec{r}_{BA}$ of the origin of frame B relative to the origin of frame A. The above expression can be differentiated with respect to time to obtain

$$\vec{v}_{PA} =$$

for the velocity of the particle in frame A in terms of the particle velocity $\vec{v}_{PB}$ in frame B and the velocity $\vec{v}_{BA}$ of frame B as measured in frame A. This expression is valid even if the two frames are accelerating relative to each other.

Take special care with the subscripts. The first names an object and the second names the coordinate frame used to measure the position or velocity of the object. You should say all the words as you read the symbols. That is, when you see $\vec{r}_{BA}$ you should say "the position vector of the origin of frame B relative to the origin of frame A". You will then get acquainted with the notation fast and won't get it mixed up later.

The expression for the velocity can be differentiated with respect to time to obtain

$$\vec{a}_{PA} =$$

for the acceleration of the particle in frame A in terms of the particle acceleration $\vec{a}_{PB}$ in frame B and the acceleration $\vec{a}_{BA}$ of frame B as measured in frame A. Specialize this equation to the case when the two frames are *not* accelerating with respect to each other: $\vec{a}_{PA} = $ _____. Now the acceleration of the particle is the same in both frames.

If a particle has a different velocity in two reference frames, then the frames must be moving relative to each other. Similarly, if a particle has a different acceleration in two frames, then the frames must be accelerating relative to each other.

To remember what each of the symbols mean, invent a specific example that you can visualize easily. Here's one you might try. Draw an airplane on a small piece of paper and a straight line across a larger piece of paper. Lay the large paper on a table and move the airplane along the line with constant speed. The large paper represents the air and what you see is the motion of the airplane relative to the air. The speed of the airplane relative to the air can be found simply by dividing the length of the line by the time the airplane takes to fly it. Now, move the large paper across the table top with constant velocity as you move the airplane along the line on the paper. You are now viewing the airplane from the ground as the wind blows. Its ground speed can be found by measuring the distance it moves on the table and dividing by the time. You might mark the starting and ending positions of the airplane with small pieces of tape on the table, then measure the distance between the pieces of tape. Convince yourself that the velocity of the airplane relative to the ground is the vector sum of its velocity relative to the air and the velocity of the air relative to the ground. Try various directions for the airplane and wind velocities.

Airplanes flying in moving air or ships sailing in moving water are often used as examples of relative motion. The airplane or ship is the particle, one coordinate frame moves with the air or water, and the other coordinate frame is fixed to Earth. The heading of the airplane or ship is in the direction of its velocity as measured relative to the air or water, *not* relative to the ground. Use your paper airplane to convince yourself of this. Its long axis is parallel to the line on the moving paper and is not necessarily parallel to the line of motion on the table.

The diagram on the right shows the velocity $\vec{v}_{PA}$ of a plane relative to the air and the velocity $\vec{v}_{AG}$ of the air relative to the ground. Draw the vector that represents the velocity $\vec{v}_{PG}$ of the plane relative to the ground. Near the midpoint of this vector draw a small plane oriented correctly; that is, with its long axis parallel to $\vec{v}_{PA}$.

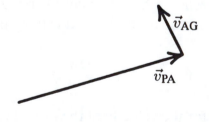

II. PROBLEM SOLVING

The problems of this chapter deal mainly with the definitions of average and instantaneous velocity and average and instantaneous acceleration, with projectile motion, with uniform circular motion, and with relative motion.

To calculate the average velocity you need to know the position at the beginning and end of a time interval. To calculate the velocity you need to know the position as a function of time. To calculate the average acceleration you need to know the velocity at the beginning and end of a time interval. To calculate the acceleration you need to know the velocity as a function of time.

When you read a projectile motion problem you should be able to identify two events, just as you did for one-dimensional problems. Take the time to be 0 for one of them, the launching of the projectile, for example. Take the y axis to be vertically upward and the x axis to be horizontal in the plane of the motion. The coordinates and velocity components are x_0, y_0, v_{0x}, and v_{0y} for the event at time 0. Let t be the time of the other event. The coordinates and velocity components are x, y, v_x, and v_y for that event. Identify the known and unknown quantities, then solve $x = x_0 + v_{0x}t$, $y = y_0 + v_{0y}t - \frac{1}{2}gt^2$, and $v_y = v_{0y} - gt$ simultaneously for the unknowns. As an alternative you might use $v_y^2 - v_{0y}^2 = -2g(y - y_0)$ instead of the equation for y or the equation for v_y. Remember $v_x = v_{0x}$.

All centripetal acceleration problems are solved using $a = v^2/r$. This equation contains three quantities. Two must be given, either directly or indirectly. Remember that the acceleration vector points toward the center of the circle if the speed is constant. Sometimes the period T of the motion is given. Remember that $v = 2\pi r/T$, where r is the radius of the circular orbit. You can use this expression to eliminate v in favor of r or r in favor of v in the expression for the centripetal acceleration.

All relative motion problems are essentially vector addition problems. Identify the two reference frames of interest. Identify the velocities in the equation $\vec{v}_{PA} = \vec{v}_{PB} + \vec{v}_{BA}$, write the equation in component form and solve for the unknown quantities. Sometimes the equation must be rewritten in terms of magnitudes and angles rather than in terms of components.

III. NOTES

Chapter 5
FORCE AND MOTION — I

I. BASIC CONCEPTS

You now start the study of how objects influence each other's motions. This is the central chapter of the mechanics section of the text. Be sure you understand the concepts of force and mass and pay particular attention to the relationship between the net force on an object and its acceleration.

You will also learn about a special property of all forces: an object cannot exert a force on a second object without the second object exerting a force on it. The magnitudes of these forces are the same but they are in opposite directions.

Before studying this chapter be sure you understand acceleration. It is defined and discussed in Chapter 4. To help in your understanding of inertial frames of reference, review the discussion of relative motion in Chapter 4. Free-fall acceleration, used in this chapter, was introduced in Chapter 2.

Dynamics. The fundamental problem of dynamics is to find the acceleration of an object, given the object and its environment. The problem is split into two parts, connected by the idea of a force: the environment of an object produces forces on the object and the net force on it causes it to accelerate. The first part of the problem is to find the net force on the object, given the relevant properties of the object and its environment. The second part of the problem is to find the acceleration of the object, given the net force. In this chapter you concentrate on the second part.

Newton's first law. The text gives two statements of Newton's first law. Write both of them here and learn them:

Statement #1: _____
_____
_____

Statement #2: _____
_____
_____

The first statement in the text is closer to Newton's words, the second is closer to the modern interpretation of the law.

The first law helps us define <u>inertial</u> <u>reference</u> <u>frames</u>. Suppose we have found a particle on which there is zero net force and we attach reference frame S to it. Clearly the acceleration of the particle, as measured in S, is zero. Describe a reference frame S' in which the acceleration of the particle is not zero: _____

Which of these frames is an inertial frame? _____
Why? _____
What is the acceleration of another inertial frame, relative to the one attached to the particle?
_____ How is the velocity of another inertial frame related to the velocity of the one attached
to the particle? _____

Newton's second law. This is the central law of classical mechanics. It gives the relationship
between the net force $\vec{F}_{net}$ on an object and the acceleration $\vec{a}$ of the object:

$$\vec{F}_{net} = m\vec{a},$$

where m is the mass of the object. To understand the law, you must understand the definitions
of force and mass.

A <u>force</u> is measured, in principle, by applying it to the standard (1 kg) mass and measuring
the _____ of the standard mass. If SI units are used, the magnitudes of these quantities
are numerically equal. Both are vectors in the same direction. That forces obey the laws of
vector addition can be checked by simultaneously applying two forces in different directions
and verifying that the result is the same as when the vector sum of the forces is applied as a
single force. All acceleration measurements must be made using an inertial reference frame.

The SI unit of force is _____ and is abbreviated _____. In terms of the SI base
units (kg, m, s) it is _____.

The <u>mass</u> of an object is measured, in principle, by comparing the accelerations of the
object and the standard mass when the same net force is applied to them. In particular, the
mass of the object is given by $m =$ _____, where a is the magnitude of the acceleration of
the object, a_0 is the magnitude of the acceleration of the mass standard, and m_0 is the mass
of the mass standard. The accelerations must be measured using an inertial frame.

Note that small masses acquire a _____ acceleration than large masses when the
same force is applied. Mass is said to measure <u>inertia</u> or resistance to changes in motion.

Mass is a scalar and is always positive. The mass of two objects in combination is the sum
of the individual masses.

Newton's second law $\vec{F}_{net} = m\vec{a}$ is a vector equation. It is equivalent to the three component equations

$$F_{net\ x} =$$

$$F_{net\ y} =$$

$$F_{net\ z} =$$

You must be aware that in these equations F_{net} is the vector sum of all the individual
forces on an object. This means that in any given situation you must identify all the forces on
the object and then sum them *vectorially*.

Note that $\vec{F}_{net} = 0$ implies $\vec{a} = 0$. If the net force vanishes, then the object does not
accelerate; its velocity as observed in an inertial reference frame is constant in both magnitude

and direction. The resultant force may vanish because there no forces on the object or because the forces sum vectorially to zero. For some situations you may know that three forces act but are given only two of them. If you also know that the acceleration vanishes, you can solve $\vec{F_1} + \vec{F_2} + \vec{F_3} = 0$ for the third force.

Newton's third law. Newton's third law tells us something about forces. If the force of object A on object B is $\vec{F}_{BA}$, then according to the third law, the force of object B on object A is given by

$$\vec{F}_{AB} =$$

Compared to the force of A on B, the force of B on A is _____ in magnitude and _____ in direction. You should also be aware that these two forces are of the same type. That is, if the force of object A on object B is a *gravitational* force, then the force of B on A is also a *gravitational* force.

   The third law is useful in solving problems involving more than one object. If two objects exert forces on each other, we immediately use the same symbol to represent their magnitudes and, in writing the second-law equations, we remember the forces are in opposite directions. In addition, we remember that the forces are on different objects. When we want to write Newton's second law for object A, one of the forces we include is the force of B on A, but emphatically *NOT* the force of A on B. The force of A on B, in addition to the other forces on B, determines the acceleration of B, not A.

Gravitational force. One force law you will use a great deal gives the gravitational force with which an object is pulled toward a nearby large astronomical body, such as Earth. Its magnitude is given by $F_g =$ _____, where m is the mass of the object and g is the magnitude of the free-fall acceleration at the position of the object. Near the surface of Earth the direction of the gravitational force is _____.

   The magnitude of the gravitational force is called the weight of the object. Be sure you understand that mass and weight are quite different concepts. Mass is a property of an object and does not change as the object is moved from place to place or even into outer space. It is a scalar. Weight, on the other hand, is the magnitude of a force. It varies as the object moves from place to place and vanishes when the object is far from all other objects, as in outer space. This is because g, not the mass, varies from place to place.

   Remember that the gravitational force on an object is mg regardless of its acceleration. If appropriate, the gravitational force is included in the vector sum of all forces on an object. This sum equals $m\vec{a}$ and if other there are other forces, then $\vec{a}$ is different from $\vec{g}$.

   The SI unit of weight is _____ .

Forces of strings. If a string with negligible mass connects two objects, it pulls on each with a force of the same magnitude, called a _____ force. You may think of the string as simply transmitting a force from one object to the other; the situation is exactly the same if the objects are in contact and exert forces on each other. Strings pull, not push, along their lengths, so a string serves to define the direction of the force. Pay careful attention to the way forces of strings are handled in the sample problems of the text.

Normal forces. When an object is in contact with a surface, the surface may push on it. If the surface is frictionless, that push must be perpendicular to the surface. Thus, it is called a normal force. Unless some adhesive is between the object and surface a normal force can only push on the object. It must be directed away from the surface and toward the interior of the object.

If the surface is at rest or moving with constant velocity, the normal force adjusts until the component of the object's acceleration perpendicular to the surface vanishes. We often use this condition to solve for the normal force. Set the sum of the perpendicular components of all the forces on the object equal to zero. Since the normal force is one of these, the resulting equation can be solved for it in terms of the perpendicular components of the other forces.

Sometimes the surface is accelerating in the direction normal to the surface. An example is the floor of an elevator that is slowing down or speeding up. Then the normal force adjusts so that the normal component of the acceleration of an object on the surface is the same as the normal component of the acceleration of the surface. Again we use Newton's second law to solve for the normal force. Set the sum of the normal components of all the forces on the object equal to the product of the mass of the object and the normal component of the acceleration of the surface. This equation can be solved for the normal force.

The normal force of a surface on an object that is resting on it depends on the directions and magnitudes of other forces on the object and on the normal component of the acceleration of the surface. Be on the lookout for these conditions as you study the sample problems and work your homework assignments.

II. PROBLEM SOLVING

Some problems deal with the definitions of force and mass. If a force is applied to the standard kilogram and, as a result, the standard kilogram has an acceleration $\vec{a}_0$, then the magnitude of the force in newtons is numerically equal to the magnitude of the acceleration in meters per second squared. Force is a vector and is in the direction of the acceleration. The net force on an object is the *vector* sum of the individual forces on it. If identical forces are applied to the standard kilogram and another object and their accelerations are $\vec{a}_0$ and $\vec{a}$, respectively, then the mass in kg of the object is given by $m = (a_0/a)m_0$, where m_0 is 1 kg.

You should know how to handle objects connected by strings. Recognize that a massless string pulls on the objects attached at each end with the same force, equal to the tension force.

Sometimes the acceleration is given indirectly by giving information that can be used in the kinematic equations to calculate it.

A definite procedure has been devised to solve dynamics problems. It ensures that you consider only one object at a time, reminds you to include all forces on the object you are considering, and guides you in writing Newton's second law in an appropriate form. Follow it closely. Use the list below as a check list until the procedure becomes automatic.

1. Identify the object to be considered. It is usually the object on which the given forces act or about which a question is posed.

2. Represent the object by a dot on a diagram or by a sketch of its outline. Do not include the environment of the object since this is replaced by the forces it exerts on the object.

3. On the diagram draw arrows to represent the forces of the environment on the object. Try to draw them in roughly the correct directions. The tail of each arrow should be at the dot or outline. Label each arrow with an algebraic symbol to represent the magnitude of the force, regardless of whether a numerical value is given in the problem statement.

The hard part is getting all the forces. If appropriate, don't forget to include the gravitational force on the object, the normal force of a surface on the object, and the forces of any strings or rods attached to the object. Carefully go over the sample problems in the text to see how to handle these forces.

Some students erroneously include forces that are not acting on the object. For each force you include you should be able to point to something in the environment that is exerting the force. This simple procedure should prevent you from erroneously including a normal force, for example, when the object you are considering is not in contact with a surface.

4. Draw a coordinate system on the diagram. In principle, the placement and orientation of the coordinate system do not matter as far as obtaining the correct answer is concerned but some choices reduce the work involved. If you can guess the direction of the acceleration, place one of the axes along that direction. The acceleration of an object sliding on a surface at rest, such as a table top or inclined plane, for example, is parallel to the surface. Once the coordinate system is drawn, label the angle each force makes with a coordinate axis. This will be helpful in writing down the components of the forces later.

The diagram, with all forces shown but without the coordinate system, is called a <u>free-body diagram</u>. We add the coordinate system to help us carry out the next step in the solution of the problem.

5. Write Newton's second law in component form: $F_{\text{net } x} = ma_x$, $F_{\text{net } y} = ma_y$, and, if necessary, $F_{\text{net } z} = ma_z$. The left sides of these equations should contain the appropriate components of the forces you drew on your diagram. You should be able to write the equations by inspection of your diagram. Use algebraic symbols to write them, not numbers; most problems give or ask for force magnitudes so you should usually write each force component as the product of a magnitude and the sine or cosine of an appropriate angle.

6. If more than one object is important, as when two objects are connected by a string, you can sometimes treat them as a single object. To do this you must know that their accelerations are the same. On the other hand, if you are asked for the force of one object on another, you must carry out the steps given above separately for each object. There is then an additional condition you must consider. Usually the condition is that the magnitudes of their accelerations are the same. You must then invoke Newton's third law: the force of the two objects on each other are equal in magnitude and opposite in direction. Use the same algebraic symbol to represent the magnitudes of these forces and draw their arrows in opposite directions on the free-body diagrams.

7. Identify the known quantities and solve for the unknowns.

III. NOTES

Chapter 6
FORCE AND MOTION — II

I. BASIC CONCEPTS

This chapter contains a great many applications of Newton's laws, with special emphasis on frictional and centripetal forces. Here's where your understanding of the fundamentals begins to pay off!

The concept of acceleration, defined and discussed in Chapters 2 (one-dimensional motion) and 4 (two- and three-dimensional motion) plays an important role in this chapter. Many of the ideas presented in Chapter 5, most notably those of force, mass, Newton's second and third laws of motion, gravitational force, friction, normal force, and tension force of a string, are used a great deal. Review all of them to help in your understanding. Also review the discussion of uniform circular motion in Chapter 4.

Friction. Two macroscopic objects in contact may exert frictional forces on each other. Friction is unavoidable when the objects are sliding on each other, although lubricants and air films may make it small. Even when the objects are stationary with respect to each other, they exert frictional forces if other forces present would otherwise cause them to slide. Explain in words the physical mechanism that gives rise to a frictional force: _____

_____

_____

_____

Although all frictional forces arise from the same fundamental phenomenon, two types are discussed in the text: static and kinetic. Tell how you can identify the situation in which each is acting.

static: _____

kinetic: _____

When the two objects are not moving relative to each other, we determine the force of static friction using the condition that their accelerations are equal. Usually, but not always, an object rests on a surface (a table top or an inclined plane, for example) that is itself at rest. Then, the force of static friction on the object is just sufficient to hold it at rest. Mathematically the frictional force is determined, via Newton's second law, by the condition that the component of acceleration parallel to the surface is zero. This condition is analogous to the condition used to determine the normal force. The difference is that the normal force of one object on another is perpendicular to the surface of contact while the frictional force is parallel to it.

The magnitude f_s of the force of static friction exerted by one surface on another must be less than a certain value, determined by the nature of the surfaces and by the magnitude of the normal force one surface exerts on the other. In particular,

$$f_s \leq$$

where μ_s is called _____ and N is the magnitude of the normal force. If the force of friction required to hold the surfaces at rest with respect to each other is greater than the maximum allowed, then the surfaces slide over each other. Once this happens the magnitude of the force of friction is given by

$$f = \mu_k N$$

where μ_k is called _____ .

List some properties of the objects in contact that determine μ_s and μ_k:

_____

_____

_____

The normal force that appears in the expressions for the force of kinetic friction and the maximum force of static friction must be computed for each situation using Newton's second law. As you know by now the magnitude of the normal force depends on the directions and magnitudes of other forces acting.

Air resistance. When an object moves in air, the air exerts a force on it. When the relative speed of the object and air is so great that the air flow around the object is turbulent, the force of the air on the object is proportional to the square of the relative speed. In fact, the magnitude of the force is given by:

$$D =$$

where v is _____ , ρ is _____ , A is _____ , and C is _____ .
The direction of the drag force is _____ .

When an object falls in air, it approaches a constant speed, called the _____ speed. You can use Newton's second law to find a value for this speed. At terminal speed the gravitational and drag forces are equal in magnitude and sum vectorially to zero. Thus,

$$v_t =$$

The larger the combination $C\rho A$ the _____ the terminal speed and the shorter the time taken to reach that speed from rest.

You should realize that when the object is dropped from rest its acceleration is g at first. As it picks up speed the drag force increases, thereby reducing the acceleration. The object continues to gain speed but at a lesser rate. At terminal speed its acceleration vanishes. From then on the acceleration remains zero, the speed does not change, and the drag force remains constant.

Look at Table 6–1 of the text to see the terminal speeds of some objects in air.

Uniform circular motion. An object in uniform circular motion has a non-zero acceleration because the direction of its velocity changes with time. A force must be applied to the object in order to produce its acceleration. If m is the mass of the object, then the applied force must have magnitude

$$F =$$

where R is the radius of the orbit and v is the speed of the object. The force is directed _____ and, because of its direction, is called a _____ force. Acquire the habit of pointing out to yourself the object in the environment that exerts the force. It might be a string, for example.

If a force $\vec{F}$ is applied to an object of mass m traveling at speed v and the force is maintained perpendicular to the velocity, then the object will travel with constant speed in a circle of radius $R =$ _____. If the magnitude of the force is decreased, the radius of the path will _____.

If you are sitting in a car without a restraining belt and the car rounds a horizontal curve, the force that pulls you around the curve with the car is provided by the _____ force between you and the car seat. If this force is not great enough, you slide toward the outside of the curve; you are going to a larger-radius path. If the force is zero (a very slippery seat), you will travel in a straight line as the seat moves out from under you around the curve.

II. PROBLEM SOLVING

Many problems of this chapter deal with frictional forces. Proceed as before: draw a free-body diagram and write down Newton's second law in component form, just as for any other second-law problem. Use an algebraic symbol, f say, for the magnitude of the frictional force. You must now decide if the frictional force is static or kinetic. If static friction is involved, f is probably an unknown but the acceleration is known or is related to other known quantities in the problem. If the object is at rest on a stationary surface, for example, its acceleration is zero. If it is at rest relative to an accelerating surface, its acceleration is the same as that of the surface. Kinetic friction is involved if one surface is sliding on the other. Then, the magnitude of the frictional force is given by $\mu_k N$.

If you do not know that the object is at rest relative to the surface, assume it is and use Newton's second law, with the acceleration of the object equal to the acceleration of the surface, to calculate both the force of static friction f_{rest} that will hold it at rest and the normal force N. Compare f_{rest} with $\mu_s N$. If $f_{rest} < \mu_s N$, the object remains at rest relative to the surface and the force of friction has the value you computed. That is, $f = f_{rest}$. If $f_{rest} > \mu_s N$, then the object moves relative to the surface. Go back to the second law equations and set $f = \mu_k N$, then solve for the acceleration.

To decide on the direction of a force of static friction, first decide which way the object would move if the frictional force were absent. The frictional force is in the opposite direction. Consider an object on an inclined plane that is tilted so the object will slide down if you do not exert a force on it. Suppose, however, you pull on it with a force F that is parallel to the plane and directed up the plane. You will find that you can apply a fairly wide range of forces without having the object move. If F is small, the force of friction is up the plane; if F is large, the force of friction is down the plane. The static frictional force can have any value from $\mu_s N$ down the plane to $\mu_s N$ up the plane (including 0), depending on the value of F.

In some situations the surface of contact is moving. Consider, for example, a crate on the bed of a moving pick-up truck. If the crate and truck move together, the force of friction acting on the crate is whatever is necessary to give the crate the same acceleration as the truck. This must be less than $\mu_s N$, where N is the normal force of the truck on the crate. If the static frictional force that is required to hold the crate on the truck is greater than $\mu_s N$, then the crate slides and the magnitude of the frictional force is given by $\mu_k N$.

Uniform circular motion problems are solved in much the same way as any other second-law problem. Carry out the set of instructions given in Chapter 5 of this Student's Companion. Draw a free-body diagram. Place the coordinate system so one of the axes is in the direction of the acceleration, pointing from the object toward the center of its orbit. For most problems you will want to substitute v^2/R for the magnitude of the acceleration. Here v is the speed of the object and R is the radius of its orbit. To see how it is done, you should carefully study the sample problems discussed in the text. Always identify the source of the centripetal force that pulls the object around the circle.

III. NOTES

Chapter 7
KINETIC ENERGY AND WORK

I. BASIC CONCEPTS

The central concept of this chapter is the idea of <u>work</u>. You should learn and understand the definition of work, you should learn how to calculate the work done by forces in various situations, and you should learn how work is related to the change in the kinetic energy of a particle (the work-kinetic energy theorem).

Here are some concepts that were discussed in earlier chapters and that you should review: free-fall acceleration from Chapter 2, scalar product from Chapter 3, displacement, velocity, speed, and uniform circular motion from Chapter 4, mass, force, Newton's second law of motion, normal force, tension force, and weight from Chapter 5, and uniform circular motion from Chapter 6.

Kinetic energy and the work-kinetic energy theorem. The significance of work is found in the work-kinetic energy theorem, which shows us that work tends to change the kinetic energy of a particle. First, the kinetic energy of a particle with mass m and speed v is defined by

$$K = $$

Kinetic energy is a scalar and so does not have a direction associated with it. Sometimes you will be given the velocity components v_x and v_y for an object moving in the xy plane and asked to compute the kinetic energy. In terms of the velocity components the kinetic energy is given by $K =$ _____ . You cannot interpret the two terms in this expression as components of a vector. They are not.

The work-kinetic energy theorem is: during any portion of a particle's motion the *net* work done by *all* forces acting on the particle equals the change in the particle's kinetic energy. Let W be the net work done on a particle during some portion of its motion. If K_i is the initial kinetic energy and K_f is the final kinetic energy for that portion, then

$$W = $$

Be sure you perform the subtraction in the correct order.

If the net work is negative, then the kinetic energy and speed of the particle both _____ _____ ; if the net work is positive, then the kinetic energy and speed _____ ; and if the net work is zero, then the kinetic energy and speed _____ . A particle speeds up when the acceleration has a vector component in the direction of the velocity and slows down when the acceleration has a vector component opposite the velocity. In the first case the net force is doing positive work and in the second it is doing negative work. If the net force is always perpendicular to the velocity, it does no work and the speed and kinetic energy of the particle do not change. If several forces act on an object but the kinetic energy of the object does not change, then you know that the works done by the forces sum to _____ .

Definition of work. Several definitions of work are given, for situations of increasing complexity. You should remember them and the situations to which they apply.

1. The particle moves through a displacement $\vec{d}$. The force $\vec{F}$ being considered is constant and makes the angle ϕ with the displacement when $\vec{F}$ and $\vec{d}$ are drawn with their tails at the same point. Then, the work done by $\vec{F}$ is given by

   $$W =$$

   Alternatively, if the particle has displacement Δx, along the x axis, and the force has a constant x component F_x, then the expression for the work can be written

   $$W =$$

2. The particle moves along a straight line (the x axis). The force $\vec{F}$ is parallel to the x axis and is not constant. Then, the work it does as the particle moves from x_1 to x_2 is given by the integral

   $$W =$$

   Be sure your definition allows for a force in the same direction as the displacement and for one in the opposite direction. To evaluate the integral, the x component of the force must be known as a function of _____. The work done by the force is the area under the graph of _____ versus _____.

3. A particle moves in the xy plane, subjected to a variable force $\vec{F}$. Write the integral definition of the work done by $\vec{F}$ as the particle moves from $\vec{r_1}$ to $\vec{r_2}$:

   $$W =$$

   Explain how the integral can be evaluated, in principle, by dividing the path into a large number of segments. Don't forget to give the quantity to be evaluated for each segment.

   _____

   _____

   This expression, generalized slightly to three dimensions, is the general definition of work. The expressions you wrote above in 1 and 2 for special situations can be derived from it.

   A person carrying a heavy box horizontally with constant velocity does no work on the box because _____.
   The normal force of a stationary surface on a sliding crate does no work, no matter what the orientation of the surface, because _____

   _____.

   A string used to whirl an object around a circle with constant speed does no work because

   _____.

   Work can be positive or negative. Each of the four diagrams below shows a block moving on a table top. On each of the first pair show how you would apply a force so it does positive work. On each of the second pair show how you would apply a force so it does negative work. In each case direct the force so it is not parallel to the velocity and assume the block continues to move in the same direction, at least for a short while after the force is applied.

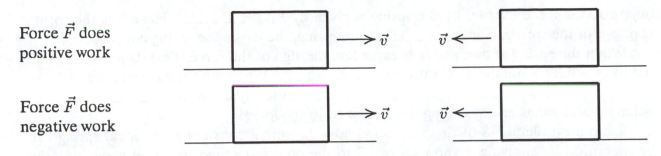

Force $\vec{F}$ does
positive work

Force $\vec{F}$ does
negative work

Work is a scalar. It does not have a direction associated with it. When several forces act on an object and you want to find the net work done, you simply add the works done by the individual forces. You must, of course, include the appropriate sign for each work. The direction of each force and the direction of the displacement are important for calculating the work done by a force but neither these directions or any others must be taken into account when the individual works are summed. The net work is the same as the work done by the net force.

The SI unit of energy and work is _____. In terms of SI base units it is _____.

Work done by a gravitational force. You should know how to calculate the work done by a gravitational force and the work done by the force of an ideal spring. In addition to being excellent examples of a constant and a variable force, respectively, they are used in many problems.

When an object of mass m falls from height y_i to height y_f near Earth's surface, the gravitational force of Earth does work $W =$ _____ on it. When the object is raised from height y_i to height y_f, the gravitational force of Earth does work $W =$ _____ on it. The two expressions you wrote should be identical. On the way down the sign of the work done by the gravitational force is _____ while on the way up it is _____. If a ball is thrown into the air, falls, and is caught at the height from which it was thrown, the work done by the gravitational force of Earth over the round trip is _____.

You should know that the work done by a gravitational force is the same no matter what path is taken between the initial and final points. In addition, all that counts is the initial and final altitudes. The two positions need not have the same horizontal coordinate.

Remember that the expression you wrote for the work done by a gravitational force is valid even if the object experiences air resistance. If air resistance is present, this expression does not, of course, give the net work done by all forces.

Work done by an ideal spring. The force exerted by an ideal spring on an object is a variable force. Its direction depends on whether the spring is extended or compressed and its magnitude depends on the amount of extension or compression.

Assume the spring is along the x axis with one end fixed and the other attached to the object, as shown. When the object is at $x = 0$, the spring is neither extended or compressed. This is the equilibrium configuration. When the object is at any coor-

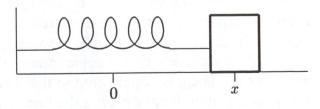

dinate x, the force exerted by the spring is given by $F_s = $ _____ . Here k is the <u>spring constant</u> of the spring. The _____ the spring, the larger the spring constant.

When the spring shown above is extended, the sign of the force it exerts is _____ and the force tends to pull the object toward _____ . When the spring is compressed, the sign of the force is _____ and the force tends to push the object toward _____ . This behavior is summarized by calling the force a <u>restoring force</u>.

Sample Problem 7–8 of the text shows how the spring constant of an ideal spring can be measured by applying a known force F to the object attached to it and measuring the elongation x of the spring when the applied force exactly matches the force of the spring in magnitude. In terms of F, m, x, and k the net force on the object is given by _____ and since the object is in equilibrium this must be zero. Thus, $k = $ _____ .

The SI units of a spring constant are _____ .

As the object moves from some initial coordinate x_i to some final coordinate x_f the work done by the spring is

$$ W_s = \int_{x_i}^{x_f} -kx \, dx = $$

Give an example in which the spring does positive work: _____

_____

Give an example in which the spring does negative work: _____

_____

In general, the spring does positive work whenever the object attached to it is moving toward the equilibrium point and does negative work whenever the object is moving away from the equilibrium point.

Suppose an object is attached to a horizontal spring and is free to move on a frictionless table top, as in the diagram above. Now, you pull on the object with a force given by $F_{ext} = +kx$. As the object goes from $x = x_i$ to $x = x_f$ the work you do is given by $W_{ext} = $ _____ . Of course, you may pull on the object with any force you like, not necessarily $+kx$. The external force $+kx$, however, has special significance because the net force on the object is then _____ and the object does not accelerate during the pulling. When this external force is applied, the net work done by the spring and your force, taken together, is _____ .

Proof of the work-kinetic energy theorem. Carefully review Section 7–6 of the text, which gives a proof of the work-kinetic energy theorem. This proof shows that the theorem is a direct consequence of Newton's second law. It also explains why the *net* work (done by *all* forces acting on the particle) enters the theorem.

Power. In words, <u>power</u> is _____ .
Suppose $W(t)$ is the work done by any force $\vec{F}$ during a time interval that ends at time t. Then, the instantaneous power delivered by the force is given by $P = $ _____ . The SI unit of power is _____ . In terms of SI base units the unit for power is _____ .

Consider a particle that at some instant of time is moving with velocity $\vec{v}$ and is acted on by a force $\vec{F}$. The power delivered to the particle by the force is given by $P = $ _____ . Be sure your equation is valid even if the force is not parallel to the velocity.

II. PROBLEM SOLVING

You should know how to calculate the work done by a force if the force is constant or if its component along the path is given as a function of the object's position. In some cases you might need to solve a Newton's second law equation to find the force. In other cases the force as a function of position might be given as a graph and you should be able to obtain the work from the area under the curve. The force in question might be the only force acting on an object or one of several.

The work-kinetic energy theorem tells us that the net work W done on a particle is equal to the change in the kinetic energy of the particle. That is, $W = \Delta K$, or since the kinetic energy is given by $\frac{1}{2}mv^2$, $W = \frac{1}{2}m(v_f^2 - v_i^2)$. Here m is the mass of the particle, v_i is its speed at the beginning of the interval, and v_f is its speed at the end of the interval.

In some problems you are asked to find the net work done on an object, given its initial and final speeds (and its mass). This is a direct application of the work-kinetic energy theorem. In other problems you will use the definition of work to calculate its value, given the force and displacement, then use the work-kinetic energy theorem to find the final speed, given the initial speed.

A spring provides a good example of a variable force. If one end is fixed and the other end is moved so the spring is either extended or compressed from its equilibrium length, then the force exerted by the spring is given by $F = -kx$, where k is the spring constant. The coordinate x of the spring end is measured with the origin at the position of the movable end when the spring has its equilibrium length. If the end of the spring is moved from x_i to x_f, the work done by the spring is $-\frac{1}{2}(x_f^2 - x_i^2)$.

Some problems involve calculations of the power delivered by forces. In some cases you evaluate $P = dW/dt$ while in others you evaluate $P = \vec{F} \cdot \vec{v}$.

III. MATHEMATICAL SKILLS

Scalar products. These are used extensively in this chapter to calculate the work done and the power delivered by a force: $W = \vec{F} \cdot \vec{d}$ for a constant force and $P = \vec{F} \cdot \vec{v}$ for any force. Be sure you know how to evaluate them. In particular know that

$$\vec{F} \cdot \vec{d} = Fd \cos \phi,$$

where ϕ is the angle between $\vec{F}$ and $\vec{d}$ when they are drawn with their tails at the same point. Also know how to evaluate a scalar product in terms of components:

$$\vec{F} \cdot \vec{d} = F_x d_x + F_y d_y + F_z d_z.$$

Review the appropriate sections of Chapter 3 of the text and this Student Companion.

IV. NOTES

Chapter 8
POTENTIAL ENERGY
AND CONSERVATION OF ENERGY

I. BASIC CONCEPTS

The closely related concepts of a conservative force and a potential energy are central to this chapter. Pay close attention to their definitions. If *all* forces exerted by objects in a system on each other are conservative and no net work is done on the objects by an outside agent, then the mechanical energy of the system (the sum of the kinetic and potential energies) does not change. When non-conservative forces act, another energy, called the thermal energy of the system, must be included in the sum for the principle of energy conservation to hold.

This chapter makes use of the ideas of free-fall acceleration from Chapter 2, displacement from Chapters 3 and 4, force, mass, and gravitational force from Chapter 5, and the work-kinetic energy theorem, friction, work done by a gravitational force, the force exerted by an ideal spring, and work done by an ideal spring from Chapter 7.

Definitions   Section 8–2 of the text gives two ways to test a force to see if it is <u>conservative</u>. One of them is: _____

_____

_____

The other is: _____

_____

_____

The two tests are equivalent to each other. If a force meets either of them, then it automatically meets the other. If it fails either of them, then it automatically fails the other.

Give some examples of conservative forces: _____

Give at least one example of a non-conservative force: _____

Consider a block attached to a horizontal spring and on a horizontal surface. Friction exists between the block and the surface. Suppose the system starts with the spring neither extended nor compressed. An external force is applied so the block moves a distance d from its initial position, thereby extending the spring. Then it moves back to its initial position. You should be able to show that the spring does zero work during this motion and that the force of friction does non-zero work.

work done by spring on outward trip: $W_s = $ _____
work done by spring on inward trip: $W_s = $ _____
sign of the work done by friction on outward trip: _____
sign of the work done by friction on inward trip: _____

Because a force of friction is actually the sum of a large number of forces, acting at the welds that form between two surfaces, and the welds move through displacements that are different from the displacement of the object, the work done by friction is *not* given by $\int \vec{f} \cdot d\vec{r}$, where $d\vec{r}$ is an infinitesimal displacement of the object. You cannot calculate the work done by friction without a detailed model of the surfaces but you should be able to argue that the work done by friction cannot be zero over a round trip. You should also be able to argue that the work done by friction on an object sliding on a stationary surface is negative.

A <u>potential</u> energy <u>function</u> can be associated with a conservative force. Here we consider forces that are given as functions of the positions of the interacting objects and not, for example, as functions of their velocities or the time. For a given force we consider the system consisting of the two objects that are interacting via the force (Earth and a projectile, a spring and an object attached to it, for example). Some configuration of the system, called the reference configuration, is selected and the potential energy is arbitrarily assigned the value zero for this configuration. The potential energy for any other configuration is taken to be the negative of the work done by the force as the system moves from the reference configuration to the configuration being considered. For the force of gravity a configuration is specified by giving the positions of the objects relative to each other. For the force of a spring a configuration is specified by giving the extension or compression of the spring or, alternatively, the position of the object attached to the spring. Later on, when you study rotational motion, a specification of configuration may also include the orientations of the objects.

The work done by a conservative force as the system goes from any initial to any final configuration is the negative of the change in the potential energy. That is,

$$W =$$

where U_i is the potential energy associated with the initial configuration and U_f is the potential energy associated with the final configuration. The path taken is immaterial since the work done by a conservative force is independent of the path. Every time the system reaches the same configuration it has the same potential energy. This property is at the very heart of the potential energy concept. Carefully note that a potential energy is associated with at least two objects, not with a single object.

To test your understanding of potential energy, explain in words why a potential energy cannot be associated with a frictional force. After all, the force does work. Why not just define the potential energy to be the negative of the work? _____

_____

_____

Potential energy is a scalar. If two or more conservative forces act, the total potential energy is simply the sum of the individual potential energies.

The SI unit for potential energy is _____ and is abbreviated _____. In terms of SI base units this unit is _____.

Only *changes* in potential energy have physical meaning. If the same arbitrary value is added to the potential energy for every configuration of the system, the motions of the objects in the system do not change. This is why almost any configuration can be selected as the reference configuration. Selection of a new reference configuration simply adds the same value to the potential energy for every configuration.

Because potential energy is a function of configuration a system may be considered to store energy as potential energy. The potential energy of the system is increased when positive work is done on the system by an external agent without changing its kinetic energy. This stored energy can be converted to kinetic energy. As you study this chapter and its sample problems or work end-of-chapter problems notice when work being done by an external agent is changing the potential energy of a system and when potential energy is being converted to kinetic energy or vice versa.

Gravitational potential energy. In this chapter we take the gravitational force on an object of mass m to be its weight; that is, we assume the free-fall acceleration $\vec{g}$ is due entirely to the interaction between Earth (or other large astronomical body) and the object. Then the gravitational potential energy of a system consisting of Earth and an object of mass m, near its surface, is given by

$$U =$$

where y is the vertical coordinate of the object, measured relative to Earth, and the reference configuration is _____ .

Gravitational potential energy is often ascribed to the object alone, but it is actually a property of the Earth-object system.

Remember that the gravitational potential energy depends only on the altitude of the object above the surface of Earth, even if the object moves horizontally as well as vertically. When the object is raised a distance Δy and is simultaneously moved horizontally by Δx, the gravitational potential energy increases by $mg\Delta y$. You should be able to show this, starting with $\Delta U = -\vec{F} \cdot \vec{d}$, where $\vec{d}$ is the displacement $\Delta x\,\hat{\imath} + \Delta y\,\hat{\jmath}$. Here the positive y direction was taken to be upward.

If the altitude of an object above Earth is increased, the sign of the work done by gravity is _____ and the sign of the potential energy change is _____ . If the altitude is decreased, the sign of the work done by gravity is _____ and the sign of the potential energy change is _____ . See Sample Problem 8–2 of the text for a calculation of gravitational potential energy and carefully note the sign of the potential energy change. The sample problem also shows that the choice of reference configuration does not influence *changes* in the potential energy.

When an object is lifted with constant speed in Earth's gravitational field, the _____ energy of the Earth-object system increases. When the object is released and falls, _____ energy is converted to _____ energy by the action of the gravitational force.

Spring potential energy. The potential energy of a system consisting of an object attached to an ideal spring with spring constant k is given by

$$U =$$

where the coordinate x of the object is measured relative to an origin at _____ .

The reference configuration is _____ .

When a spring is elongated by pulling the object outward from its equilibrium position, the sign of the work done by the spring is _____ and the sign of the change in the spring potential energy is _____ . When the spring is compressed by pushing the object inward from its equilibrium position, the sign of the work done by the spring is _____ and the sign of the change in the spring potential energy is _____ . An external agent can increase the spring potential energy by compressing or extending the spring. This stored energy is converted to kinetic energy when the spring is released.

Conservation of mechanical energy.   Consider a system of objects that interact with each other via conservative forces and suppose no net work is done on objects of the system by outside agents. The change in the total potential energy as the system changes configuration is the negative of the total work done by all forces. According to the work-kinetic energy theorem, the total work is also equal to the change in _____ . Thus, if all forces are conservative, the sum of the _____ and _____ energies does not change as the system changes configuration. In symbols, $\Delta K + \Delta U = 0$.

The <u>mechanical energy</u> of a system is defined by E_{mec} = _____ + _____ and if all forces are conservative, then $\Delta E_{mec} = 0$. You should recognize that, in general, K is the sum of the kinetic energies of all objects in the system. When, for example, an object moves in Earth's gravitational field, K is strictly the sum of the kinetic energies of the object and Earth. If, however, the object is much less massive than Earth, then the kinetic energy of Earth does not change significantly and can be omitted from the energy equation. Similarly, an ideal spring is considered to be massless and so does not contribute to the kinetic energy of any system of which it is a part.

Conservation of mechanical energy can be used to solve some problems. If you know the potential and kinetic energies for one configuration of the system, then you can calculate the total mechanical energy for that configuration. If mechanical energy is conserved, then it has the same value for all configurations. Now, suppose you are given or can calculate the potential energy for a second configuration. Then, the conservation of energy principle allows you to compute the kinetic energy for the second configuration. Likewise, if you know the kinetic energy for the second configuration, then the conservation of mechanical energy principle allows you to compute the potential energy.

For example, suppose an object of mass m is attached to a spring with spring constant k. It moves on a horizontal frictionless surface. Initially it is pulled out so the spring is elongated by x_i and it is given a speed v_i to start its motion. In terms of these quantities the mechanical energy of the spring-object system is E_{mec} = _____ . Suppose the speed of the object is v_f when it is at x_f. In terms of these quantities the mechanical energy is E_{mec} = _____ . Equate the two expressions for the mechanical energy to obtain $\frac{1}{2}mv_i^2 + \frac{1}{2}kx_i^2 = \frac{1}{2}mv_f^2 + \frac{1}{2}kx_f^2$. This equation can be solved for any one of the quantities that appear in it.

Suppose a ball of mass m is thrown into the air with an initial speed v_0. Take the zero of gravitational potential energy to be at the release point and neglect the motion of Earth. The mechanical energy of the Earth-ball system is then E_{mec} = _____ . What is the speed of the ball when it is a distance y above the release point? The potential energy is then

$U(y) =$ _____ and if we let v be the speed of the ball, its kinetic energy is $K =$ _____.
Equate the two expressions for the mechanical energy to obtain $\frac{1}{2}mv_0^2 = \frac{1}{2}mv^2 + mgy$. This
can be solved for v. The speed is the same whether it is going up or coming down. You can
also find how high the ball goes before falling back again. Put $v = 0$ and solve for y.

Calculation of force from potential energy. You have learned how to compute the potential
energy associated with a given conservative force. The reverse calculation can also be carried
out. If the potential energy function is known, the force can be computed by evaluating its
derivatives with respect to coordinates. Consider an object moving along the x axis and acted
on by a conservative force F. If $U(x)$ is the potential energy as a function of the object's
coordinate, then

$$F =$$

If you are given a graph of the potential energy as a function of x and asked for the force on
the particle when its coordinate has a given value, you measure the _____ of the line
that is _____ to the curve at that point. Be careful about the signs here: a positive slope
indicates a force in the _____ x direction and a negative slope indicates a force in the
_____ x direction.

Potential energy curves. You should know how to obtain information about the motion of
an object from a potential energy curve, the potential energy plotted as a function of its coordinate. In thinking about these curves you may assume that the agent exerting the force is
essentially stationary. Only the object whose motion is being considered has kinetic energy.

Consider the potential energy function shown below for an object that moves along the
x axis. Suppose the object has mechanical energy E_{mec}, represented by a dotted line, and
suppose further that mechanical energy is conserved. Indicate on the graph the potential and
kinetic energies of the object when it is at x_1. Use x_{min} to label the minimum coordinate
of the object in its motion and use x_{max} to label the maximum coordinate of the object in
its motion. Look at the slope of the potential energy function and mark with an arrow the
directions of the force when the object is at x_{min} and at x_{max}. x_{min} and x_{max} are called the
_____ of the motion.

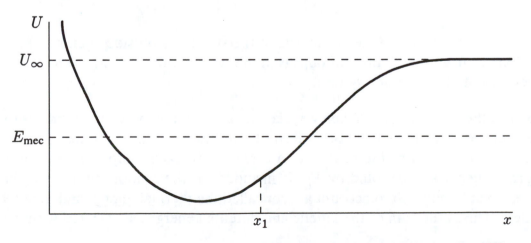

Suppose the object starts at x_1 and travels in the negative x direction. Qualitatively describe its subsequent motion, telling where it is speeding up, where it is slowing down, where it momentarily stops, and what it does after it stops: _____

_____

_____

_____

If the mechanical energy of the object is increased slightly, what happens to the minimum and maximum coordinates? The minimum coordinate _____ and the maximum coordinate _____ .

Suppose now that the object is far to the right and is traveling in the negative x direction with a mechanical energy that is slightly greater than U_∞ on the graph. Describe its subsequent motion: _____

_____

Equilibrium points are coordinates for which the _____ vanishes. The slope of the potential energy curve is _____ at these points. When an object is released near a point of stable equilibrium, its subsequent motion is _____. When an object is released near a point of unstable equilibrium, its subsequent motion is _____. When an object is released near a point of neutral equilibrium, its subsequent motion is _____.

The graph above shows an equilibrium point just to the left of x_1. Is it a point of stable, unstable, or neutral equilibrium? _____

External work. Consider a system of objects that may interact with each other and with other objects outside the system. Its mechanical energy is not conserved if an external force does work on any object in the system. Suppose, for example, a crate is carried up some stairs. The force of the person doing the carrying is an external force if the system is taken to consist of Earth and the crate. The work done by this force increases the mechanical energy of the Earth-crate system. If the speed of the crate does not change, the increase in energy appears as potential energy; the crate is further away from Earth than previously. If the crate's speed does increase, the increase in energy is both potential and kinetic.

When external forces do work on objects in the system, the energy equation is written

$$\Delta K + \Delta U = W,$$

where W is the total work done *on* objects of the system by outside agents. The potential energy U arises from interactions of objects in the system with each other and K is the total kinetic energy of objects in the system.

Internal and thermal energy. When we take into account the individual particles that make up an object, we may need to add other terms to the energy. If the motions of large internal parts of an object, such as the pistons of a car, must be taken into account a term called the internal energy and denoted by E_{int} is included. If the random motions of individual molecules must be taken into account a term called the thermal energy and denoted by E_{th} is included. This is the sum of the kinetic and potential energies associated with motions of the particles and with their mutual interactions, as distinct from the motion of the object as a whole, motions of large internal parts, or interactions of the object with other objects. Strictly speaking, thermal energy is a form of internal energy but it is treated separately in this text.

When a gas is compressed, for example, its thermal energy changes. Non-conservative forces, such as friction, always change the thermal energy of the object on which they act because the welds between the surfaces deform and the microscopic surface structure changes as time goes on. Thus, both the macroscopic kinetic energies and the thermal energy of the system change. Similarly, air resistance results from collisions of many air molecules with particles in an object. These change both the macroscopic kinetic energy of the object and its thermal energy.

When internal and thermal energies are included, the energy equation becomes

$$\Delta K + \Delta U + \Delta E_{th} + \Delta E_{int} = W.$$

Here K is the sum of the kinetic energies of objects in the system, U is the sum of the potential energies associated with interactions between objects in the system, E_{th} is the total thermal energy of the system, E_{int} is the total internal energy of the system (exclusive of thermal energy), and W is the total work done on particles of the system by outside agents.

Carefully note that if no net work is done on objects of the system by outside agents, then $\Delta K + \Delta U + \Delta E_{th} + \Delta E_{int} = 0$. Energy, in the form $K + U + E_{th} + E_{int}$, is conserved. Forces between particles of the system may change the kinetic energies of the objects, the potential energies of their interactions, the thermal energy, or the internal energy, but the sum retains the same value. Energy may change form but the total does not change.

Consider a block sliding on a horizontal table top. Because friction is present it slows and stops. No external work is done on the system composed of the block and table so its total energy is conserved. All the kinetic energy originally possessed by the block ends up as _____ energy. If $\vec{f}$ is the force of friction acting on the block and $\vec{d}$ is the displacement of the block, then $-\vec{f} \cdot \vec{d}$ gives the increase in the thermal energy of the block and the surface on which it slides.

An increase in the thermal energy of an object may result in an increase in the temperature of that object. When the temperature of an object is different from the temperature of its surroundings, energy is transferred as heat and still another term must be included in the energy equation given above. You will learn more about this when you study Chapter 19.

II. PROBLEM SOLVING

You should know how to compute the potential energies of an object-Earth system and an object-spring system and know how changes in the potential energies are related to the gravitational and spring forces.

Some problems of this chapter involve the calculation of a potential energy change, given the force and the end points of the path. Since the change in potential energy is just the negative of the work done by the force, the calculation proceeds like the calculations of work in the last chapter. If you are asked for the potential energy itself, the reference configuration (for which the potential energy is zero) must be identified, either in the problem statement or by you. You then calculate the negative of the work done on the system as the system changes from the reference configuration to the configuration of interest.

Problems involving the conservation of mechanical energy are relatively straightforward. You must first select a system of objects. Normally you will pick the system so that no outside agents do work on objects in the system. You must then decide if all the forces of interaction between objects of the system are conservative. The force of gravity and the force of an ideal spring are; the force of friction and drag forces are not. In one dimension forces that are uniform over all possible paths are conservative. For other forces you may have to apply one of the tests for a conservative force. See Section 8–2 of the text or the Basic Concepts section of this chapter of *A Student's Companion*.

If all forces that do work are conservative, then conservation of mechanical energy yields $K_1 + U_1 = K_2 + U_2$, where K_1 and U_1 are the kinetic and potential energies for one configuration and K_2 and U_2 are the kinetic and potential energies for another. You must identify the two configurations from the problem statement. You will be given enough information to calculate three of the four energies that appear in the energy equation and can use the conservation of mechanical energy to compute the fourth.

In many problems you are asked to find some parameter that occurs in the expression for the kinetic or potential energy at the second configuration. For example, you may be asked for the compression of a spring, a spring constant, the speed of an object, or its mass. You first find the relevant energy (kinetic or potential), then solve for the parameter.

You should be able to interpret a potential energy graph for a particle. This tells you the potential energy $U(x)$ for every particle coordinate. If you know the mechanical energy E_{mec}, you can use $K = E_{\text{mec}} - U$ to find the kinetic energy of the particle when it is at any given coordinate. Sometimes the speed is given for one coordinate and requested for another. Calculate the value of K for the first coordinate and read the value of U for that coordinate from the graph, then calculate E_{mec}. Read the value of U for the second coordinate from the graph, then use $K = E_{\text{mec}} - U$ to calculate the kinetic energy for that coordinate. Finally, solve $K = \frac{1}{2}mv^2$ for the speed at the second coordinate.

You should also know how to obtain the force from the graph. At any coordinate it is the negative of the slope of the curve at that coordinate.

Some problems deal with the change in thermal energy through friction or air resistance. Sometimes you can use $\Delta E_{\text{th}} = -\vec{f} \cdot \vec{d}$. Sometimes you must calculate the decrease in mechanical energy. Other problems deal with the influence of an external force on the mechanical energy of a system. The work done by the net external force tells you the change in the mechanical energy of the system.

III. NOTES

Chapter 9
SYSTEMS OF PARTICLES

I. BASIC CONCEPTS

In this chapter you consider systems of more than one particle. The center of mass is important for describing the motion of the system as a whole and its velocity is closely related to the total linear momentum of the system. External forces acting on the system accelerate the center of mass and when the net external force vanishes the velocity of the center of mass is constant. The total linear momentum of the system is then conserved.

The following concepts from Chapter 4 are used in this chapter: position vector, displacement, velocity, and acceleration. Newton's second and third laws of motion, force, and mass, all from Chapter 5, are also used as are the ideas of work, kinetic energy, and the work-kinetic energy theorem from Chapter 7.

Coordinates of the center of mass. You should know how to calculate the coordinates of the center of mass of a collection of particles. If particle i has mass m_i and coordinates x_i, y_i, and z_i, then the coordinates of the center of mass are given by

$$x_{\text{com}} =$$

$$y_{\text{com}} =$$

$$z_{\text{com}} =$$

where M is the total mass of the system. In vector notation, the position vector of the center of mass is given by

$$\vec{r}_{\text{com}} =$$

where $\vec{r}_i$ is the position vector of particle i. The center of mass is not necessarily at the position of any particle in the system. Sample Problem 9–1 of the text shows you how to calculate the coordinates of the center of mass of a small collection of particles.

For a *continuous* distribution of mass the coordinates of the center of mass are given by the integrals

$$x_{\text{com}} =$$

$$y_{\text{com}} =$$

$$z_{\text{com}} =$$

If the mass is distributed symmetrically about some point or line, then the center of mass is at that point or on that line. For example, the center of mass of a uniform spherical shell is at its _____, the center of mass of a uniform sphere is at its _____, the center of mass of a uniform cylinder is at the midpoint of its _____, the center of mass of a uniform rectangular plate is at its _____, and the center of mass of a uniform triangular plate is at the intersection of _____ and halfway through its thickness.

Sometimes an object with a complicated shape can be thought of as a group of parts such that the center of mass of each part can be found easily. Then, each part is replaced by a particle with mass equal to the mass of the part, positioned at the center of mass of the part. The center of mass of the whole object is at the position of the center of mass of these particles. This idea can also be used to find the center of mass of an object with a hole. See Sample Problem 9–2 of the text.

Velocity and acceleration of the center of mass. As the particles that comprise the system move, the coordinates of the center of mass might change. Consider a system of discrete particles and differentiate the expression for the center of mass position vector to find the velocity of the center of mass in terms of the individual particle velocities:

$$\vec{v}_{\text{com}} =$$

where m_i is the mass and $\vec{v}_i$ is the velocity of particle i.

Consider two objects moving along the x axis, one with mass m_A and coordinate given by $x_A(t) = x_{A0} + v_A t$ and the other with mass m_B and coordinate given by $x_B(t) = x_{B0} + v_B t$, where v_A and v_B are constant velocities. Use the definition of the center of mass to find an expression for the coordinate of the center of mass as a function of time: $x_{\text{com}}(t) =$ _____ _____. Now, differentiate this expression to find an expression for the velocity of the center of mass: $v_{\text{com}} =$ _____. How should the velocities of the objects be related for the center of mass to be at rest? _____

_____

The center of mass might accelerate as the particles accelerate. Differentiate the general expression for the velocity of the center of mass to find the acceleration of the center of mass in terms of the accelerations of the individual particles:

$$\vec{a}_{\text{com}} =$$

where m_i is the mass and $\vec{a}_i$ is the acceleration of particle i.

Newton's second law for the center of mass. Since each individual particle obeys Newton's second law, $\vec{F}_{\text{net}} = M\vec{a}_{\text{com}}$, where $\vec{F}_{\text{net}}$ is the vector sum of all forces on all the particles of the system, some of which might be due to particles in the system and some of which might be due to the environment of the system. Actually, the sum reduces to the sum over all *external* forces acting on particles of the system, i.e. those due to the environment of the system. Explain in words why these two sums are equal: _____

_____

_____

Thus, the center of mass obeys a Newton's second law: $\vec{F}_{net} = M\vec{a}_{com}$. The mass that appears in the law is the total mass of the system (the sum of the masses of the individual particles) and the force that appears is the vector sum of all *external* forces acting on all particles of the system.

If the total external force acting on a system is zero, then the acceleration of the center of mass is _____ and the velocity of the center of mass is _____ . If the center of mass of the system is initially at rest and the total external force acting on the system is zero, then the velocity of the center of mass is always _____ and the center of mass remains at its initial position.

These statements are true no matter what forces the particles of the system exert on each other. The particles might, for example, be fragments that are blown apart in an explosion or they might be objects that collide violently. On the other hand, they might interact with each other from afar via gravitational or electrical forces and never touch.

Linear momentum. The linear momentum of a particle of mass m, moving with a velocity $\vec{v}$ is given by $\vec{p} = $ _____ . Linear momentum is a vector. Its SI units are _____ . In terms of linear momentum, Newton's second law for a particle is

$$\vec{F}_{net} = $$

where $\vec{F}_{net}$ is _____ . You should be able to prove this formulation is exactly the same as $\vec{F}_{net} = m\vec{a}$.

If the particle's speed is close to the speed of light, relativistically correct expressions must be used. The linear momentum of a particle of mass m moving with velocity $\vec{v}$ is then given by

$$\vec{p} = $$

The total linear momentum $\vec{P}$ of a system of particles is the vector sum of the individual momenta. Since, for non-relativistic particles, $\vec{P} = m_1\vec{v}_1 + m_2\vec{v}_2 + \ldots$, the total linear momentum is given by $\vec{P} = M\vec{v}_{com}$, where M is the total mass of the system. That is, the total linear momentum of the system is identical to the linear momentum of a single particle with mass equal to the total mass of the system, moving with a velocity equal to the velocity of the center of mass. Furthermore, the net external force changes the total linear momentum of the system according to

$$\vec{F}_{net} = $$

Conservation of linear momentum. State in words the principle of conservation of total linear momentum for a system of particles. Your statement should have the form "If ... , then ... ." _____

Since linear momentum is a vector this principle is equivalent to the three equations: $P_x =$ constant if $F_{\text{net }x} = 0$, $P_y =$ constant if $F_{\text{net }y} = 0$, and $P_z =$ constant if $F_{\text{net }z} = 0$. One component of linear momentum may be conserved even if the others are not.

If the linear momentum of a system is conserved, then the acceleration of the center of mass of the system is _____ and the velocity of the center of mass is _____ .

Variable mass systems. $\vec{F}_{\text{net}} = M\vec{a}_{\text{com}}$ and $\vec{F}_{\text{net}} = d\vec{P}/dt$ hold only for systems of constant mass. Recall that when they were derived, the number of particles in the system was held constant. In Section 9–7 of the text the derivation of an expression for $d\vec{P}/dt$ is carefully carried out for a system that loses mass. To understand it, you should carefully go over the steps.

Suppose that at time t a rocket with mass M is moving with velocity $\vec{v}$. It will eject fuel of mass $-\Delta M$ (a positive quantity) in time Δt. Originally the fuel is traveling with the rocket at velocity $\vec{v}$ but after it is ejected it has velocity $\vec{U}$ and the rocket has velocity $\vec{v} + \Delta\vec{v}$. Carefully note that $\vec{v}$, $\vec{v} + \Delta\vec{v}$, and $\vec{U}$ are all measured relative to an inertial frame and that ΔM is taken to be negative, so the mass of the rocket after ejection is $M + \Delta M$. Also note that the rocket exerts a force on the fuel to eject it and the fuel exerts a force on the rocket of the same magnitude but in the opposite direction. Suppose no external forces act on the rocket-fuel system.

The rocket and the ejected fuel form a constant mass system, so $\vec{F}_{\text{net}} = d\vec{P}/dt$ is valid for it. The total linear momentum of the system at the initial time t, in terms of M and $\vec{v}$, is given by $\vec{P}_i =$ _____ . The total linear momentum of the system at time $t + \Delta t$, in terms of M, ΔM, $\vec{v}$, $\Delta\vec{v}$, and $\vec{U}$, is given by $\vec{P}_f =$ _____ . .

Use the expressions you gave above to write $(\vec{P}_f - \vec{P}_i)/\Delta t$ in terms of M, $\vec{v}$, $\vec{U}$, $\Delta\vec{v}$, and ΔM, then evaluate the expression in the limit as $\Delta t \to 0$. The term $\Delta M\Delta\vec{v}/\Delta t$ vanishes in this limit since it is proportional to an infinitesimal. Because the net external force on the rocket-fuel system is zero, the change in the total linear momentum of the system is zero. In the space below carry out the steps to show that $M d\vec{v}/dt - (\vec{U} - \vec{v})dM/dt = 0$:

Now, $\vec{U} - \vec{v}$ is the velocity of the fuel relative to the rocket. This is usually constant during firing so we write $\vec{U} - \vec{v} = \vec{v}_{\text{rel}}$. The text also makes the substitution $R = -dM/dt$ for the rate at which mass is ejected. This is a positive quantity. Then, the acceleration of the rocket is given by

$$M\vec{a} = -R\vec{v}_{\text{rel}}.$$

The term on the right is called the <u>thrust</u> of the rocket. Notice that if there are no external forces acting, the acceleration is opposite in direction to the relative velocity of the fuel. To accelerate forward, a rocket ejects fuel toward the rear.

External forces and changes in internal energy. The total kinetic energy of a system of particles can be divided into two parts. One is associated with motion of the center of mass and is called the <u>kinetic energy of the center of mass</u>. The other is associated with the motions of the various particles relative to the center of mass. It, along with the potential energy associated with mutual interactions of particles in the system, make up the <u>internal energy</u>.

For a system of particles with total mass M and center of mass velocity $\vec{v}_{com}$, the kinetic energy associated with the motion of the center of mass of the system is given by $K_{com} =$ _____. Changes in this kinetic energy are related to the net external force acting on the system and the displacement of the center of mass. For purposes of calculating ΔK_{com} the system is replaced by a single (fictitious) particle of mass M, located at the position of the center of mass and acted on by the net external force $\vec{F}_{net}$. The change in the kinetic energy of the center of mass then equals the work W_{com} done by this force on this particle.

Suppose the center of mass of a system moves from x_i to x_f along the x axis, during which motion the net external force is F_{net}, a constant. The work done by this force on the fictitious particle that replaces the system is given by $W_{com} =$ _____ and the change in the kinetic energy of the center of mass of the system is given by $\Delta K_{com} =$ _____.

Be sure you realize that W_{com} is not necessarily the actual work done by the net external force. In some cases the actual work done by $\vec{F}_{net}$ may change the internal energy of the system as well as change the kinetic energy of the center of mass. In other cases $\vec{F}_{net}$ may not actually do any work but W_{com} is not zero and the kinetic energy of the center of mass changes. This occurs, for example when the point of application of an external force does not move but the center of mass does. Then, the actual work that changes the kinetic energy is done by internal forces. If you are asked for a change in energy, use the energy equation $\Delta K_{com} + \Delta U + \Delta E_{int} = W$, where W is the actual work done by external forces. If you are asked about a change in the kinetic energy of the center of mass, use $W_{com} = \Delta K_{com}$.

Be sure you understand the motion of a skater pushing off from a railing, as discussed in the text. Assume the skater's hands do not leave the railing. When filling in the blanks below, consider what happens from the point of view of an observer at rest with respect to the railing.

The force that accelerates the skater is: _____

Does this force do any work? _____

The forces that do work are: _____

Does the kinetic energy of the center of mass of the skater change? _____

II. PROBLEM SOLVING

When asked to calculate the coordinates of the center of mass, first decide if the system consists of discrete particles, with the masses and coordinates of each given. If it does, the calculation proceeds by straightforward evaluation of $\vec{r}_{com} = (1/M) \sum m_i \vec{r}_i$.

If, on the other hand, the system can be considered to be a continuous distribution of mass, next decide if the mass is uniformly distributed. If it is, symmetry considerations are often sufficient to locate the center of mass. As a last resort you may need to evaluate the integral for the center of mass coordinates.

Perhaps you can partition the object into several smaller objects, each with an easily identifiable center of mass. Replace each of these with a single particle, with mass equal to the mass of the object it replaces and located at the center of mass of the object. Then, find the center of mass of the particles.

Some problems deal with two-body systems for which the center of mass is initially at rest and the net external force vanishes. One of the bodies changes position and the problem asks for the change in position of the other. Since the net external force is zero the velocity of the center of mass is a constant and since it is initially zero it remains zero. The center of mass does not move. Write an expression for the center of mass in terms of the initial coordinates of the bodies, write a second expression in terms of the final coordinates, and equate the two expressions. Then, solve for the final coordinates of the second body. In a variant of this type problem the two objects end up at nearly the same place. This place must be at their center of mass.

Some problems deal with the definition of linear momentum and the equation of motion $\vec{F}_{net} = d\vec{p}/dt$ for a particle. Solve this as you would any second-law problem.

Many problems of this chapter can be worked using the principle of linear momentum conservation. If you suspect the principle can be used, first check for external forces: if there are none or if they add to zero, total linear momentum is conserved. For some problems only one component of linear momentum is conserved. If the problem asks about motion in that direction, then the principle can be used. Write the momentum at the beginning of some interval in terms of the velocities and masses of the particles involved. Do the same for the momentum at the end of the interval. Equate the two expressions and solve for the unknown quantities.

Some problems deal with the energy of a system of particles. Most of these involve the relationship between the net external force and the change in the kinetic energy of the center of mass. Use $\int F_{net} \, dx = K_{com\,f} - K_{com\,i}$, in one dimension.

III. NOTES

Chapter 10
COLLISIONS

I. BASIC CONCEPTS

Here you will apply the principle of linear momentum conservation to collisions between objects. You should pay special attention to the role played by the impulses the objects exert on each other. You should also take notice of when energy conservation can be invoked to solve a problem and when it cannot.

This chapter makes use of the ideas of velocity (Chapter 4), force (Chapter 5), Newton's second and third laws of motion (Chapter 5), kinetic energy (Chapter 7), linear momentum (Chapter 9), and center of mass (Chapter 9).

Definition. As succinctly as you can, tell in words what a collision is. Be sure to include the important characteristics. _____

_____

_____

Impulse. During a collision two objects exert forces on each other. What is important is not the force alone or its duration alone but a combination called the <u>impulse</u> of the force. If one body acts on the other with a force $\vec{F}(t)$ for a time interval from t_i to t_f, the impulse of the force is given by the integral

$$\vec{J} = $$

Don't confuse impulse and work. Impulse is an integral of force with respect to _____ and work is an integral of force with respect to _____. Impulse is a vector, work is a scalar. The SI unit of impulse is _____; the SI unit of work is _____.

In terms of the average force $\vec{F}_{avg}$ that acts during the collision, the impulse is given by $\vec{J} = $ _____, where Δt is _____. This expression is often useful for calculating the average force, which in turn is useful as an estimate of the strength of the interaction.

Because Newton's second law is valid, the total impulse acting on an object gives the change in the _____ of the object. Because Newton's third law is valid the impulse of one object on the other is the negative of the impulse of the second object on the first. If external impulses can be neglected, the total _____ of the two objects is conserved during a collision.

For most collisions the impulse of one colliding object on the other is usually much greater than any external impulse and we may neglect impulses exerted by the environment of the colliding bodies. We may, for example, neglect the effects of gravity and air resistance during the time a baseball is in contact with a bat.

Collisions. Total linear momentum is conserved during the collisions we consider. Since linear momentum is conserved during a collision, the velocity of the center of mass of the colliding bodies is _____ .

Kinetic energy may or may not be conserved in a collision. If it is, the collision is said to be _____ . If it is not, the collision is said to be _____ . The distinguishing characteristic of a two-body <u>completely</u> <u>inelastic</u> collision is _____
_____ .

During a completely inelastic collision the loss in total kinetic energy is as large as conservation of linear momentum allows. That is, it is impossible for the bodies to lose a larger fraction of their original kinetic energy and together still retain all the original total linear momentum.

Explosions, in which an object splits into two or more parts as a result of internal forces, can be handled in exactly the same manner as an inelastic collision. During an explosion the total linear momentum is conserved but the total kinetic energy increases.

Series of collisions. Here you consider a stream of moving objects that collide, one after the other, with another object. Examples are a rapid succession of bullets hitting a target or the molecules of a gas hitting the walls of a container. Suppose each of the "bullets" has mass m and each suffers the same change in velocity $\Delta \vec{v}$. If n "bullets" hit in time Δt, then during that interval the change in linear momentum of the "target" is given by

$$\Delta \vec{P} =$$

and the average force exerted by the "bullets" on the "target" is given by

$$\vec{F}_{avg} =$$

This is an average over a time interval that includes many strikes. It is not the average force of any bullet during its contact with the object.

Completely inelastic one-dimensional collisions. Consider a completely inelastic two-body collision in one dimension. Both objects move along the x axis. Object 1 has mass m_1 and moves with velocity v_{1i} before the collision. Object 2 has mass m_2 and moves with velocity v_{2i} before the collision. Both objects move with velocity v_f after the collision. The equation that expresses conservation of linear momentum during the collision is

If the masses and initial velocities are known, conservation of linear momentum is sufficient to determine the common final velocity. This velocity is given by

$$v_f =$$

An example of this type collision is _____
_____ .

If object 2 is initially at rest, then

$$v_f =$$

Notice that the final speed of the combination *must* be less than the initial speed of object 1. Explain why in words: _____

_____

Kinetic energy is lost during a completely inelastic collision, transformed to internal energy, radiation, etc. Consider a completely inelastic collision with object 2 initially at rest. In terms of the masses and velocities, the initial total kinetic energy is $K_i = $ _____ and the final total kinetic energy is $K_f = $ _____ . In the space below make the substitutions to express the kinetic energies in terms of the velocities, carry out the algebra, and obtain the result given below for the fractional energy loss of the two-body system:

$$[\text{ans: } 1 - [(m_1 + m_2)/m_1](v_f^2/v_{1i}^2)]$$

Use conservation of linear momentum to substitute for v_f^2 in terms of v_{1i}^2 and find an expression for the fractional energy loss in terms of the masses alone:

$$[\text{ans: } m_2/(m_1 + m_2)]$$

To make the fractional energy loss large, the mass of object _____ should be made much larger than the mass of the other object. To make the loss small, the mass of _____ should be made much larger than the mass of the other object.

Elastic one-dimensional collisions.   Consider an elastic two-body collision in one dimension and suppose both objects move along the x axis. Object 1 has mass m_1, moves with velocity v_{1i} before the collision, and moves with velocity v_{1f} after the collision. Object 2 has mass m_2, moves with velocity v_{2i} before the collision, and moves with velocity v_{2f} after the collision. The equation that expresses conservation of linear momentum during the collision is

The equation that expresses conservation of kinetic energy during the collision is

Using the equations of linear momentum and energy conservation you should be able to show that the relative velocity with which the objects approach each other before the collision is the same as the relative velocity with which they separate after the collision. In symbols, _____. The proof follows immediately when you divide the kinetic energy conservation equation, in the form $m_1(v_{1f}^2 - v_{1i}^2) = -m_2(v_{2f}^2 - v_{2i}^2)$, by the linear momentum conservation equation, in the form $m_1(v_{1f} - v_{1i}) = -m_2(v_{2f} - v_{2i})$. You must recognize that $A^2 - B^2 = (A - B)(A + B)$ for any quantities A and B. Carry out the steps in the space below:

In the space below solve $m_1 v_{1i} + m_2 v_{2i} = m_1 v_{1f} + m_2 v_{2f}$ and $(v_{1f} - v_{2f}) = -(v_{1i} - v_{2i})$ for v_{1f} and v_{2f} in terms of the other quantities. Draw a box around your answers and check them with the text.

Using these general expressions, you should be able to obtain the results for some special elastic collisions. Write the results below without looking them up and explain what has happened during the collision. Unless specifically stated otherwise do not assume either object is initially at rest.

a. If the two masses are the same, then $v_{1f} =$ _____ and $v_{2f} =$ _____.
   In words, what has happened is _____
   _____

b. If the two masses are the same and object 2 is initially at rest, then $v_{1f} =$ _____ and
   $v_{2f} =$ _____.
   In words, what has happened is _____
   _____

c. If the target object, object 2 say, is very massive compared to the incident object, object 1, then $v_{1f} =$ _____ and $v_{2f} =$ _____.
   In words, what has happened is _____

d. If the target object, object 2, is very massive compared to the incident object, object 1, and is initially at rest, then $v_{1f} =$ _____ and $v_{2f} =$ _____ .
In words, what has happened is _____

e. If object 1 is very massive compared to object 2 and object 2 is initially at rest, then $v_{1f} =$ _____ and $v_{2f} =$ _____ .
In words, what has happened is _____

For all these collisions the total kinetic energy (the sum of the individual kinetic energies of the colliding objects) is the same after the collision as before. However, kinetic energy is nearly always transferred from one object to the other. Sometimes you will want to know how much. The fractional loss of kinetic energy by object 1, for example, is given by $(K_{1i} - K_{1f})/K_{1i}$ and, since $K = \frac{1}{2}mv^2$, this is $1 - v_{1f}^2/v_{1i}^2$.

Two-dimensional collisions. Consider the two-dimensional elastic collision diagramed below.

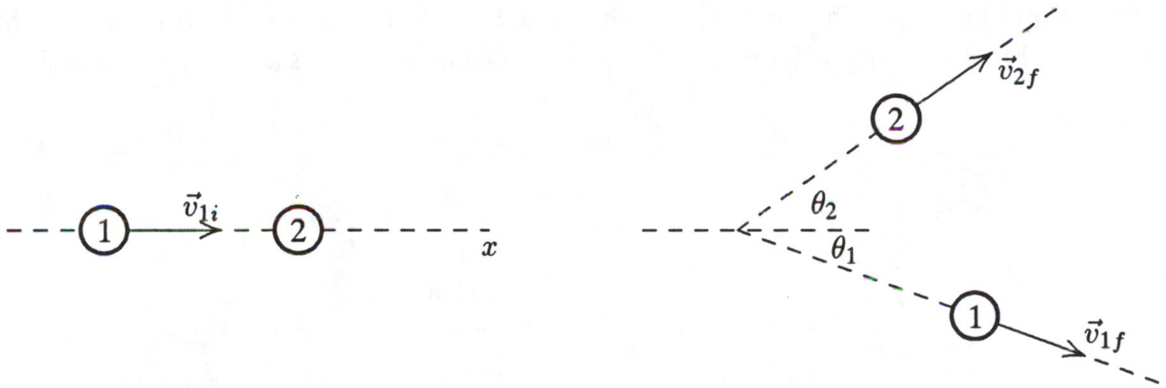

Before Collision   After Collision

Object 2 is initially at rest and object 1 impinges on it with speed v_{1i} along the x axis. Object 1 leaves the collision with speed v_{1f} along a line that is below the x axis and makes the angle θ_1 with that axis. Object 2 leaves the collision with speed v_{2f} along a line that is above the x axis and makes the angle θ_2 with that axis. Take the y axis to be upward in the diagram. In terms of the masses, speeds, and angles between the velocities and the x axis, conservation of total linear momentum $\vec{P}$ leads to the equations:

conservation of P_x:

conservation of P_y:

Suppose the collision is elastic and write the equation that expresses conservation of kinetic energy in terms of the masses and speeds:

conservation of K:

These equations can be solved for three unknowns. If the masses (or their ratio) are given, then, of the five other quantities (v_{1i}, v_{1f}, v_{2f}, θ_1, and θ_2) that enter the equation, two must be given.

Note that the masses and the initial velocity of object 1 alone do not determine the outcome of a two-dimensional elastic collision. The magnitude and direction of the impulse acting on each object during the collision also play important roles. Often these are not known, so the outcome cannot be predicted. Experimenters usually measure the speed or the orientation of the line of motion of one of the objects after the collision then use the equations to calculate other quantities. You should know how to solve for any three of the quantities, given the others. Some hints are given in the Problem Solving section below.

Here's an important result for two-dimensional collisions that you should remember: if a particle collides elastically with another particle of the same mass, initially at rest, then the two velocity vectors after the collision are perpendicular to each other. In this case, $\vec{v}_{1i} = \vec{v}_{1f} + \vec{v}_{2f}$ and $v_{1i}^2 = v_{1f}^2 + v_{2f}^2$. The three vectors form a triangle and the square of one side equals the sum of the squares of the other two sides. The triangle must be a right triangle with $\vec{v}_{1i}$ as the hypotenuse.

If a two-dimensional collision is completely inelastic, the objects stick together after the collision and kinetic energy is not conserved. We now consider collisions for which both objects are initially moving. The geometry is shown below. Before the collision mass m_1 has velocity $\vec{v}_{1i}$ and mass m_2 has velocity $\vec{v}_{2i}$. After the collision both masses have velocity $\vec{v}_f$.

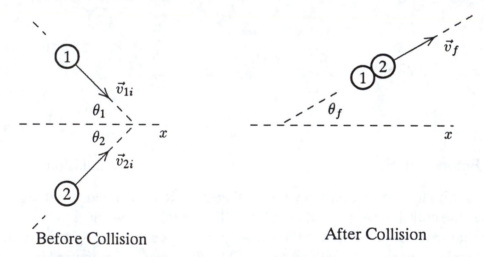

Before Collision   After Collision

Since $v_{1f} = v_{2f}$, we write v_f for both of speeds. Take the y axis to be upward in the diagram. The conservation of linear momentum equations become

conservation of P_x:

conservation of P_y:

These two equations contain eight quantities (m_1, m_2, v_{1i}, v_{2i}, θ_1, θ_2, v_f, and θ_f) and can be solved for 2 unknowns. If the masses (or their ratio) are given, then 4 additional quantities must be given. Unlike an elastic collision, the masses and initial conditions completely determine the outcome. That is, knowledge of m_1, m_2, v_{1i}, v_{2i}, θ_1, and θ_2 allow us to solve for

v_f and θ_f. Notice that the six given quantities allow us to find the velocity of the center of mass. Since linear momentum is conserved and the objects stick together this is the same as the final velocity of both objects.

II. PROBLEM SOLVING

Most calculations of impulse are rather straightforward. You might be given the force acting on an object as a function of time and asked to find the impulse over a given time interval. You simply evaluate the integral that defines impulse. Remember that impulse is a vector quantity and you must, for example, use the x component of the force to find the x component of the impulse. For one-dimensional motion the force might be given graphically as a function of time. You must then recognize that the impulse is the area under the curve. You might also be given sufficient information to calculate the initial and final linear momentum of a particle. Then, use $\vec{J} = \vec{p}_f - \vec{p}_i$ to calculate the impulse.

Collision problems are more complicated. First decide if the collision is one- or two-dimensional. A head-on collision is always one-dimensional. So is a completely inelastic two-body collision with one object initially at rest. If the objects move along different lines, whether initially or finally, the collision is two-dimensional.

First consider one-dimensional collisions. Since total linear momentum is always conserved in collisions, nearly every problem solution begins by writing the equation for conservation of linear momentum. There is only one such equation for a one-dimensional collision. To write it, use the component, not the magnitude, of the linear momentum of each object and write the component as the product of the appropriate mass and velocity component. Always use symbols, not numbers, even for given quantities.

Make a list of the quantities given in the problem statement and a list of the unknowns. If there is only one unknown, the linear momentum conservation equation can be solved immediately for it. Once the equation has been solved, the result can be used in other calculations. You might be asked to test for the conservation of kinetic energy, for example, and classify the collision as elastic or inelastic. You might also be asked to calculate the impulse exerted by one object on the other.

If more than one quantity is unknown, look for more information in the problem statement. If it specifies that the collision is elastic, write the equation for the conservation of kinetic energy in terms of the masses and speeds of the objects. If it specifies that the collision is completely inelastic, equate the final velocities of the objects to each other and use a single symbol to denote them.

For two-dimensional collisions there are two linear momentum conservation equations, one for each component. The amount of algebraic manipulation can usually be reduced significantly if one of the coordinate axes is placed along the direction of the total linear momentum. For some problems, however, this direction is not immediately evident. In any event, select a coordinate system and write the linear momentum conservation equations in terms of the masses, speeds, and angles between the velocities and a coordinate axis.

If the line of motion of one of the objects, either before or after the collision, makes the angle θ with a coordinate axis, then you will find $\sin\theta$ in one of the components of the linear momentum conservation equation and $\cos \theta$ in the other. To eliminate θ, use the trigonometric identity $\cos^2 \theta + \sin^2 \theta = 1$. That is, solve one equation for $\sin \theta$ and the other for $\cos \theta$, square both results, add, and set the sum equal to 1.

Sometimes the angle of the line of motion is sought. You then solve the linear momentum conservation equations for the sine or cosine of the angle and evaluate the inverse trigonometric function. Sometimes the equation for $\sin \theta$ or $\cos \theta$ contains an unknown and you must eliminate it before solving for the angle. An unknown can often be eliminated by dividing the expression for $\sin \theta$ by the expression for $\cos \theta$. The resulting equation for $\tan \theta$ is then solved.

Suppose the collision is elastic and you wish to eliminate one of the speeds. Solve one of the linear momentum conservation equations for the speed to be eliminated and substitute the resulting expression in the energy equation, rather than vice versa. If you solve the energy equation first, the resulting expression will contain a square root and probably is difficult to manipulate.

III. NOTES

Chapter 11
ROTATION

I. BASIC CONCEPTS

You now begin the study of rotational motion. The pattern is similar to that used for the study of translation: you first study kinematics, the description of the motion, and then the dynamics of the motion. Important new concepts in this chapter include angular displacement, angular velocity, angular acceleration, torque, and rotational inertia. Take special care to understand their definitions and the roles they play in rotational motion. Concentrate on Newton's second law for rotation. Also concentrate on rotational kinetic energy and the work-kinetic energy theorem for rotation.

Some important concepts from previous chapters that are used in this chapter are: displacement, position, velocity, speed, and acceleration from Chapter 2, Newton's second law of motion, force, and mass from Chapter 5, and work, kinetic energy, and the work-kinetic energy theorem from Chapter 7. You might also review uniform circular motion, discussed in Chapters 4 and 6.

Definitions. If a rigid body undergoes pure rotation about a fixed axis, then the path followed by each point on the body is a _____, centered on the _____.

The angular position of the body is described by giving the angle between a reference line, fixed to the body, and a non-rotating coordinate axis. Suppose the body rotates around the z axis and describe how a reference line might be chosen. In particular, identify the plane it is parallel to and the angle it makes with the axis of rotation. _____

_____

Illustrate the idea in the space to the right by drawing the outline of a rotating body, with the axis of rotation, the z axis, out of the page. Mark the rotation axis with a circled dot ($\odot$). Show x and y axes and a reference line, all in the plane of the page. Usually the angle θ that gives the angular position is measured counterclockwise from a coordinate axis. Show θ on your diagram.

The angular position θ is often measured in radians. See the Mathematical Skills section of Chapter 3 of this *Student's Companion* for a discussion of radian measure.

If the body rotates from θ_1 to θ_2, then its angular displacement is $\Delta\theta = $ _____. If θ_2 is greater than θ_1, then the angular displacement is positive. According to the convention stated above for measuring θ, the body rotated in the _____ direction. If θ_2 is less than θ_1, then the angular displacement is negative and the body rotated in the _____ direction.

If the body rotates through more than 1 revolution, θ continues to increase beyond 2π rad. Be sure you understand that an angular displacement of 2π rad is NOT equivalent to 0. A rotation from $\theta = 0$ to $\theta = 2\pi$ rad represents an angular displacement of 2π rad, not 0. An angular displacement of 2 revolutions represents an angular displacement of _____ rad.

If the body undergoes an angular displacement $\Delta\theta$ in time Δt, its <u>average angular velocity</u> during the interval is $\omega_{avg} = $ _____ . Its <u>instantaneous angular velocity</u> at any time is given by the derivative $\omega = $ _____ . According to the convention used above to measure the angular position, a positive value for ω indicates rotation in the _____ direction and a negative value for ω indicates rotation in the _____ direction. Instantaneous angular velocity is usually called simply angular velocity.

<u>Angular speed</u> is the magnitude of the angular velocity and is also denoted by ω.

If the angular velocity of the body changes with time, then the body has a non-vanishing angular acceleration. If the angular velocity changes from ω_1 to ω_2 in the time interval Δt, then the <u>average angular acceleration</u> in the interval is $\alpha_{avg} = $ _____ . The <u>instantaneous angular acceleration</u> at any time t is given by the derivative $\alpha = $ _____ . Instantaneous angular acceleration is usually called simply angular acceleration.

Common units of angular velocity are deg/s, rad/s, rev/s, and rev/min. Corresponding units of angular acceleration are _____ , _____ , _____ , and _____ .

You should understand that a positive angular acceleration does NOT necessarily mean the angular speed is increasing and a negative angular acceleration does NOT necessarily mean the angular speed is decreasing. Remember that the angular speed decreases if ω and α have opposite signs and increases if they have the same sign, no matter what the sign of α. If ω is negative and α is positive, for example, then ω becomes less negative as time goes on and its magnitude becomes smaller. If the positive acceleration continues beyond $\omega = 0$, of course, the magnitude of ω starts increasing with time.

The values of $\Delta\theta$, ω, and α are the same for every point in a rigid body. The angular positions of different points may be different, of course, but when one point rotates through any angle, all points rotate through the same angle. All points rotate through the same angle in the same time and their angular velocities change at the same rate.

Angular velocity and angular acceleration, but NOT angular position or displacement, can be considered to be vectors along the axis of rotation. To determine the direction of $\vec{\omega}$, use the right hand rule: _____

The diagram on the right shows a rigid rod rotating around the z axis, with points to the right of the axis moving into the page and points to the left of the axis moving out of the page. Show the angular velocity vector $\vec{\omega}$ on the diagram. If the body were rotating in the opposite direction, the vector angular velocity would be in the _____ direction.

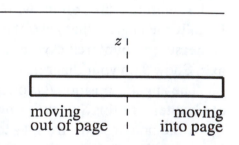

If the body rotates faster as time goes on, then $\vec{\alpha}$ and $\vec{\omega}$ are in the same direction. If it rotates more slowly, then $\vec{\alpha}$ and $\vec{\omega}$ are in opposite directions. Consider the situation illustrated above (right side moving into page, left side moving out of page) and suppose the body rotates faster as time goes on. Then, the angular acceleration vector $\vec{\alpha}$ is in the _____ direction. If the body rotates more slowly as time goes on, $\vec{\alpha}$ would be in the _____ direction.

Although directions can be assigned arbitrarily to angular displacements, they do not add as vectors. The result of two successive angular displacements around different axes that are not parallel, for example, depends on the order in which the angular displacements are carried out. This means angular displacements are not vectors.

Rotation with constant angular acceleration. If a body is rotating around a fixed axis with constant angular acceleration α, then as functions of time its angular velocity is given by

$$\omega(t) =$$

and its angular position is given by

$$\theta(t) =$$

Here ω_0 is the angular velocity at time $t =$ _____ and θ_0 is the angular position at time $t =$ _____.   The reference line can always be selected so $\theta_0 = 0$. We suppose that has been done.

These two equations are solved simultaneously to find answers to constant angular acceleration problems. You should memorize them or, better yet, learn to derive them quickly using techniques of the integral calculus. Another useful equation can be obtained by using $\omega = \omega_0 + \alpha t$ to eliminate t from $\theta = \omega_0 t + \frac{1}{2}\alpha t^2$. It is

$$\omega^2 - \omega_0^2 = 2\alpha\theta,$$

where θ_0 was taken to be 0.

Table 11–1 of the text lists all possible results when one of the equations given above is used to eliminate a variable from the other. You may use these equations to solve rotational problems or else simultaneously solve the two you wrote above.

Any consistent set of units can be used. Thus, θ in degrees, ω in degrees/s, and α in degrees/s^2 as well as θ in radians, ω in radians/s, and α in radians/s^2 or θ in revolutions, ω in revolutions/s, and α in revolutions/s^2 can be used. Be careful not to mix units, however.

Relationships between linear and angular quantities. Each point in a rotating body has a velocity and acceleration and these are related to the angular variables. Suppose the body is rotating around a fixed axis and consider a point in the body a perpendicular distance r from the rotation axis. If the body turns through the angle $\Delta\theta$ (in radians), the point moves a distance $\Delta s =$ _____ along its circular path. The speed of the point is $v = ds/dt$ and the angular velocity of the body is $\omega = d\theta/dt$, so v is related to ω by $v =$ _____, where the units of ω MUST be _____. The tangential component of the acceleration of the point is $a_t = dv/dt$ and the angular acceleration is $\alpha = d\omega/dt$ so a_t is related to α by $a_t =$ _____, where the units of α MUST be _____.

Because the point is moving in a circular path its acceleration also has a radial component. You learned in Chapter 4 that $a_r = v^2/r$. Now, write the relationship in terms of the angular speed instead of the speed: $a_r =$ _____. In terms of the components a_r and a_t, the magnitude of the total acceleration is given by $a =$ _____ and, in terms of ω, α, and r, is given by $a =$ _____. Recall that an angle in radians is really unitless and check the units on the two sides of each equation you just wrote to be sure they are consistent.

For a point in a body rotating with constant angular acceleration, the tangential component of the acceleration is also constant. If s represents the distance traveled by the point along its circular path from time 0 to time t, then $s(t) = v_0 t + \frac{1}{2} a_t t^2$. The speed v of the point is given by $v(t) = v_0 + a_t t$. Divide these equations by the radius r of the orbit to obtain the equations for $\theta(t)$ and $\omega(t)$.

Be sure you understand the relationships between the angular and linear quantities. Every point on a rotating body turns through the same angle in the same time, has the same angular velocity, and has the same angular acceleration. This means that compared to a point on the rim of a rotating wheel, for example, a point halfway out travels _____ the distance, has _____ the speed, and has _____ the tangential acceleration.

Rotational inertia.   Rotational inertia is a property of a rotating body that depends on the distribution of mass in the body and on the axis of rotation. It determines the angular acceleration and change in kinetic energy produced by the net torque acting. That is, the same torque acting on different bodies produces different angular accelerations and different changes in kinetic energy because the bodies have different rotational inertias.

The rotational inertia I of a rigid body consisting of N particles, rotating about a given fixed axis, is defined by the sum over all particles:

$$I =$$

where m_i is the mass of particle i and r_i is the distance of particle i from the axis of rotation. This distance is measured along a line that is perpendicular to the axis of rotation, from the rotation axis to the particle.

The sum is difficult to carry out for a body with more than a few particles. Some bodies, however, can be approximated by a continuous distribution of mass and techniques of the integral calculus can be used to find the rotational inertia. For a body with a continuous distribution of mass the rotational inertia is given by the integral

$$I =$$

Table 11–2 of the text gives the rotational inertias of several bodies, some for two different axes. You will need to refer to this figure as you work through problems.

The equation that defines the rotational inertia as a sum over the particles of a body is very useful when we need to think about the rotational inertia of an object. For example, the sum shows us that the rotational inertia depends on the square of the distance of each particle from the rotation axis as well as on the mass of each particle. A particle that is far from the rotation axis contributes more to the rotational inertia than a particle of equal mass close to the rotation axis.

Look at the expressions given in Table 11–2 for the rotational inertias of a hoop about the cylinder axis and for a solid cylinder, also about the cylinder axis. The rotational inertia of a solid cylinder is less than the rotational inertia of a hoop with the same mass and radius because _____

_____.

The defining equation also shows that the rotational inertia depends on the orientation and position of the axis of rotation. For example, the rotational inertia of a thin rod rotating about an axis that is perpendicular to the rod and through its center is less than the rotational inertia of the same rod but rotating about a perpendicular axis through one end. The rotational inertia is smaller when the rotation axis is through the center because _____

_____

_____.

The defining equation for the rotational inertia is a sum over all particles in the body. If we like, we can consider the body to be composed of two or more parts and calculate the rotational inertia of each part about the same rotation axis, then add the results to obtain the rotational inertia of the complete body. As an example, suppose two identical solid uniform balls, each of mass m and radius r, are attached to each other at a point on their surfaces and the composite body rotates around the line that joins their centers. Both balls rotate about a diameter so each contributes $2mr^2/5$ to the rotational inertia of the complete body. Thus, for the composite, $I = 4mr^2/5 = 2Mr^2/5$, where $M = 2m$ is the total mass.

The defining equation for the rotational inertia is used to prove the parallel-axis theorem. Here we consider two identical bodies. One is rotating about an axis through the center of mass and the other is rotating about an axis that is parallel to the first axis but is displaced from the center of mass by a distance h (measured along a line that is perpendicular to the axis). The rotational inertia I for the second body is related to the rotational inertia I_{com} for the first by $I =$ _____, where M is the total mass of the body.

Consider a uniform sphere of radius R and mass M, rotating about an axis that is tangent to its surface. The distance between this axis and a parallel axis through the center is _____. Look up the expression for the rotational inertia of a uniform solid sphere about an axis through its center and apply the parallel-axis theorem. The rotational inertia about the axis tangent to the surface is given by $I =$ _____ + _____ = _____. You should get $I = 7MR^2/5$.

The parallel-axis theorem tells us that of all possible parallel axes the one that leads to the smallest rotational inertia is the one through the center of mass and that the rotational inertia increases as the rotation axis moves away from the center of mass (remaining parallel to its original orientation, of course).

Rotational kinetic energy. Suppose a rigid body, rotating about a fixed axis, is made up of N particles. The kinetic energy of the particles is given by $K = \sum \frac{1}{2}m_i v_i^2$, where m_i is the mass of particle i and v_i is its speed. Substitute $v_i = r_i \omega$, where r_i is the distance of particle i from the rotation axis, and obtain $K =$ _____. Notice that the sum over particles for the rotational inertia appears in the expression for K. For a rigid body with rotational inertia I, rotating with angular velocity ω about a fixed axis, the total kinetic energy of all the particles in the body is given by $K =$ _____. For this expression to be valid ω must be in radians per second.

A particle far from the axis of rotation contributes more to the rotational kinetic energy than a particle of the same mass closer to the rotation axis. Use $v = r\omega$ to explain: _____

Torque. The net torque acting on a rotating body gives the body an angular acceleration. It does work and changes the rotational kinetic energy of the body.

If a force is applied to a body free to rotate about a fixed axis, the torque (about the axis) associated with the force is intimately related to the force component tangent to the circle through the point of application and centered at the axis. In fact, $\tau =$ _____, where r is the radius of the circle (the distance from the origin to the point of application) and F_t is the tangential component of the force.

The diagram on the right shows a particle traveling counterclockwise around a circle of radius r. A force $\vec{F}$ acting on it at some instant of time is also shown. On the diagram show the tangential and radial components of the force. The radial component holds it in the circle. The tangential component produces an angular acceleration. In terms of the magnitude F of the force and the angle ϕ shown, the radial component is $F_r =$ _____ and the tangential component is $F_t =$ _____. The magnitude of the torque is _____.

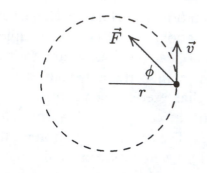

The diagram on the right shows a body subjected to three forces. The axis of rotation is through O. The forces and their points of application are all in the plane of the page. For each force draw the line from the rotation axis to the point of application and label it r_1, r_2, or r_3, as appropriate. For each, label the angle ϕ that appears in $\tau = Fr \sin \phi$.

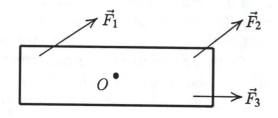

Notice that the magnitude of a torque depends on the distance from the rotation axis to the point of application. A tangential force produces a greater angular acceleration if it is applied far from the rotation axis than if it applied closer.

For rotation about a fixed axis we may take a torque to be positive if it tends to turn the object counterclockwise and negative if it tends to turn the body clockwise. This convention is consistent with the one introduced in the last chapter for the signs of the angular velocity and acceleration. For the situation illustrated in the diagram above, the sign of τ_1 is _____, the sign of τ_2 is _____, and the sign of τ_3 is _____.

Newton's second law for rotation about a fixed axis. This law relates the net torque τ_{net} acting on a rigid body to the angular acceleration α of the body. Quantitatively the law is

$$\tau_{net} =$$

where I is the rotational inertia of the body about the axis of rotation. It is as important to the study of rotational motion about a fixed axis as $F_{net} = ma$ is to the study of one-dimensional translational motion. Concentrate on learning to use it.

The law is easy to derive for a particle moving in a circle and subjected to a force $\vec{F}$. The tangential component of the force produces a tangential acceleration a_t according to $F_t =$ _____ , where m is the mass of the particle. Since the tangential and angular accelerations are related by $a_t = r\alpha$, you can write $F_t =$ _____ . Now, multiply both sides by r to obtain $\tau =$ _____ and notice that the rotational inertia mr^2 appears as a factor. For a body composed of many particles, acted on by more than one torque, the derivation is slightly more complicated but the result is the same: $\tau_{net} = I\alpha$.

Be careful that you use the net torque in this equation. You must identify and sum all the torques that are acting and you must be careful about signs. You must also be careful to use radian measure for the angular acceleration α.

Work done by a torque.   Consider a particle traveling counterclockwise in a circular orbit, subjected to a force with tangential component F_t. As it moves through an infinitesimal arc length ds, the magnitude of the work done by the force is d$W =$ _____ . Since d$s = r\,d\theta$, where dθ is the infinitesimal angular displacement, and $\tau = F_t r$, the expression for the work becomes d$W =$ _____ . As the particle travels from θ_1 to θ_2 the work done by the torque is given by the integral

$$W =$$

In terms of the torque τ and angular velocity ω the power supplied by the torque is given by $P = dW/dt =$ _____ .

The work done by a torque may be positive or negative. If the torque and angular displacement are in the same direction, both clockwise or both counterclockwise, then the work is _____ ; if they are in opposite directions, one clockwise and one counterclockwise, then the work is _____ .

On the left-hand diagram below, show a force that does positive work on the particle and give the signs of the torque and the angular velocity, using the convention described above (positive for counterclockwise, negative for clockwise). On the right-hand diagram, do the same for a force that does negative work.

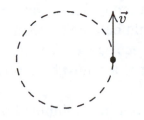

Force does positive work
Sign of torque: _____
Sign of angular velocity: _____

Force does negative work
Sign of torque: _____
Sign of angular velocity: _____

The same results hold for an extended body. If more than one torque acts, sum the works done by the individual torques to find the net work. This is also the work done by the net torque.

A work-kinetic energy theorem is valid for rotational motion: the net work done by all torques acting on a body equals the change in its rotational kinetic energy. Suppose the net torque τ_{net} is constant. Further suppose that in a certain time interval the body turns through the angular displacement $\Delta\theta$ and its angular velocity changes from ω_0 to ω. In terms of these quantities the work-kinetic energy theorem becomes: _____ = _____, where I is the rotational inertia of the body.

The rotational kinetic energy of a rotating body must be included in the mechanical energy of a system that contains the body. For example, suppose a string is wrapped around a drum that is free to rotate on a horizontal axis and the other end is attached to a crate, say. The string unwraps from the drum as the crate falls. The mechanical energy of the system consisting of Earth, drum, and crate is $E_{mech} = \frac{1}{2}mv^2 + \frac{1}{2}I\omega^2 + mgy$, where v is the speed of the falling crate, I is the rotational inertial of the drum, and y is the vertical coordinate of the crate. If there is no friction this quantity is conserved. If the string does not slip on the drum, then $v = R\omega$, where R is radius of the drum. You can use these considerations to find how the speed of the crate and the angular speed of the drum vary as the crate falls.

II. PROBLEM SOLVING

Many problems of this chapter are mathematically identical to those of one-dimensional linear kinematics, discussed in Chapter 2: θ replaces x, ω replaces v, and α replaces a.

If a problem involves rotation with constant angular acceleration, first write the kinematic equations $\theta(t) = \theta_0 + \omega_0 t + \frac{1}{2}\alpha t^2$ and $\omega(t) = \omega_0 + \alpha t$, then solve them simultaneously for the unknowns. As an alternative, search Table 11–1 for the equation containing the quantities that are given in the problem statement.

Recall that when you solve one-dimensional problems you can always place the origin at any point you choose. In particular you might select the origin to be at the position of the particle when time $t = 0$ and thus take x_0 to be zero. Similarly, to describe rotational motion you can always orient the fixed reference frame and select the reference line in the body so the initial angular position θ_0 is zero. This is usually helpful because it reduces the number of algebraic symbols you must carry in the calculation.

Some problems deal with relationships between linear and angular variables. Given the radius of the orbit of a point in a rotating body you should be able to obtain values for the linear variables from values for the angular variables and vice versa. Use $\Delta s = r\Delta\theta$, $v = r\omega$, $a_t = r\alpha$, and $a_r = r\omega^2$. Don't forget that the angular variables must be expressed in radian measure. This may require a change in units.

If you are asked to calculate the acceleration of a point in a spinning body, remember it has both tangential and radial components. You need to know ω to compute the radial component ($a_r = r\omega^2$) and α to compute the tangential component ($a_t = r\alpha$). These might be given or might be solutions to another problem. If you are asked for the magnitude of the acceleration, use $a = \sqrt{a_r^2 + a_t^2}$.

You should know how to compute the rotational inertia of a system composed of a small number of discrete particles. Given the masses and positions of the particles you must evaluate the sum $I = \sum m_i r_i^2$, where r_i is the perpendicular distance of particle i from the axis of rotation. You may need to use some geometry to obtain r_i. If, for example, the z coordinate axis is the axis of rotation, then $r_i^2 = x_i^2 + y_i^2$. For many problems you will want to make use of Table 11–2. For some you will need to use the parallel-axis theorem.

Some problems ask you to compute the torque associated with a given force. Simply find the force component tangent to the circular orbit at the application point, then multiply by the perpendicular distance from the rotation axis to the application point. For fixed axis rotation you may treat the torque as a scalar. Assign a sign to it according to whether it tends to cause rotation in the counterclockwise $(+)$ or clockwise $(-)$ direction when it is applied to an object at rest.

Many problems deal with the rotational motion of a rigid body about a fixed axis and can be solved using $\tau_{\text{net}} = I\alpha$. Draw a force diagram. This is a sketch of the rotating object with the rotation axis perpendicular to the page. Draw arrows for all forces acting on the object, with their tails at the points of application. Label the arrows with symbols for the force magnitudes. Also label the distances from the rotation axis to the points of application. Write the sum of the torques in terms of the force magnitudes and the distances to the rotation axis. Be sure to include the signs of the individual torques (positive for counterclockwise and negative for clockwise, for example). Set the sum of torques equal to $I\alpha$, identify the known quantities and solve for the unknown quantities.

Some problems involve several bodies, some in translation and some in rotation. The accelerations of the translating bodies and the angular accelerations of rotating bodies are usually related. Perhaps a string runs from a translating body and around a rotating body. Then, if the string does not slip or stretch, the magnitude of the tangential acceleration of a point on the rim of the rotating body must be the same as the magnitude of the acceleration of the translating body. You write $a = \pm\alpha r$. Which sign you use depends on which directions you chose for positive acceleration and positive angular acceleration. If you use the $+$ sign, you are saying that a positive acceleration of the translating body is consistent with a positive angular acceleration of the rotating body. If you use the $-$ sign, you are saying a positive acceleration of the translating body is consistent with a negative angular acceleration of the rotating body.

To solve this type problem, draw a force diagram for each body. For a rotating body, choose the direction of positive rotation (the counterclockwise direction, say) and write the sum of the torques, taking their directions into account. Set the sum equal to $I\alpha$. For a translating body, choose a coordinate system with one axis in the direction of the acceleration, if possible. Write the sum of the force components for each coordinate direction and set each sum equal to the product of the mass of the body and the appropriate acceleration component. Write all equations in terms of the magnitudes of the forces and appropriate angles. Use algebraic symbols, not numbers.

Don't forget tension forces of strings, gravitational forces, normal forces, and frictional forces, if they act. If a string is wrapped around a disk that is free to rotate and if the string does not slip on the disk, then the force exerted by the string on the disk is the tension force

T. Since the string must be tangent to the disk, the torque it exerts has magnitude TR, where R is the radius of the disk. If a string passes over a pulley, then the tension force of the string might be different on different sides. If T_1 is the tension force on one side and T_2 is the tension force on the other side, then $\pm(T_1 - T_2)R$ is the net torque exerted by the string on the pulley. The sign, of course, depends on which direction of rotation is taken to be positive.

The second-law equations (for translation and rotation) and the equation or equations that link the accelerations of translating bodies and the angular accelerations of rotating bodies are solved simultaneously for the unknowns.

Some problems can be solved using the work-kinetic energy theorem for rotation or the principle of energy conservation. Consider using an energy technique if angular displacements and angular velocities are given or requested. On the other hand, use Newton's second law for rotation if angular accelerations and torques or forces are given or requested. If you use an energy method, write the initial and final kinetic energies for some time interval in terms of given or requested quantities. These are usually the rotational inertia and angular velocities. If the system also includes a translating object, such as a falling body that exerts a torque (via a string) on a rotating object, you must include the kinetic energy of that body. Then write an expression for the total work done by external forces and set the expression equal to the change in the total kinetic energy. Identify the known quantities and solve for the unknown quantities.

III. NOTES

Chapter 12
ROLLING, TORQUE,
AND ANGULAR MOMENTUM

I. BASIC CONCEPTS

This chapter starts with a discussion of rolling objects; they simultaneously translate and rotate. Pay careful attention to the role played by friction between the object and the surface on which it rolls. Learn the relationship between the velocity of the center of mass and the angular velocity when an object rolls without sliding. Next you study rotational motion using a new concept, that of angular momentum. Learn the definition for a single particle and learn how to calculate the total angular momentum of a collection of particles. Pay particular attention to the result for a rigid body rotating about a fixed axis. You will also study the change in the angular momentum of a system brought about by the total external torque applied to the system. You will need to know how to find the direction of a vector torque. Angular momentum is at the heart of one of the great conservation laws of mechanics. Pay attention to the conditions for which this law is valid.

The important concepts from other chapters that are used in this chapter are: velocity and acceleration from Chapter 4, force, mass, Newton's second and third laws of motion, weight, and tension force of a string from Chapter 5, friction from Chapter 6, kinetic energy from Chapter 7, center of mass and linear momentum from chapter 9, angular velocity, angular acceleration, rotational inertia, rotational kinetic energy, torque, and Newton's second law for rotation from Chapter 11. Be sure you understand these concepts.

Combined rotational and translational motion. In this section you study an object, such as a rolling wheel, which rotates as its center of mass moves along a line. The center of mass obeys Newton's second law: _____, where $\vec{F}_{net}$ is the total force acting on the object, M is the total mass of the object, and $\vec{a}_{com}$ is the acceleration of the center of mass. Rotation about an axis through the center of mass is governed by Newton's second law for rotation: _____, where τ_{net} is the total torque acting on the object, I is the rotational inertia of the object, and α is the angular acceleration of the object.

You should understand that these two motions are related by the forces that act on the wheel. The total force accelerates the center of mass and the total torque (derived from the forces) produces an angular acceleration. If, in addition to the mass and rotational inertia of the body, you know all the forces acting and their points of application you can calculate the acceleration of the center of mass and the angular acceleration about the center of mass.

Rolling without sliding is a special case. We usually do not know the force of friction exerted by the surface on the body at the point of contact but we do know what that force does. If the object rolls without sliding, it causes the point on the wheel in contact with the surface to have a velocity of _____. This means that the speed v_{com} of the center of mass and the angular speed ω around the center of mass are related by $v_{com} = $ _____, where R is the radius of the wheel. Furthermore, since the wheel accelerates without sliding, the magnitude of the acceleration of the center of mass a_{com} is related to the magnitude of the angular acceleration around the center of mass by $a_{com} = $ _____.

You should understand the relationship between v_{com} and ωR for a rolling wheel. The velocity of the point in contact with the ground is given by the vector sum of the velocity due to the motion of the center of mass and the velocity due to rotation: $v = v_{com} - \omega R$, where the forward direction was taken to be positive. If the wheel slides, then $v \neq 0$ and v_{com} is different from ωR. If it does not slide, then $v = 0$ and $v_{com} = \omega R$.

For a wheel rolling without sliding along a horizontal surface there are three basic equations containing the forces, torques, and accelerations. They are:

_____ for translation of the center of mass

_____ for rotation around the center of mass

_____ for the condition of no sliding.

These can be solved, for example, for the linear and angular accelerations and for the force of friction if the other quantities are known.

To test if the wheel slides or not, you will also need the vertical component of Newton's second law for the center of mass. This is used to solve for the normal force of the surface on the wheel. Imagine that the wheel does not slide and calculate the force of friction required to prevent sliding. Compare the magnitude of this force with $\mu_s N$, where μ_s is the coefficient of static friction. If it is less, the wheel does not slide and the force of friction is as computed. If it is greater, the wheel does slide and the force of friction is $\mu_k N$, where μ_k is the coefficient of kinetic friction.

A rolling wheel, whether sliding or not, has both translational and rotational kinetic energy. If the center of mass has velocity v_{com} and the wheel is rotating with angular velocity ω about an axis through the center of mass, then the total kinetic energy is given by

$$K = $$

If the wheel is not sliding, then both terms of the kinetic energy can be written in terms of v_{com} or in terms of ω. Use $v_{com} = \omega R$ to eliminate ω in favor of v_{com} or use the same expression to eliminate v_{com} in favor of ω.

If an object is rolling without sliding, the point in contact with the surface is instantaneously at rest. As a result, the frictional force acting there does zero work.

An object that is rolling without sliding on a surface may be considered to be in pure rotation about an axis that is at each instant through the point of contact with the surface. Consider a wheel of radius R, rolling on a horizontal surface. If its center of mass is at its center and its rotational inertia about an axis through that point is I_{com}, then, according to the parallel axis theorem, its rotational inertia about the point of contact is given by

$$I = $$

The total kinetic energy of the wheel is now just its rotational kinetic energy $\frac{1}{2}I\omega^2$. Substitute for I in terms of I_{com} and obtain the result you wrote in the last paragraph, $K =$

_____.

Torque. When a force $\vec{F}$ is applied to an object at a point with position vector $\vec{r}$, relative to some origin, then the torque $\vec{\tau}$ associated with the force is given by the vector product

$$\vec{\tau} = $$

Review the parts of Chapter 3 that deal with vector products. The magnitude of the torque is given by $\tau =$ _____, where ϕ is the angle between _____ and _____ when they are drawn with their tails at the same point. The direction of the torque is given by the right-hand rule: _____

_____.

You should be able to make a connection with the definition used in the last chapter. The diagram shows a particle going around a circle of radius R in a plane perpendicular to the plane of the page. The horizontal dotted line is a diameter. Suppose a force, tangent to the circle and into the page, is applied to the particle. Take the origin to be at O on the axis and draw a vector to indicate the torque. Since $\vec{r}$ and $\vec{F}$ are perpendicular to each other, the magnitude of the torque is given by

$$\tau = $$

_____.

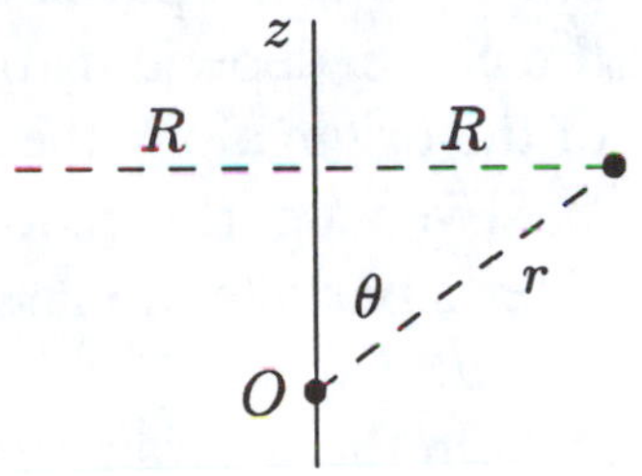

In terms of the angle θ shown the z component of τ is $\tau_z =$ _____. Notice that $r\sin\phi = R$, the radius of the circle. Thus, in terms of R, $\tau_z =$ _____. In the last chapter we were interested in motion about a fixed axis and we needed only the component of the torque along that axis. We used $\tau = F_t R$. Now you see this is only one component of the vector torque. If the origin is at the center of the circle, it is the only non-vanishing component.

Angular momentum. If a particle has linear momentum $\vec{p}$ and position vector $\vec{r}$ relative to some origin, then its <u>angular momentum</u> is defined by the vector product

$$\vec{\ell} = $$

Notice that the angular momentum depends on the choice of origin. In particular, if the origin is picked so $\vec{r}$ and $\vec{p}$ are parallel at some instant of time, then $\vec{\ell} =$ _____ at that instant.

You should be familiar with some frequently occurring special cases. If a particle of mass m is traveling around a circle of radius R with speed v and the origin is placed at the center of the circle, then the magnitude of its angular momentum is $\ell =$ _____. If the orbit is in the plane of the page and the particle is traveling in the counterclockwise direction, the direction of its angular momentum is _____. If it is traveling in the clockwise direction, the direction is _____.

For rotation about a fixed axis the component of the angular momentum along the axis of rotation is, in most cases, the component of greatest interest. Nevertheless you should be able to compute the angular momentum vector and its component along any axis.

Suppose a particle of mass m is traveling around a circle of radius R parallel to the xy plane and centered on the z axis above the origin. The diagram on the right shows the xz plane as the particle crosses the plane going into the page. Draw the position vector $\vec{r}$ and the angular momentum vector $\vec{\ell}$. In terms of m, v, and r (the distance from the origin), the magnitude of $\vec{\ell}$ is $\ell =$ _____.

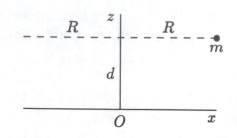

Let θ be the angle between $\vec{r}$ and the positive x axis. In terms of θ the z component of $\vec{\ell}$ is $\ell_z =$ _____ and the x component is $\ell_x =$ _____. In terms of m, R, and v the z component of $\vec{\ell}$ is given by $\ell_z =$ _____. In terms of m, d, and v the x component of $\vec{\ell}$ is given by $\ell_x =$ _____. The component of $\vec{\ell}$ in the plane of the orbit is radial and follows the particle around the circle, always pointing toward or away from the center of the circle, depending on the direction of travel. Notice that $\vec{\ell}$ and ℓ_x depend on the position of the origin along the z axis but ℓ_z does not.

Even a particle moving in a straight line may have a non-zero angular momentum. Consider a particle of mass m traveling with velocity $\vec{v}$ in the negative y direction along the line $x = d$.

On the diagram draw the vector $\vec{r}$ from the origin to the particle. In terms of m, v, and d, the magnitude of its angular momentum about the origin is $\ell =$ _____ and the direction of its angular momentum is _____. Show the angular momentum vector on the diagram by drawing an × in a small circle ($\otimes$) if it is into the page and a dot in a small circle ($\odot$) if it is out of the page. Label it $\vec{\ell}$. If the particle is traveling along the same line but in the positive y direction, the direction of its angular momentum is _____. If the particle is traveling along a line through the origin, its angular momentum is _____.

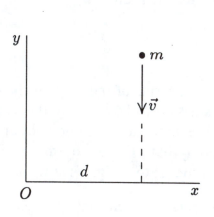

The total angular momentum $\vec{L}$ of a system of particles is the vector sum of the individual angular momenta of all the particles in the system. For an object rotating with angular velocity ω about the z axis, the z component of the total angular momentum is the sum

$$L_z =$$

where m_i is the mass of particle i and r_i is the distance from particle i to the rotation axis. In terms of the rotational inertia I of the object it is

$$L_z =$$

Notice that this equation is also valid for a single particle traveling in a circular orbit. Then, the rotational inertia is mR^2.

If the mass is distributed symmetrically about the axis of rotation, the total angular momentum is along the axis of rotation and the magnitude of L_z is also the magnitude of $\vec{L}$. The direction of $\vec{L}$ is given by the right hand rule: curl the fingers in the direction of _____, then the thumb will point in the direction of _____.

Newton's second law for rotation. If a net torque acts on a particle, then its angular momentum changes with time. Recall that, in terms of the linear momentum of a particle, Newton's second law can be written $\vec{F} = d\vec{p}/dt$, where $\vec{F}$ is the net force acting on the particle. Take the vector product of this equation with $\vec{r}$ to obtain $\vec{r} \times \vec{F} = \vec{r} \times d\vec{p}/dt$. Note that $\vec{r} \times \vec{F} = \vec{\tau}$, the net torque acting on the particle. With a small amount of mathematical manipulation you can show that $\vec{r} \times (d\vec{p}/dt) = d(\vec{r} \times \vec{p})/dt - (d\vec{r}/dt) \times \vec{p} = d\vec{\ell}/dt$. The second term, $(d\vec{r}/dt) \times \vec{p}$, vanishes because _____ is in the same direction as _____. Thus, in terms of torque and angular momentum, Newton's second law becomes $\vec{\tau} = $ _____.

For a system of particles the time rate of change of the angular momentum of each particle equals the net torque acting on the particle. This means that the time rate of change of the total angular momentum for the system is the vector sum of all torques acting on all particles. The sum includes the torque exerted by any particle in the system on any other particle in the system as well as external torques, exerted on particles in the system by objects in the environment of the system.

According to Newton's third law, the sum of the torques exerted by particles of the system on each other is _____. (This result follows from the third law if the force of each object on each of the others is along the line joining the particles.) Only torques exerted by objects outside the system can change the total angular momentum of the system. Be sure you understand this. One particle in the system can change the angular momentum of another particle in the system but the second particle simultaneously changes the angular momentum of the first and the two changes are equal in magnitude and opposite in direction.

Newton's second law for the angular momentum of the system is

$$\vec{\tau}_{net} = $$

where $\vec{\tau}_{net}$ is the total *external* torque exerted on all particles of the system and $\vec{L}$ is the total angular momentum of the system.

This is the fundamental equation for a system of particles. You should learn to apply it to as many situations as you can. Remember it is a three-dimensional equation. In component form $\tau_{net\,x} = $ _____, $\tau_{net\,y} = $ _____, and $\tau_{net\,z} = $ _____. You must use the same origin to compute the angular momentum and torques but the result is the same for any origin you choose.

For a body rotating about the z coordinate axis, we are usually interested in only the z components of the torque and angular momentum. If an external torque with other components is applied, the bearings must apply whatever additional torque is required to hold the axis fixed. If the spinning body is symmetric about the axis of rotation, then $\vec{L}$ is along that axis and the bearings do not need to exert a torque to hold the axis fixed.

Since $L_z = I\omega$ for a body rotating about the z axis, Newton's second law for fixed axis rotation becomes $\tau_{\text{net } z} = $ _____. In some cases the body is rigid and I does not change with time. Then, Newton's second law for fixed axis rotation becomes

$$\tau_{\text{net } z} =$$

where α is the angular acceleration of the body. This relationship was explored in the last chapter. In this chapter you will encounter some situations for which I does change.

Conservation of angular momentum. If the total external torque acting on a body is zero, then the change over any time interval of the angular momentum of the body is _____. Angular momentum is then said to be conserved. This is a vector law. One component of the angular momentum may be conserved while other components are not. To see if angular momentum is conserved or not, calculate the total external torque acting on the body.

If the z component of the total external torque acting on a body rotating about the z axis vanishes, then $I\omega$ is a constant. Consider an ice skater, initially rotating on the points of her skates with her arms extended. When she drops her arms to her side, her angular velocity increases. By dropping her arms she decreases her _____ and her angular velocity must increase for _____ to remain the same.

Suppose the skater is spinning in the counterclockwise direction when viewed from above, so her angular velocity vector points upward. Consider her torso and arms as two individual objects. As she drops her arms the angular momentum of her torso increases, so you know the torque exerted by her arms on her torso is in the _____ direction. Her torso exerts a torque of equal magnitude but opposite direction on her arms. This means the magnitude of the angular momentum of her arms _____ as they drop.

II. PROBLEM SOLVING

Dynamics problems for a rotating body are based on the vector equation $\vec{\tau} = \mathrm{d}\vec{L}/\mathrm{d}t$. Some problems simply ask for a calculation of a torque or angular momentum using the general definitions. Identify the point to be used as the origin, then evaluate the appropriate vector product. To find a torque, take $\vec{r}$ to be the vector from the origin to the point of application of the force and evaluate $\vec{r} \times \vec{F}$. To find the angular momentum of a particle, take $\vec{r}$ to be the vector from the origin to the particle, then evaluate $\vec{r} \times \vec{p}$ or $m\vec{r} \times \vec{v}$. You must be given the linear momentum of the particle or else its mass and velocity. To find the angular momentum of a rigid body rotating about a fixed axis, use $L_z = I\omega$, where I is the rotational inertia of the body and ω is its angular speed. The component of $\vec{L}$ along the axis of rotation is the most important component and its direction is given by a right-hand rule.

Problems involving rolling objects can be solved by considering the forces acting on the object. Many can be also solved using the energy method. In either case, don't forget the condition of no-sliding: the point on the rolling object in contact with the surface is instantaneously at rest with respect to the surface.

A dynamical solution starts with a force diagram. See Section 12–3 of the text for an illustrative example, a wheel rolling down a ramp. Show all forces and label them with algebraic symbols. Since the application points of forces are important for rotation you cannot represent the body as a point. Instead, sketch the body and place the tail of each force vector at the point where the force is applied. Choose a coordinate system to describe the motion of the center of mass, with one axis in the direction of its acceleration, if possible. Sum the force components for each coordinate direction, again taking care with the signs. Equate the sum to the product of the mass and the corresponding component of the acceleration of the center of mass. Then choose a direction for positive rotation. For each force write an expression that gives the torque about the axis of rotation. Be careful about the signs. Sum the torques and equate the sum to $I\alpha$.

When a wheel accelerates without sliding, a frictional force is present at the point of contact between the body and the surface on which it rolls. Don't forget to include it, even if it is not mentioned in the problem statement. You then write an additional equation relating the acceleration of the center of mass to the angular acceleration about the center of mass: $a_{com} = \pm \alpha R$. The sign depends on the choices made for the directions of positive translation and rotation.

If the body is sliding, the magnitude of the force of friction is given by $f = \mu_k N$, where μ_k is the coefficient of kinetic friction for the body and surface and N is the magnitude of the normal force exerted by the surface on the body. In some cases you can determine the direction of the force before solving the problem and can take it into account in drawing the force diagram and writing Newton's second law. In other cases you must arbitrarily select a direction, then examine the solution to see if the direction of rotation is consistent with your selection. If it is not, rework the problem using the opposite direction.

The energy method makes use of either the work-kinetic energy theorem or conservation of mechanical energy. If external forces do work on the object, use $W_{net} = \Delta K$, where W_{net} is the total work done by all forces and ΔK is the change in the kinetic energy. If all forces that do work are internal and conservative, use $U_i + K_i = U_f + K_f$, where U_i and U_f are the initial and final potential energies and K_i and K_f are the initial and final kinetic energies. In either case don't forget to include both the kinetic energy of translation and the kinetic energy of rotation. If the object is not sliding, include the auxiliary equation $v_{com} = \pm \omega R$. Remember that gravitational potential energy is given by Mgy, where y is the height of the center of mass above the reference height.

Notice that the energy method requires fewer equations than the dynamical method. Less information is required and fewer unknowns can be evaluated. Directions of forces and torques enter only insofar as they determine the work or potential energy. The energy method cannot be used to solve for directions except in special cases.

Some problems can be solved using the principle of angular momentum conservation. To see if the principle can be used, you must select a system and examine the net external torque acting on it to see if it vanishes. If it does, then angular momentum is conserved. If one component of the total torque vanishes, then that component of the angular momentum is conserved, regardless of whether the other components change.

Some problems deal with two objects that exert torques on each other, thereby changing each other's angular momentum but are subject to no external torques. Write the total angular momentum in terms of the quantities that describe the motions before the objects interact (their initial velocities or angular velocities); write the angular momentum in terms of the quantities that describe the motions after the interaction (their final velocities or angular velocities). Equate the two expressions and solve for the unknown. In some problems one of the objects is moving along a straight line, either before or after the interaction. Don't forget to include its angular momentum.

Some problems deal with an object whose rotational inertia changes at some time while it is spinning about a fixed axis. The problem statement gives information that can be used to calculate the initial values I_0 and ω_0, before the rotational inertia changes, and asks for either I or ω at a later time, after they change. If no external torques act, angular momentum is conserved and $I_0\omega_0 = I\omega$. This can be solved for one of the quantities in terms of the others.

III. MATHEMATICAL SKILLS

You should know how to find the magnitude and direction of vector products, such as $\vec{a} \times \vec{b}$. Recall that the magnitude is $ab\sin\phi$, where ϕ is the smallest angle between $\vec{a}$ and $\vec{b}$ when they are drawn with their tails at the same point. You should also remember the right hand rule for finding the direction. The product is perpendicular to the plane of $\vec{a}$ and $\vec{b}$. Curl the fingers of your right hand so they rotate $\vec{a}$ toward $\vec{b}$ through the angle ϕ. Then, the thumb will point in the direction of the vector product.

You also need to know how to compute the vector product in terms of the components of the factors:

$$(\vec{a} \times \vec{b})_x = a_y b_z - a_z b_y \qquad (\vec{a} \times \vec{b})_y = a_z b_x - a_x b_z \qquad (\vec{a} \times \vec{b})_z = a_x b_y - a_y b_x$$

IV. NOTES

Chapter 13
EQUILIBRIUM AND ELASTICITY

I. BASIC CONCEPTS

This chapter consists largely of applications of Newton's second laws for center of mass motion and for rotation, specialized to situations in which the total force and total torque both vanish. However, several new ideas are introduced to deal with the torque exerted by gravity and with elastic forces: the center of gravity and the elastic moduli. Pay careful attention to what they are and how they are used.

Here are some of the important concepts covered in previous chapters that you will use while studying this chapter: acceleration from Chapter 4, force, mass, and weight from Chapter 5, friction from Chapter 6, center of mass and Newton's second law for a system from Chapter 9, angular velocity, angular acceleration, and torque from Chapter 11, angular momentum and Newton's second law for rotation from Chapter 12.

Equilibrium conditions.   A rigid body is said to be in static equilibrium if its center of mass is not moving and if it is not rotating. Its total momentum and its total angular momentum both vanish. In terms of the external forces and torques acting on it, equilibrium means

$$\vec{F}_{net} =$$

and

$$\vec{\tau}_{net} =$$

where $\vec{F}_{net}$ is the vector sum of all external forces and $\vec{\tau}_{net}$ is the vector sum of all external torques. You will use these equations in applications to determine the external forces and torques on static objects. In some cases, you will be interested in the force or torque exerted by one part of an object on another part. In that event, simply consider the system to be the part of the object on which the force or torque acts and treat the force or torque as an external force or torque.

In nearly all examples and problems in this chapter, the forces considered are in the same plane. If you pick an origin in this plane, every torque is perpendicular to the plane and the number of equations to be solved is reduced to three. If the plane of the forces is the xy plane, then the three equations are $F_{net,\,x} =$ _____, $F_{net,\,y} =$ _____, and $\tau_{net,\,z} =$ _____. They can be solved for three unknowns. Usually the last equation is written using τ_{net} to stand for $\tau_{net,\,z}$.

Recall that the value of a torque depends on the origin used to specify the point of application of the force. When an object is in equilibrium, however, the torques sum to zero no matter which point is used as the origin. The choice is a matter of convenience, not necessity, but the same origin must be used to compute all torques. If an origin is not specified by the problem statement, place it in the plane of the forces. Carefully study the sample problems of Section 13–4 to see how equilibrium problems are solved.

Center of gravity. The force of gravity is often one of the forces acting on an object. Although each particle of the object experiences a gravitational force, for purposes of determining equilibrium these forces can be replaced by a single force acting at a point called _____. If the acceleration $\vec{g}$ due to gravity is uniform throughout the object, the magnitude of the replacement force is _____, where M is the total mass of the object, and the direction of the replacement force is _____. In addition, the point where the replacement force is applied then coincides with _____.
In this case, the position of the center of gravity does not depend on the orientation of the object. As you read the sample problems of Section 13–4 in the text, carefully note how gravitational forces are taken into account. In each case, their sum is replaced by a single force of magnitude Mg, directed downward and applied at the center of mass.

Even if the acceleration due to gravity is not uniform over the object, the sum of the gravitational forces can still be replaced by a single force applied at the center of gravity. Then, however, the center of gravity does not coincide with the center of mass. Furthermore, the position of the center of gravity depends on the orientation of the object. Practically speaking, it is not as useful a concept then.

Elasticity. In many cases, the conditions that the net force and torque both vanish are not sufficient to determine the forces acting on an object. From a mathematical viewpoint there are too many unknowns for the number of equations. As a result, the equations have an infinite number of solutions and additional information is needed to decide which one is physically correct.

Contact forces, like the normal force of a surface on an object, are <u>elastic</u> in nature. That is, they arise from slight deformations of the objects in contact. In many circumstances, the additional information you need to understand equilibrium is how elastic forces depend on deformation.

You should know the special terminology that is used in the study of elasticity. A force per unit area is called a _____ and a fractional deformation is called a _____.

There are two regimes of deformation. If the stress is less than the <u>yield strength</u>, then, when the stress is removed, _____

_____.

If the stress is greater than the yield strength but less than the <u>ultimate strength</u>, then, when the stress is removed, _____

_____.

If the stress is greater than the ultimate strength, then _____

_____.

For stresses well below the yield strength, the strain is proportional to the stress and the material is said to be elastic. Elastic materials behave like springs.

Suppose forces that are equal in magnitude and opposite in direction are applied to the opposite ends of a rod. If the forces are perpendicular to the rod faces, the rod will be elongated or compressed, depending on the directions of the forces. Suppose the forces are pulling outward and let ΔL be the elongation of the rod. Also suppose the magnitude of each force is F, the area of each face is A, and the undeformed rod length is L. Then the stress is given by ＿＿＿＿ and the strain is given by ＿＿＿＿. If the object remains elastic (small stress), then the stress is proportional to the strain. The proportionality is written

$$\frac{F}{A} = E \frac{\Delta L}{L},$$

where the modulus of elasticity E is called ＿＿＿＿＿＿＿＿＿＿. E is a property of the material and does not depend on F, A, L, or ΔL.

Values of E for some materials are listed in Table 13–1 of the text. A large value of E means a large force is required to compress or elongate a given length of sample by a given amount. If E is small and the length and area are the same, only a small force is required.

If the forces are opposite in direction but parallel to the rod faces, then shear occurs. Suppose one face moves a distance Δx relative to the other, along a line that is parallel to the faces. Also suppose the magnitude of each force is F, the area of each face is A, and the rod length is L. The stress is given by ＿＿＿＿ and the strain is given by ＿＿＿＿. Stress is proportional to strain and the proportionality is written

$$\frac{F}{A} = G \frac{\Delta x}{L}.$$

The modulus of elasticity G is called ＿＿＿＿＿＿＿＿＿. G is a property of the material. Material with a small modulus of elasticity shears more easily than material with a large modulus of elasticity, provided of course, the lengths and areas of the samples are the same.

If the object is placed in a fluid and pressure is applied uniformly on its surface, then hydraulic compression occurs. In this case, the stress is the ＿＿＿＿＿ in the fluid. This is the same as the force per unit ＿＿＿ acting on the object. The strain is ＿＿＿ where V is the original volume and ΔV is the change in volume. The modulus of elasticity is called the ＿＿＿＿ modulus and is denoted by B. In symbols, $B =$ ＿＿＿＿ where p is the pressure in the fluid.

II. PROBLEM SOLVING

Except for the ideas of center of gravity and of elasticity, this chapter chiefly covers applications of concepts discussed in earlier chapters and the problem solving techniques you will need are not new to you.

In most situations all forces acting on the object under consideration are in the xy plane. Select the origin for the computation of torques to be in that plane. All torques are then along the z axis, perpendicular to the plane. The method can easily be extended to situations in which forces with z components and torques with x and y components act.

Here are the steps you should use to solve problems. First select the object to be considered. Draw a force diagram showing as arrows all external forces acting on the object. Don't forget normal and frictional forces exerted by any objects in contact with the selected object. Use an algebraic symbol to designate the magnitude of each force, known or unknown. Use either Mg or W to label the weight, depending on whether the weight or mass of the object is given.

Draw each force arrow in the correct direction, if you know it. Gravity acts downward, a string pulls along its length, a frictionless surface pushes along the perpendicular to the surface. Sometimes you will know the line of a force but not its direction. You might know, for example, that a rigid rod acts along its length but you may not know if it is pushing or pulling. In that event, draw the arrow in either direction. If, after you solve the problem, the value turns out to be positive, then you picked the correct direction; if it turns out to be negative, the force is actually opposite the direction you picked. In other cases, you may know only that the force acts in the xy plane. Pick an arbitrary direction, not along one of the coordinate axes. You will then solve for both components.

Because you need to compute torques, be sure you indicate the point of application of each force by placing the tail or head of the arrow at the appropriate place on the diagram. Gravity acts at the center of gravity. For the situations considered here this is the same as the center of mass. A string or rod acts at its point of attachment. In many cases, an object makes contact only at a point and that is the point of application of the force it exerts.

Choose a coordinate system for calculating force components. If you know the direction of one of the unknown forces, you can usually simplify the algebra by selecting one of the coordinate axes to be in that direction. Write $F_{net, x} = 0$ and $F_{net, y} = 0$ in terms of the individual forces.

Choose an origin for calculating torques. The choice is completely arbitrary but you can usually reduce the algebra considerably by choosing an origin so that one or more torques vanish. Try to place the origin on the line of action of a torque associated with an unknown force. Write $\tau_{net} = 0$ in terms of the forces and their lever arms. Pay special attention to the signs of the torques.

Some elasticity problems are simply direct applications of Eqs. 13–25, 13–26, or 13–27 in the text. You may be asked to calculate the stress or the strain, given the other. In variants, you may be asked for the change in length under compression or the relative displacement of the faces under shear. The appropriate modulus must be given or else listed in Table 13–1. In some cases, you may be asked to calculate the modulus, given the other quantities in the equations. These problems are all straightforward. In some situations, the elastic properties of objects determine the forces at equilibrium. To solve these, more complicated problems, you must simultaneously apply an equation of elasticity and the equations of equilibrium.

III. NOTES

Chapter 14
GRAVITATION

I. BASIC CONCEPTS

You will learn about the force law that describes the gravitational attraction of two particles for each other. By considering the attraction of every particle in one object for every particle in a second object the law is extended to deal with objects containing many particles. Then, the law is applied to the motion of planets and satellites. Here you will bring to bear some of the concepts you studied in earlier chapters.

Here are some topics you should review before studying this chapter: free-fall acceleration (Chapter 2), force, mass, Newton's second law, Newton's third law (Chapter 5), uniform circular motion (Chapter 6), kinetic energy (Chapter 7), potential energy, mechanical energy, conservation of mechanical energy (Chapter 8), linear momentum, conservation of linear momentum (Chapter 9), angular momentum, Newton's second law for rotation, and conservation of angular momentum (Chapter 12).

Newton's law of gravity. According to Newton's law of gravity, two particles with masses m_1 and m_2, separated by a distance r, each attract the other with a force of magnitude

$$F = $$

where the universal gravitational constant G has the value $G =$ _____ N·m^2/kg^2.

Gravitational forces, like all forces, are vectors. The force that particle 1 exerts on particle 2 is toward _____, along the line that joins the particles. The force that particle 2 exerts on particle 1 is toward _____, along the same line. Note that the force obeys Newton's third law: the force of particle 1 on particle 2 has the same magnitude as the force of particle 2 on particle 1 but is in the opposite direction. Also note that the magnitude of the force is inversely proportional to the square of the particle separation. If that distance is doubled the gravitational force is _____ as great.

Experimental evidence indicates that the masses in the law of gravity are identical to those in Newton's second law of motion. If the gravitational force exerted by particle 1 is the only force acting on particle 2 then the acceleration of particle 2 is given by $a =$ _____, toward _____. Notice that the acceleration of particle 2 depends on m_1 but is independent of m_2.

When more than two particles are present, the force on any one of them is the <u>vector</u> sum of the individual forces the other particles exert on it. If an object can be considered a continuous distribution of mass, the gravitational force it exerts on a particle is computed as the integral $\vec{F} =$ _____, where $d\vec{F}$ is the force exerted by an infinitesimal volume element of the distribution. The integral represents the vector sum of _____

Two very important consequences of Newton's law of gravity are stated in Sections 14–2 and 14–5. Both are concerned with the force of gravity exerted by a uniform spherical shell of mass M on a particle of mass m. In the first case, the particle is *outside* the shell, a distance r from the shell center. Then, the magnitude of the force of the shell on the particle is given by

$$F = $$

The direction of the force is _____ . In the second case, the particle is *inside* the shell. Then, the magnitude of the force of the shell on the particle is given by $F =$ _____ , no matter where the particle is located inside.

The first of these theorems can be used to show that the force of attraction of two objects with spherically symmetric mass distributions is given by an equation that is identical to Newton's law of gravity for particles, provided we interpret r as _____

_____ .

If the force of gravity exerted by Earth is the only force acting on a particle of mass m at its surface, we may write $F = ma_g$, where a_g is the acceleration due to gravity at Earth's surface. Assume Earth is a sphere with a spherically symmetric mass distribution and equate F to the force given by Newton's gravitational law to obtain an expression for a_g in terms of the mass M_e and radius R_e of Earth:

$$a_g = $$

This expression can be used to calculate the mass of Earth, if a_g, R_e, and G are known. A similar expression (with R_e replaced by the distance between the center of Earth and a particle) can be used to estimate the acceleration due to gravity at any given distance above Earth's surface. This calculation is an estimate because we have made the assumption that Earth has a spherically symmetric mass distribution, an assumption that is not precisely valid.

Even if deviations from a spherical mass distribution can be ignored, the acceleration due to gravity is not the same as the acceleration g of an object in free-fall because Earth is spinning. At Earth's surface the difference between the two is $a_g - g = \omega^2 R$, where ω is the angular velocity of _____ and R is the radius of the circle traversed by the object as it spins around with Earth. At the equator R is the radius of _____ and numerically $a_g - g =$ _____ m/s^2. The measured weight of the object, as read by a spring balance, is

_____ .

The shell theorems can also be used to find an expression for the gravitational force on a particle located *within* a spherically symmetric mass distribution. Consider a particle somewhere inside Earth, a distance r from the center. If Earth's mass distribution were spherically symmetric, we could calculate the force of Earth on the particle by considering only that part of Earth's mass that is less than r from the center. All parts of Earth attract the particle but the forces due to parts further from the center than the particle sum to zero.

To test your understanding, assume Earth's mass is uniformly distributed and use the space below to show that the magnitude of the force on the particle is given by $(GM_em/R_e^3)r$, where M_e is the mass of Earth, m is the mass of particle, and r is the distance from the center of Earth to the particle. First show that the mass inside a spherical surface of radius r is $M_e r^3 / R_e^3$, then substitute this expression for one of the masses in Newton's law of gravity.

Notice that the force on a particle at the center of Earth is zero.

In many cases, the dimensions of an extended body are small enough that the acceleration due to gravity is very nearly uniform over the body. Then, for purposes of studying the motion of the body, we may treat it as a particle located at its center of mass. Satellites in Earth's gravitational field and planets in the Sun's gravitational field are discussed in the text as if they were particles.

Gravitational potential energy. The force of gravity is a conservative force. Go back to Chapter 8 and review the tests for conservative forces. In general, when a system changes configuration how does the work done by a conservative force depend on the paths taken by parts of the system? _____

Since the gravitational force is conservative a potential energy function is associated with it. If two point masses m and M are separated by a distance r, their gravitational potential energy is given by

$$U(r) =$$

where the potential energy is taken to be zero for infinite separation. This expression also gives the potential energy of two non-overlapping bodies with spherically symmetric mass distributions. Then, r is interpreted as the separation of their _____. It is important to recognize that this energy is associated with the *pair* of objects, NOT with either object alone.

As you know, potential energy is a scalar. To calculate the gravitational potential energy of a system of particles, sum the potential energies of each *pair* of particles in the system. The total potential energy of a collection of particles is the work done by an external agent to assemble the particles from the reference configuration. The particles start from rest and are placed at rest in their final positions. If the particles are brought from infinite separation the external agent must do negative work since the mass already in place attracts any new mass being brought in and the agent must pull back on it. Also remember that the potential energy is the *negative* of the work done by the gravitational forces of the particles on each other as the system is assembled.

If the particles of a system interact only via gravitational forces and no external forces act, then the mechanical energy of the system is conserved. The sum of the kinetic energies of all the particles and the total gravitational potential energy remains the same as the particles move.

For a two-particle system the mechanical energy consists of three terms, corresponding to the kinetic energy of each particle and the potential energy of their interaction. Suppose that at some instant particle 1 (with mass m_1) has speed v_1, particle 2 (with mass m_2) has speed v_2, and the particles are a distance r apart. Then, the mechanical energy of the system is given by

$$E =$$

At a later time v_1, v_2, and r may have different values but the sum of the three terms will have the same value. You can equate the expressions for the mechanical energy in terms of the speeds and separation at two different times and use the resulting equation to solve for one of the quantities that appear in it.

If we consider a satellite in orbit around Earth or a planet in orbit around the Sun, then one body is much more massive than the other. We may usually place the origin at the center of the more massive body and take its velocity to be zero. Then, the mechanical energy is the sum of two terms, the kinetic energy of the less massive body and the gravitational potential energy.

Suppose an Earth satellite with mass m (much less than that of Earth) has an initial speed v_0 when it is a distance r_0 from Earth's center and has speed v when it is a distance r from Earth's center. Write the conservation of mechanical energy equation in terms of m, M_e, r_0, v_0, r, and v:

This equation can be used to solve for any one of the quantities in it, given the others.

The escape speed is the initial speed that an object must be given at the surface of Earth (or other object with large mass) in order to _____

_____.

To use the conservation of mechanical energy equation to calculate the escape speed for an object on Earth's surface, set $r_0 =$ _____ , $r =$ _____ , and $v =$ _____ , then solve for v_0. The result is: $v_0 =$ _____ .

What other conservation principles are valid for the gravitational interaction of two spherically symmetric bodies if no other objects exert forces on them? _____

_____

Planetary motion. The motions of planets are controlled by gravity, due chiefly to the Sun. Here we consider only the gravitational force of the Sun and neglect the influence of other planets. The motion of a planet is then at least partially described by Kepler's three laws. State the laws in words:

1. Law of orbits: _____

_____

2. Law of areas: _____

_____

This law is a direct consequence of the principle of _____ conservation.

3. Law of periods: _____

_____

A planetary orbit can be described by its semimajor axis, which is _____

_____

and its eccentricity, which is _____

_____.

The point of closest approach to the Sun is called the _____ and, in terms of the semimajor axis a and eccentricity e, the distance of this point from the Sun is given by R_p = _____. The point of maximum distance from the Sun is called the _____ and this distance is given by R_a = _____. For circular orbits e = _____ and $R_a = R_p = a$. An eccentricity of nearly 1 corresponds to an ellipse that is much longer than it is wide.

An elliptical orbit for a planet is shown to the right. Label the position of the Sun, the semimajor axis, aphelion, and perihelion.

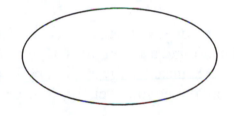

The law of areas provides us with a means of relating a planet's speed at one point to its speed at another point. The simplest relationship holds for points that are the greatest and least distances from the Sun because at these points the velocity is perpendicular to the position vector from the Sun.

In terms of the mass m of the planet, distance r from the Sun, and speed v of the planet, the magnitude of the angular momentum at one of these points is given by ℓ = _____, where the origin was placed at the Sun. Let R_p and v_p be the distance from the Sun and speed at perihelion and let R_a and v_a be the distance from the Sun and speed at aphelion. Then, conservation of angular momentum leads to the equality: _____.

For a circular orbit the period T is related to the radius r by

$$T^2 =$$

where M is the mass of the central body (the Sun). This expression can be used, for example, to calculate the mass of the Sun. The equation is also valid for elliptical orbits if r is replaced by the semimajor axis a.

You should recognize that asteroids and recurring comets in orbit around the Sun and satellites (including the Moon) in orbit around Earth or another planet also obey Kepler's laws. When the two bodies have comparable mass, as for example the two stars in a binary star system, each travels in an elliptical orbit around the _____.

Consider a body of mass m in a circular orbit about a much more massive sun (mass M), essentially at rest. The speed of the body and the mechanical energy of the system are closely related to the radius of the orbit. Gravity produces a centripetal force of magnitude $F = GmM/r^2$ and, according to Newton's second law, this must equal mv^2/r, so $v =$ _____ . You can use this equation to find the speed of a body in a given circular orbit.

As a function of the orbit radius r the kinetic energy is $K = \frac{1}{2}mv^2 =$ _____ . If we take the potential energy to be zero for infinite separation, then the potential energy for an orbit of radius r is

$$U = $$

and the mechanical energy, as a function of r, is

$$E = $$

The negative value of the mechanical energy indicates that the system is bound. The orbiting body does not have enough kinetic energy to escape. This expression for the mechanical energy is valid for elliptical orbits if we replace r with a, the semimajor axis. You should be aware that the kinetic energy is *not* given by $GmM/2a$ for non-circular orbits but the mechanical energy *is* given by

$$E = -\frac{GmM}{2a}.$$

To emphasize these ideas, consider a planet of mass m in an elliptical orbit with semimajor axis a, around a sun of mass M, essentially at rest. No matter where the planet is in its orbit, the mechanical energy is given by $E =$ _____ . When it is a distance r from the sun, the potential energy is given by $U =$ _____ , and the kinetic energy is given by

$$K = E - U = GmM \left(\frac{1}{r} - \frac{1}{2a} \right).$$

This expression can be used to find the speed of the planet if M, r, and a are given. As this discussion indicates, the speed of the orbiting body cannot be changed without changing the semimajor axis.

II. PROBLEM SOLVING

Some problems are straightforward applications of Newton's law of gravity. The law relates the gravitational force between two particles to their masses and separation. In some situations, you are asked for the force of two or more particles on another. Then, you must evaluate the vector sum of the individual forces. In some variants, you are given the force and asked for the position of one of the particles. If one of the objects is not a particle but is an extended body and can be approximated by a continuous mass distribution, then you may need to evaluate an integral. Look for any simplification brought about by symmetry.

Some problems deal with the force of gravity at Earth's surface or the variation of that force with altitude near Earth's surface. Don't forget that the r in the gravitational force law for uniform spheres is their center-to-center distance. Other related problems deal with the measured weight of an object. This differs from the force of gravity because Earth is spinning and the object is spinning with it. Be sure to distinguish between the acceleration due to gravity and the free-fall acceleration.

Some problems ask you to use the shell theorems to find the gravitational force exerted on a point particle by a spherically symmetric distribution of mass. You use Newton's law of gravity, but for one of the masses you substitute the mass that is inside an imaginary spherical surface that passes through the point particle. You may need to carry out a side calculation to find that mass.

You might be asked for the initial speed that an object must be given so it barely escapes from Earth (or other large astronomical body). After the object is removed, it is infinitely far from Earth and its kinetic energy is zero. The initial mechanical energy is the potential energy of the Earth-object system plus the kinetic energy of the object. The final mechanical energy is simply the potential energy of the Earth-object system with the object removed to far away. Use conservation of mechanical energy to find the initial kinetic energy of the object.

Some problems might require you to find the potential energy of a collection of point particles. The problem might be phrased in terms of the work required to assemble the particles from infinite separation or to remove them to infinite separation. In any case, identify all the possible *pairs* of particles and sum the potential energies of the pairs. If there are four particles, for example, the pairs are 1–2, 1–3, 1–4, 2–3, 2–4, and 3–4. For each pair find the separation r_{ij} of the two particles, then evaluate $-Gm_im_j/r_{ij}$ for the potential energy of the pair. Finally, sum all the contributions to find the total potential energy U. As the system is assembled gravity does work $-U$ and an external agent does work $+U$. As the system is disassembled gravity does work $+U$ and the agent does work $-U$. U itself is negative.

Many planetary motion problems can be solved using the relationship between the mechanical energy E and the semimajor axis a: $E = -GmM/2a$, where m is the mass of the satellite and M is the mass of the central body. In terms of the distance r from the central body and the speed v, $E = -GmM/r + \frac{1}{2}mv^2$. If the mechanical energy and mass are known you can calculate the semimajor axis. If, in addition, the distance from the central body is known you can calculate the speed.

To solve some problems, you will also need to know the relationship between the aphelion distance, the perihelion distance, and the semimajor axis: $a = (R_a + R_p)/2$. Sometimes the eccentricity e is given or requested. Then, you will need $R_a = a(1 + e)$ and $R_p = a(1 - e)$. When the period is given or requested, use $T^2 = (4\pi^2/GM_e)a^3$.

III. MATHEMATICAL SKILLS

Gravitational force near Earth's surface. The magnitude of the gravitational force exerted by Earth on a body a distance h above its surface is given by $GmM/(R + h)^2$, where M is the mass of Earth, m is the mass of the body, and R is the radius of Earth. You may need to calculate the force for h small compared to R. Use the binomial expansion:

$$(R + h)^{-2} = R^{-2} - 2R^{-3}h + 3R^{-4}h^2 + \ldots .$$

This expression finds practical use in some calculations. Suppose you wished to find the difference in the gravitational field at opposite ends of a vertical rod. In principle, you could use Newton's law for gravity directly, substituting $R + h_1$ to find the force at one end and $R + h_2$ to find the field at the other end. R, however, is generally so much greater than h_1 or h_2 that your calculator truncates both numbers to R. On the other hand, if you use the first two terms of the binomial expansion to write an expression for each force, then subtract the expressions, the first terms cancel and you are left with the terms that are proportional to h_1 and h_2. These can be calculated easily.

IV. NOTES

Chapter 15
FLUIDS

I. BASIC CONCEPTS

In this chapter you will study gases and liquids, first at rest and then in motion. The most important concepts for the study of static fluids are those of pressure and density. Learn their definitions well and learn to calculate the pressure in various situations, paying particular attention to the variation of pressure with depth in a fluid. Then, use the concepts to understand two of the most basic principles of fluid statics: Archimedes' and Pascal's principles. Two equations are fundamental for understanding fluids in motion: the continuity and Bernoulli equations. They express the relationship between pressure, velocity, density, and height at points in a moving fluid. To understand these equations, you must first understand the ideas of streamlines and tubes of flow.

This chapter uses the following topics from previous chapters: velocity and speed from chapter 4, mass, force, and weight from Chapter 5, work, kinetic energy, and the work-kinetic energy theorem from Chapter 7, gravitational potential energy from Chapter 8, and equilibrium from Chapter 13. You should review any that you do not fully understand.

Definitions. Describe some properties of solids, liquids, and gases that can be used to distinguish them from each other:

Solids: _____

_____

Liquids: _____

_____

Gases: _____

_____

Both liquids and gases are fluids.

If a small volume ΔV of fluid has mass Δm, then its <u>density</u> ρ at that place is given by

$$\rho =$$

The definition includes a limiting process in which the volume being considered shrinks to a point. Thus, density is defined at each point in a fluid (or solid) and may vary from point to point. Note that density is a scalar. In many cases, the density of a fluid is uniform and we may write $\rho = M/V$, where M is the mass and V is the volume of the fluid.

Normally a fluid exerts an outward force on the inner surface of the container holding it. The force on any small surface area ΔA is proportional to the area and is perpendicular to the surface. If ΔF is the magnitude of the force on the area ΔA, then the <u>pressure</u> p exerted by the fluid at that place is defined by the scalar relationship

$$p =$$

You must take the limit as the area ΔA tends toward zero. Thus, pressure is defined at each *point* and may vary from point to point on the surface. In many cases, the surface being considered is a plane and the pressure is uniform over it. Then, we may write $p = F/A$, where F is the force exerted by the fluid on the surface and A is the area of the surface.

Pressure ultimately arises from forces that must be exerted on particles of the fluid in order to contain them. Container walls must exert forces on particles that impinge on them in order to reverse the normal components of their velocities and keep the particles within the container. The particles, of course, exert forces with equal magnitude and opposite direction on the walls. If the container top is open, the atmosphere above exerts forces on fluid particles at the fluid surface.

Pressure also exists in the interior of a fluid. Fluid within any volume exerts an outward force on the fluid around it and fluid outside a volume exerts an inward force on the fluid it surrounds. The force is normal to the imaginary surface that bounds the volume and the pressure is defined in the same way as at the container walls.

The SI unit of pressure (N/m^2) is called _____ and is abbreviated _____ . Other units are: 1 atmosphere (atm) = _____ Pa, 1 bar = _____ Pa, 1 mm of Hg = _____ Pa, and 1 torr = _____ Pa. Pressure is a scalar quantity.

Table 15–1 of the text lists the densities of some materials and Table 15–2 lists various pressures that exist in nature. Note the wide range of values.

The density of a fluid depends on the pressure. If the pressure at any point is increased (by squeezing the container, for example), the density at that point increases. Most liquids are not readily compressible; great pressure is required to change their densities. Gases, on the other hand, are readily compressible.

Variation of pressure with depth in a fluid. Pressure varies with depth in a fluid subjected to gravitational forces. Consider an element of fluid at height y in a larger body of fluid. Suppose the upper and lower faces of the element each have area A and the element has thickness Δy, as shown. If the fluid has density ρ, the mass of the element is _____ and the force of gravity on it is _____ . If the pressure at the upper face is p_1, the downward force of the fluid there is _____ . If the pressure at the lower face is p_2, the upward force of the fluid there is _____ . Since the element is in equilibrium the net force must vanish. This means

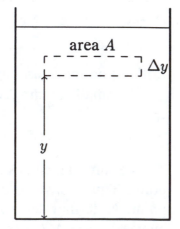

$$p_2 - p_1 =$$

The pressure is _____ at points higher in the fluid than at lower points. In the limit as Δy becomes infinitesimal, $p_2 - p_1$ becomes the infinitesimal dp and the equation above becomes

$$\frac{dp}{dy} =$$

If the fluid is incompressible, then the density is the same everywhere and the pressure difference between any two points in the fluid, at heights y_1 and y_2, respectively, is given by

$$p_2 - p_1 =$$

If p_0 is the pressure at the upper surface of an incompressible fluid, then $p =$ _____. is the pressure a distance h below the surface. Notice that the pressure is the same at any points that are at the same height in a homogeneous fluid.

To derive an expression for the pressure as a function of height in a fluid, no matter how the density depends on pressure, start with $dp/dy = -\rho g$. This leads to $g\,dy = -dp/\rho$ and then to

$$g(y_2 - y_1) = -\int_{p_1}^{p_2} \frac{dp}{\rho},$$

where p_1 is the pressure at y_1 and p_2 is the pressure at y_2. If the density is a known function of pressure, the integration can be carried out, at least in principle.

Assume the density does not depend on pressure, evaluate the integral, and show that $p_2 - p_1 = -\rho g(y_2 - y_1)$:

If the fluid is inhomogeneous, as it is if it consists of layers of immiscible fluids with different densities, you must apply the expression for $p(y)$ to each layer separately. For practice, consider the stack of two incompressible fluids shown on the right. The upper surface is open to the atmosphere and the pressure there is atmospheric pressure p_0. Each layer has thickness $2d$. Write an expression for the pressure at each of the labelled points:

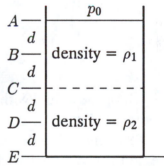

$$p_A =$$
$$p_B =$$
$$p_C =$$
$$p_D =$$
$$p_E =$$

To work some problems, you must also know when the pressure is the same at two points. Although atmospheric pressure varies with height, the variation is negligible over distances on the order of meters. The surfaces of two portions of a fluid that are both exposed to the atmosphere may be taken to be at the same pressure (atmospheric pressure). Thus, the upper surface in each arm of an open U-tube may be taken to be at atmospheric pressure, even if the surfaces are at different heights.

If two points are at the same height *and* they can be joined by a line such that the fluid density does not change along the line, then the pressure is the same at the points. Thus, the pressure is the same at any two points at the same height in the water shown in Fig. 15–xx of the text, even if the points are in different arms of the U-tube. A line can be drawn from one to the other through the liquid in the bottom of the tube. Note that the pressure is not the same at points at the same height if one point is in the water on the right and the other is in the oil on the left. A line joining them must pass through two liquids of different density.

Measurement of pressure. This section serves two purposes. It describes two instruments used to measure pressure: the barometer and the open-tube manometer. It also provides some excellent applications of ideas discussed earlier in the chapter. You should study it with both these purposes in mind.

In the space on the right, draw a diagram of a mercury barometer. Clearly mark the mercury, the region of near vacuum, and the region where the pressure has the value indicated by the height of the mercury column.

You should be able to use what you know about the variation of pressure with height in an incompressible fluid to show that the pressure at the top of the mercury pool is given by $\rho g h$, where ρ is the density of mercury and h is the height of the top of the mercury column above the pool. Carry out the derivation in the space below:

Suppose a fluid with half the density of mercury is used in a barometer. The height of the column will be _____ the height of a mercury column.

You should understand the importance of the region above the mercury column. If this region is not a near vacuum but instead is filled with a gas, then the pressure at the surface of the mercury pool is NOT proportional to the height of the mercury column. In fact, if the pressure above the column is p_0 and the height of the column is h, then the pressure above the pool is given by $p =$ _____ .

In the space to the right, draw a diagram of an open-tube manometer. Mark the fluid. Use p_0 to label the fluid surface where the pressure is atmospheric and p to label the fluid surface where the pressure has the value being measured.

Describe exactly what characteristic of the fluid is measured and tell how this is related to the pressure to be found:

Of these two instruments one measures <u>absolute</u> <u>pressure</u> and the other measures <u>gauge</u> <u>pressure</u>. Distinguish between absolute and gauge pressure and tell which instrument is used to measure each:_____

_____

_____

Pascal's principle.   Suppose the external pressure applied to the surface of a fluid is changed by Δp_{ext}. Pascal's <u>principle</u> states that the pressure everywhere in the fluid changes by _____.

In the space to the right, draw a diagram of a simple hydraulic system that can be used to lift a heavy object by applying a force that is much less than its weight. Clearly identify the input and output forces F_i and F_o. Suppose the input force is exerted over an area A_i and the output force is exerted over an area A_o. According to Pascal's principle these quantities are related by

$$\frac{F_i}{A_i} =$$

Suppose the fluid surface at the input force moves a distance d_i and the object moves a distance d_o. If the fluid is incompressible, then the volume of fluid does not change and $d_i A_i = d_o A_o$. Substitute $A_o = d_i A_i / d_o$ into the equation for Pascal's principle to show that $F_i d_i = F_o d_o$:

This means that the work done by the external force on the fluid is the same as the work done by the fluid on the object. If the fluid is compressible, the work done by _____ is larger than the work done by _____. The difference results in an increase in the internal energy of the compressed fluid.

An automobile braking system uses hydraulics to transmit the force you apply at the brake pedal to the brake shoes. You may think of a small diameter hose containing brake fluid running from the pedal to a shoe. Since a small force applied by you results in a huge force applied to the shoe, you know that the surface area of fluid being pushed on by the pedal is much _____ than the area of fluid pushing on the shoe. Furthermore, since the fluid is essentially incompressible, the distance the pedal moves is much _____ than the distance the shoe moves.

Archimedes' principle.   State Archimedes' principle in your own words: _____

_____

_____

The buoyant force on an object wholly or partially immersed in a fluid is a direct result of the pressure exerted on it by the fluid and depends on the variation of pressure with height. The pressure at the bottom of the object is greater than the pressure at the top and the net buoyant force is upward.

You know that Archimedes' principle is valid because, if the submerged portion of an object is replaced by an equal volume of fluid, the fluid would be in equilibrium and the net force associated with the pressure of surrounding fluid would equal the _____ of the fluid that replaced the object.

To calculate the buoyant force acting on a given object, first find the submerged volume. If the object is completely surrounded by fluid, this is the volume of the object. If the object is floating on the surface, it is the volume that is beneath the surface. Suppose the submerged volume is V_s. Then, the magnitude of the buoyant force is given by $F_b =$ _____, where ρ_f is the density of _____.

To predict if an object floats or sinks, first assume it is totally submerged and compare the buoyant force with the weight. If the weight is greater, the object _____. If the buoyant force is greater, it _____. For an object with uniform density in an incompressible fluid this procedure is the same as comparing the density of the object with the density of the fluid. If the density of the fluid is greater than the density of the object, the object _____. If the density of the object is greater than the density of the fluid, the object _____.

For purposes of calculating the torque on a floating object, the buoyant force of the fluid on the object can be treated as a single force acting at the center of mass of _____. This point is called the center of _____. The force of gravity, on the other hand, can be treated as a single force acting at the center of mass of _____. If buoyancy and gravity together produce a net torque, the object tilts.

Buoyant torque is used to test the stability of a boat. Assume the boat is tilted slightly and calculate the buoyant torque about the center of gravity. If it tends to right the boat, the boat is stable. If it tends to tilt the boat more, the boat is unstable. Three cases are shown below, with the centers of buoyancy marked B and the centers of gravity marked G. For each, draw vectors to indicate the directions of the force of gravity and the buoyant force, give the direction of the buoyant torque about the center of gravity (into or out of the page), and label the boats "stable" and "unstable", as appropriate.

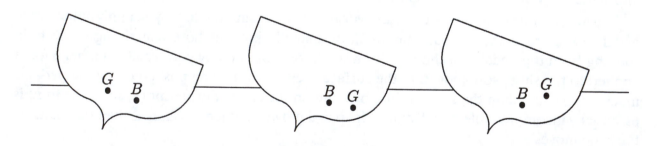

Moving fluids. Fluid flow can be categorized according to whether it is steady or nonsteady, compressible or incompressible, viscous or nonviscous, and rotational or irrotational. This chapter deals chiefly with steady, incompressible, nonviscous, irrotational flow.

The velocity and density of a fluid in <u>steady</u> <u>flow</u> do NOT depend on _____
_____ .

If we follow all the particles that eventually get to any selected point, we find they all have the same velocity as they pass that point, regardless of their velocities when they are elsewhere. In addition, particles flow into and out of any volume in such a way that the mass in the volume at any time is the same as at any other time. Be very careful here. Steady flow does NOT imply that the velocity of any fluid particle is constant nor does it imply that the density is the same everywhere.

If the flow is <u>incompressible</u>, the density does not depend on either the _____ or the
_____ . It is the same everywhere in the fluid and retains the same value through time. Steady flow does NOT imply incompressible flow.

You should also know the meanings of the other terms. Qualitatively distinguish viscous from nonviscous flow: _____
_____

Qualitatively distinguish rotational from irrotational flow: _____
_____

<u>Streamlines</u> are a convenient and effective means of visualizing fluid flow. They are also important for several derivations carried out in this chapter. Define the term streamline:
_____
_____

In steady flow, streamlines are fixed curves in space. Every fluid particle on the same streamline passes through the same sequence of points and at each point has the same velocity as other particles when they are at that point. At any point the fluid velocity is _____ to the streamline through the point.

In steady flow, can the velocity of a particle be different when it is at different points on the same streamline? _____ Do different streamlines ever cross? _____ Give an argument to substantiate your claim: _____
_____

Define the term <u>tube</u> <u>of</u> <u>flow</u>: _____
_____

In steady flow, do particles ever cross the boundary of a tube of flow? _____ Justify your answer: _____
_____

Does a streamline ever cross the boundary of a tube of flow? _____ Why or why not? _____
_____

We conclude from these statements that streamlines crowd closer together in narrow portions of tubes of flow than in wide portions. Consider a narrow tube of flow at a place where it has cross-sectional area A. If the fluid there has density ρ and speed v, then the mass of fluid that passes the cross section per unit time is given by _____ and the volume of fluid that passes the cross section per unit time is given by _____ . The former quantity is called the <u>mass</u> <u>flow</u> <u>rate</u> and the latter is called the <u>volume</u> <u>flow</u> <u>rate</u> and is denoted by R. The SI units of mass flow rate are_____ and the SI units of volume flow rate are _____ . Other commonly used units for the volume flow rate are L/s, gal/s, and gal/min.

Equation of continuity. The equation of continuity is derived from the condition that _____ is conserved in fluid flow. If there are no sources or sinks of fluid, the mass contained in any given volume can change only because the mass that flows into the volume differs from the mass that flows out at any time. In steady flow, the mass in any volume does not change, so in any time interval the mass flowing in equals the mass flowing out.

For steady flow of an incompressible fluid the equation of continuity is a statement that the _____ is the same for every cross section along a tube of flow. Let the subscripts 1 and 2 label two places along a tube of flow with cross-sectional areas A_1 at 1 and A_2 at 2. Suppose the fluid speed is v_1 at 1 and v_2 at 2. Then, for steady flow the equation of continuity is

$$A_1 v_1 =$$

For incompressible steady flow the volume flow rate is uniform along a tube of flow. This means that fluid particles have greater speed in a _____ portion of a tube than in a _____ portion. Because streamlines crowd closer together in narrow portions of tubes of flow than in wider portions we conclude that a high concentration of streamlines corresponds to a _____ fluid speed and a low concentration corresponds to a _____ fluid speed.

Bernoulli's equation. Bernoulli's equation for steady, incompressible, nonviscous, irrotational flow tells us that the quantity _____ has the same value at every point along a streamline. You should realize, however, that the value of the quantity may be different for different streamlines.

This equation arises directly from the work-energy theorem, applied to the fluid in a narrow tube of flow, in the limit as the tube becomes a streamline. The term p appears because the _____ exerted by neighboring portions of fluid do work, the term $\rho g y$ appears because _____ does work if the height of the tube above Earth varies, and the term $\frac{1}{2}\rho v^2$ appears because the _____ of the fluid changes if work is done on it.

If the tube does not change the height above Earth, the gravitational term is not needed and Bernoulli's equation becomes _____ = constant along a streamline. This equation indicates that the pressure is large where the fluid speed is _____ and is small where the fluid speed is _____. Combining this result with the continuity equation, we can conclude that for steady, incompressible, horizontal flow the pressure is _____ where a tube of flow is narrow and _____ where it is wide.

To fix these ideas, consider the horizontal tube of flow shown on the right. Draw several streamlines within the tube, clearly illustrating where they crowd close together and where they spread apart. Label one end of the tube "high fluid velocity" and the other end "low fluid velocity", as appropriate. Label one end "high pressure" and the other end "low pressure", as appropriate.

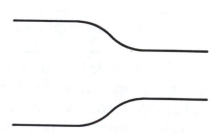

II. PROBLEM SOLVING

Some problems require you to know the definitions of pressure and density. Remember that if the pressure is uniform and the surface is a plane, then $p = F/A$. If there are several surfaces, you may need to sum the forces vectorially to obtain the net force. Remember that each force is perpendicular to the surface on which it acts. In some cases, you may need to use Newton's second law to find the force.

To calculate the variation of pressure with depth in a static incompressible fluid, use $p = p_0 + \rho g h$, where p is the pressure at depth h, p_0 is the pressure at the top of the fluid, and ρ is the density of the fluid. A variation on this type problem concerns a fluid that is accelerating. Then, the pressure differential across any fluid element provides some or all of the force that accelerates the element.

Some problems involve immiscible, incompressible fluids in U-tubes. Usually you know the pressure at the top surface of the fluid in each arm of the tube. If the end is open to the atmosphere, it is atmospheric pressure. If the end is closed and a vacuum exists above the fluid, it is zero. You can find a relationship between values of the pressure at the various fluid interfaces. Start at the fluid surface in one arm and follow a line in the tube to the surface in the other arm. Calculate the pressure at each interface and finally the pressure at the top of the second arm in terms of the pressure at the top of the first arm. Lastly, set the expression you obtain equal to the known value of the pressure at the top of the second arm. This gives you an equation to solve for the unknown in the problem.

All Pascal's law problems are much the same. In every case, a tube containing an incompressible fluid has a different cross section at different ends. A force F_1 is applied uniformly to an end with area A_1 and the fluid at the other end exerts a force F_2 on an object. If the area at the second end is A_2, then $F_1/A_1 = F_2/A_2$. Three of these quantities must be given, then the fourth can be evaluated.

Some problems ask you to calculate the work done by each of the forces. These are $W_1 = F_1 d_1$ and $W_2 = F_2 d_2$, where d_1 and d_2 are the distances moved. If the fluid is incompressible, then $d_1 A_1 = d_2 A_2$ and $W_1 = W_2$.

Two fundamental Archimedes' principle problems have been covered in the Basic Concepts section. They involve finding the buoyant force on an object, either floating or completely submersed in an incompressible fluid, and deciding if an object floats or sinks. These and many other Archimedes' law problems start with the equations $F_g = \rho g V$ for the force of gravity and $F_b = \rho_f g V_s$ for the buoyancy, where ρ is the density of the object, ρ_f is the density of the fluid in which it is wholly or partially immersed, V is the volume of the object, and V_s is the submerged volume. If the object is floating with no other forces acting, then $\rho V = \rho_f V_s$.

In some cases, other forces are present. For example, a string may be tied to the object, either holding it up from above or pulling it down from below, or weights may be placed on the object. In these instances, simply include the additional force F in the equation for equilibrium: $\rho_f g V_s - \rho g V + F = 0$, where F is positive if the force is directed upward. In other cases, the object has an acceleration. Use Newton's second law: set the net force, including the buoyant force, equal to the product of the mass and acceleration.

For a fluid in motion, you should be able to relate the volume and mass flow rates to the dimensions of a tube of flow and to the fluid velocity. The volume flow rate gives the volume of fluid that passes a cross section per unit time and is given by Av, where A is the cross-sectional area of the tube and v is the fluid speed. The mass flow rate is the mass of fluid that passes a cross section per unit time and is given by $\rho A v$, where ρ is the fluid density. Notice that the mass flow rate is the product of the density and the volume flow rate.

For steady flow in a single tube of flow the mass flow rate is the same in all parts of the tube. If the fluid is incompressible, the volume flow rate is also the same in all parts. If two tubes merge, the sum of the two mass flow rates in the merging tubes must equal to the mass flow rate in the final tube. If the fluid is incompressible, the same statement is true for the volume flow rates.

Bernoulli's equation is used to solve some problems. It relates conditions (density, fluid speed, pressure, and height above Earth) at one point on a streamline to conditions at another point. If you are given all but one of these quantities you can use Bernoulli's equation to solve for the unknown quantity. Note that the vertical coordinates of the two points enter the equation as their difference, so only the difference need be given, not the individual coordinates. Many problems give the cross-sectional areas at the two points along the tube of flow. Then the equation of continuity and Bernoulli's equation should be solved simultaneously.

III. NOTES

Chapter 16
OSCILLATIONS

I. BASIC CONCEPTS

This chapter is about periodic motion: the motion of an object that moves back and forth between two points, its motion being the same during every cycle. Pay attention to the meanings of the terms used to describe simple harmonic motion: amplitude, period, frequency, angular frequency, and phase constant; pay attention to the transfer of energy from kinetic to potential and back again as the object moves; and also pay attention to the form of the force law that leads to this type of motion.

Concepts discussed in previous chapters that are used in this chapter include displacement, velocity, acceleration, and free-fall acceleration from Chapter 2, force, mass, and Newton's second law of motion from Chapter 5, uniform circular motion from Chapter 6, kinetic energy and the force exerted by an ideal spring from Chapter 7, potential energy, mechanical energy, and conservation of mechanical energy from Chapter 8, and rotational inertia, angular acceleration, and Newton's second law for rotation from Chapter 11.

Simple harmonic motion. Concentrate on simple harmonic motion (SHM), for which the position variable (angle or coordinate) is a sinusoidal function of time. As a model we take an object of mass m attached to a spring with spring constant k, moving along the x axis with its equilibrium position at $x = 0$. Take the potential energy to be zero when the spring is neither stretched nor compressed. Then, as functions of the coordinate, the force on the object is given by

$$F(x) =$$

and the potential energy of the system is given by

$$U(x) =$$

You should recognize that these equations are valid only if x = 0 is the equilibrium point. If the origin is placed elsewhere, other terms must be added to the right sides.

The force and potential energy for the spring-object system have certain distinguishing characteristics in common with all systems executing simple harmonic motion. The object, of course, does not move along a line for all SHM systems. For example, some objects have rotational motions and for them position is described by an angle rather than by a linear coordinate.

If we place the origin at the point for which the force or torque is zero, then for SHM to occur the force or torque must always be proportional to _____ and if, in addition, we choose the potential energy to be zero when the position variable is zero then this energy is proportional to _____ .

The sign in the force law is important. When the coordinate is positive, the force is _____ and when the coordinate is negative, the force is _____. A force that always pushes an object toward its equilibrium position is called a _____ force.

The force law is substituted into Newton's second law to obtain the equation of motion. For the spring-object system it is

$$\frac{d^2 x}{dt^2} =$$

The most general solution to this equation is

$$x(t) = x_m \cos(\omega t + \phi),$$

where x_m, ω, and ϕ are constants. If the function given here is to obey the differential equation, then the constant ω, in terms of m and k, must be $\omega =$ _____.

The maximum value of the coordinate x is denoted by _____. The object moves back and forth between $x =$ _____ and $x =$ _____. x_m is called the _____ of the oscillation.

The constant ω is called the _____ of the oscillation. Since ωt is measured in radians, the units of ω are _____.

The angular frequency is related to the frequency f of the oscillation: $\omega =$ _____. The physical significance of the frequency is: _____

_____

It has the dimension of _____ and its SI unit is: _____.

The angular frequency and frequency are both related to the period T of the motion: $\omega =$ _____ and $f =$ _____. The physical significance of the period is _____

_____

It has the dimension of _____ and its SI unit is _____.

As a fraction of the period T, the time taken by the object to go directly from $x = 0$ to $x = x_m$ for the first time is _____ and the time taken to go from $x = -x_m$ to $x = +x_m$ for the first time is _____.

Suppose an oscillation has a period of 5.0 s. Then, its frequency is _____ Hz and its angular frequency is _____ rad/s. The motion repeats every _____ s or, in other words, it repeats _____ times per s.

You should remember that the angular frequency is determined by the ratio of k to m. Consider two springs, one with spring constant k and the other with spring constant $2k$. An object of mass m is attached to the first spring and set into oscillation. We can obtain an oscillation of exactly the same angular frequency by attaching an object of mass _____ to the second spring. These two systems take the same time to complete a cycle of their motions.

An expression for the velocity of the object as a function of time can be found by differentiating the expression for $x(t)$ with respect to time. The result is

$$v(t) =$$

In terms of x_m and ω, the maximum speed of the object is $v_m =$ _____. The object has maximum speed when its coordinate is $x =$ _____. The velocity of the object is zero when its coordinate is $x =$ _____ and also when its coordinate is $x =$ _____.

An expression for the acceleration of the object as a function of time can be found by differentiating $v(t)$ with respect to time. It is

$$a(t) =$$

In terms of x_m and ω, the magnitude of the maximum acceleration is $a_m =$ _____. The object has maximum acceleration when its coordinate is $x =$ _____ and also when its coordinate is $x =$ _____. This is when the force has maximum magnitude and the velocity vanishes. The acceleration is zero when the coordinate is $x =$ _____. This is when the force vanishes and the speed is a maximum.

The argument of the trigonometric function, $\omega t + \phi$, is called the _____ of the oscillation and ϕ is called the _____.

For a given system, x_m and ϕ are determined by the initial conditions (at $t = 0$). Since $x(t) = x_m \cos(\omega t + \phi)$ the initial coordinate of the object is given by $x_0 =$_____ and since $v(t) = -\omega y_m \sin(\omega t + \phi)$ the initial velocity of the object is given by $v_0 =$ _____.

x_0 is positive for ϕ between _____ and _____.

x_0 is negative for ϕ between _____ and _____.

v_0 is positive for ϕ between _____ and _____.

v_0 is negative for ϕ between _____ and _____.

The equations $x_0 = x_m \cos \phi$ and $v_0 = -\omega x_m \sin \phi$ can be solved for x_m and ϕ. To obtain an expression for x_m, solve the first equation for $\cos \phi$ and the second for $\sin \phi$, then use $\cos^2 \phi + \sin^2 \phi = 1$. To obtain an expression for ϕ, divide the second equation by the first and solve for $\tan \phi$. The results are

$$x_m =$$

$$\phi =$$

Be careful when you evaluate the expression for ϕ. There are always two angles that are the inverse tangent of any quantity but your calculator only gives the one closest to zero. The other is 180° or π radians away. Always check to be sure the values you obtain for x_m and ϕ give the correct initial coordinate and velocity. That is, $x_m \cos \phi$ must give the correct initial coordinate and $-\omega x_m \sin \phi$ must give the correct initial velocity. If they do not, add π to your answer for ϕ and check again.

On the axes below draw a graph of $x(t) = x_m \cos(\omega t + \phi)$. Take ϕ to be 0 and x_m to have the value marked on the vertical axis. Draw the function so the period is T, as marked on the time axis. On your graph label the times for which magnitude of the velocity is a maximum, the times for which it is a minimum, the times for which the magnitude of the acceleration is a maximum, and the times for which it is a minimum.

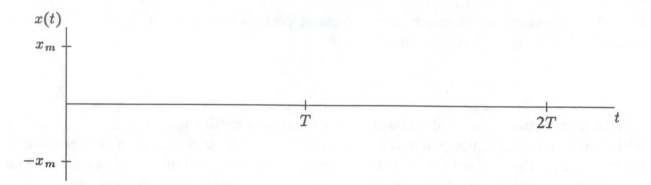

Energy considerations. An expression for the potential energy as a function of time can be found by substituting $x(t) = x_m \cos(\omega t + \phi)$ into $U = \frac{1}{2}kx^2$. The result is

$$U(t) =$$

An expression for the kinetic energy as a function of time can be found by substituting $v = -\omega x_m \sin(\omega t + \phi)$ into $K = \frac{1}{2}mv^2$. The result is

$$K(t) =$$

or, if $\omega^2 = k/m$ is used,

$$K(t) =$$

Both the potential and kinetic energies vary with time. As the object moves away from the equilibrium point it slows down; kinetic energy is converted to potential energy and stored in the extended spring. As the object moves toward the equilibrium point the stored potential energy is converted to kinetic energy and the object speeds up.

The potential energy is a maximum when the coordinate is $x =$ _____ and also when it is $x =$ _____. Then, the speed of the object is $v =$ _____ and the kinetic energy vanishes. The kinetic energy is a maximum when the coordinate is $x =$ _____. Then, the potential energy vanishes. Notice that the maximum kinetic energy has exactly the same value as the maximum potential energy.

Although both the potential and kinetic energies vary with time, the mechanical energy $E_{\text{mec}} = K + U$ is constant, as you can see by adding the two expressions you wrote above. If you know that the object has speed v when its coordinate is x, then you can use $E_{\text{mec}} =$ _____ to compute the mechanical energy. If you know the amplitude x_m of the oscillation, then you can compute the mechanical energy using $E_{\text{mec}} =$ _____. If you know the maximum speed v_m of the object, then you can compute the mechanical energy using $E_{\text{mec}} =$ _____. Since mechanical energy is conserved these three expressions must have the same value.

Applications. Several other systems that oscillate in simple harmonic motion are discussed in the text. For each of them use one of the tables below to describe the displacement variable (linear or angular coordinate) that is oscillating and give its symbol. Then, give an expression for the force or torque, as appropriate. For the torsional pendulum the torque is proportional to the angular displacement. For the other pendulums the force or torque is not proportional to the displacement unless the amplitude of the oscillation is small. Give both the exact expression and the small amplitude approximation. In all cases tell the physical meaning of the symbols that appear in the constant of proportionality that relates the force or torque and the displacement. Give expressions for the angular frequency ω and the period T in terms of properties of the system.

a. Torsional pendulum
   displacement variable: _____

   _____

   torque law: _____
   meaning of symbols: _____

   _____

   $\omega =$ _____ $T =$ _____

b. Simple pendulum
   displacement variable: _____

   _____

   force law (exact): _____
   force law (small amplitude): _____
   meaning of symbols: _____

   _____

   $\omega =$ _____ $T =$ _____

c. Physical pendulum
   displacement variable: _____

   _____

   torque law (exact): _____
   torque law (small amplitude): _____
   meaning of symbols: _____

   _____

   $\omega =$ _____ $T =$ _____

Remember that the small angle approximation $\sin\theta \approx \theta$ is valid only if θ is measured in radians. See the Mathematical Skills section for details.

SHM and uniform circular motion. When a particle moves around a circle with constant speed, the projections of its position vector on the x and y axes perform simple harmonic motions. Suppose the angular speed of the particle is ω and the radius of the circle is R. Put the origin of the coordinate system at the center of the circle and measure angles counterclockwise from the x axis. If the initial angular position of the particle is ϕ, then the x component of its position vector is given by $x(t) =$ _____. You can identify the angular speed ω of the particle with the _____ of the oscillation and the radius of the circle with the _____ of the oscillation.

Damped and forced harmonic motions. Many oscillating systems in nature are damped by a force that is proportional to the velocity. Take the damping force to be $-b(dx/dt)$, where b is a constant, and write the equation of motion obeyed by the displacement x of an object with mass m on a spring with spring constant k:

$$\frac{d^2x}{dt^2} =$$

If the damping coefficient b is small, you may think of the motion as simple harmonic with an exponentially decreasing amplitude. The angular frequency is not quite the same as in the absence of damping and is now denoted by ω'. Give the solution in terms of the angular frequency ω' and the damping constant b:

$$x(t) =$$

Give an expression for the angular frequency ω' in terms of k, m, and b:

$$\omega_d =$$

Note that if b is small, the angular frequency is nearly the same as the angular frequency for undamped motion, namely $\sqrt{k/m}$.

If $(b/2m)^2 < k/m$, then the object oscillates with decreasing amplitude. If $(b/2m)^2 \geq k/m$, then the object does not oscillate, but instead simply returns to its equilibrium point. The solution is then usually written in another form.

Sketch $x(t)$ on the graph below for $(b/2m)^2 < k/m$. Take the phase constant to be zero. Be sure to show both the oscillations and the decay of the amplitude.

$x(t)$

t

The mechanical energy of a damped oscillator is not constant. As time goes on, mechanical energy is dissipated by _____

_____.

Oscillations can be driven by applying an external sinusoidal force. Suppose the external force is given by $F_m \cos \omega_d t$ and write the equation of motion for the displacement variable x, including a damping term proportional to dx/dt:

$$\frac{d^2x}{dt^2} =$$

The angular frequency of a forced oscillation is the same as the angular frequency of the driving force and is NOT necessarily the natural angular frequency of the oscillator. Also know that the amplitude of the oscillation depends on the driving frequency, as well as on the driving amplitude F_m. The driving frequency for which the velocity amplitude is a maximum is called the _____ frequency. This is the same as the natural frequency of the oscillator. It is nearly the same as the frequency for which the amplitude is a maximum.

The amplitude of the motion does not decay with time even though a resistive force is present and mechanical energy is being dissipated. The mechanism that drives the oscillator and provides the external force does work on the system and supplies the energy required to keep it going at constant amplitude.

Study Fig. 16–xx of the text to see how the amplitude depends on the driving frequency. For a given value of the damping coefficient b, the amplitude decreases as the driving frequency moves away from the resonance frequency in either direction. As b increases, the amplitude at resonance _____. Also the width of the curve, measured at an amplitude that is half the resonance amplitude, increases. Resonance becomes less sharp.

II. PROBLEM SOLVING

Some problems make use of the relationships among angular frequency, frequency, and period for simple harmonic motion: $\omega = 2\pi f$, $f = 1/T$, and $\omega = 2\pi/T$. Occasionally the period is given indirectly by describing a time interval. You must then know, for example, that the time the oscillator takes to go from maximum displacement in one direction to maximum displacement in the other direction is $T/2$ or the time it takes to go from maximum displacement to zero displacement is $T/4$. If these time intervals or others are given, you should be able to calculate the period, frequency, and angular frequency. You should also know how to find the maximum speed and maximum acceleration in terms of the angular frequency and amplitude: $v_m = \omega x_m$ and $a_m = \omega^2 x_m$. Some problems require you to know the relationship between the angular frequency and the appropriate physical properties of the oscillating system: $\omega = \sqrt{k/m}$ for an undamped spring-object system.

A problem statement might give you an expression for the displacement of an oscillating object as a function of time and ask for the amplitude, angular frequency, and phase constant (or related quantities). Simply identify the various constants in the given expression. It might also ask for the coordinate, velocity, and acceleration at some specific time. Simply evaluate the expression and its first and second derivatives for that value of the time.

Some problems can be solved using the principle of mechanical energy conservation. For an spring-object system the mechanical energy E_{mec}, the speed v, and the coordinate x are related by $E_{mec} = \frac{1}{2}mv^2 + \frac{1}{2}kx^2$. If the object has speed v_1 when it is at x_1 and speed v_2 when it is at x_2, then conservation of mechanical energy yields $\frac{1}{2}mv_2^2 + \frac{1}{2}kx_2^2 = \frac{1}{2}mv_1^2 + \frac{1}{2}kx_1^2$. Either of these equations can be solved for one of the quantities that appear in them.

Some problems deal with the other oscillating systems discussed in the text: the torsional pendulum, the simple pendulum, and the physical pendulum. In each case, you should know how the angular frequency depends on properties of the oscillating body: $\sqrt{\kappa/I}$ for a torsional pendulum, $\sqrt{g/\ell}$ for a simple pendulum, and $\sqrt{mgd/I}$ for a physical pendulum.

Some problems deal with damped oscillations. The position of the oscillating object is then given by $x = x_m e^{-bt/2m} \cos(\omega' t + \phi)$, where $(\omega')^2 = (k/m) - (b^2/4m^2)$. Here the damping force is given by $F_d = -bv$. Other problems deal with forced oscillations. You should remember that the amplitude is a maximum when the frequency of the applied force is roughly the natural frequency of oscillation.

III. MATHEMATICAL SKILLS

Derivatives of sinusoidal functions.   You need to know how to differentiate $\sin(\omega t + \phi)$ and $\cos(\omega t + \phi)$ to verify the solutions to several of the equations of motion in the chapter and to calculate the velocity and acceleration of an oscillating body. Remember how to use the chain rule. Let $\omega t + \phi = u$. Then, $d\sin(\omega t + \phi)/dt = (d\sin u/du)(du/dt) = (\cos u)(\omega) = \omega\cos(\omega t + \phi)$. Similarly, $d\cos(\omega t + \phi)/dt = -\omega\sin(\omega t + \phi)$.

Small angle approximation.   Several of the oscillators discussed in this chapter are harmonic only if the amplitude is small. Simple and physical pendulums are examples. For the motion to be considered harmonic the angle θ of swing must be sufficiently small that $\sin\theta$ may be replaced by θ in radians without generating unacceptable error.

The Maclaurin series for $\sin\theta$ is

$$\sin\theta = \sum_{n=0}^{\infty}(-1)^n\frac{\theta^{2n+1}}{(2n+1)!},$$

and its first three terms are

$$\sin\theta = \theta - \frac{\theta^3}{6} + \frac{\theta^5}{120} - \dots.$$

The series is valid only if θ is in radians. Notice that if θ is small, each term in the series is less in magnitude than the previous term. The small angle approximation amounts to using only the first term of the series.

If θ is small, the error generated by the small angle approximation is nearly the second term $\theta^3/6$ and the fractional error is roughly $\theta^2/6$. For example, the error is less than 1 per cent if $\theta^2/6 < 0.01$ or $\theta < 0.2$ radians. In fact the error for $\theta = 0.2$ radians is 0.7 per cent.

IV. NOTES

Chapter 17
WAVES — I

I. BASIC CONCEPTS

In this chapter you study wave motion, a mechanism by which a disturbance (or distortion) created at one place in a medium propagates to other places. A wave may carry energy and momentum with it, but it does not carry matter.

The general ideas are specialized to a mechanical wave on a taut string. Pay attention to those characteristics of a string that determine the speed of a wave. You will also learn about sinusoidal waves, for which the disturbance has the shape of a sine or cosine function. This type wave has a special terminology associated with it; you will need to know the meaning of the terms amplitude, frequency, period, angular frequency, wavelength, and angular wave number. You will also need to understand what determines the values of each of these quantities.

Two or more waves present at the same time and place give rise to what are called interference effects. Special combinations of traveling waves result in standing waves. Learn what these phenomena are and how to analyze them.

Here are the main concepts that were discussed in previous chapters and are used in this chapter: displacement, velocity, speed, and acceleration (Chapter 2), Newton's second law of motion and tension in a string (Chapter 5), uniform circular motion (Chapter 6), kinetic energy, and power (Chapter 7), elastic potential energy and conservation of mechanical energy (Chapter 8), and amplitude, frequency, period, and angular frequency (Chapter 16). Review any you do not fully understand.

General characteristics of traveling waves. A distortion in one region of space causes a disturbance in neighboring regions, so the disturbance moves from place to place. After a taut string is distorted and released, for example, the distorted section pulls on neighboring sections to distort them and the neighboring sections pull the originally distorted portion back to its undistorted position.

Give some examples of mechanical waves: _____

_____

When a mechanical wave is present, both the waveform (the disturbance) and the medium (string, water, air) move. These motions are related to each other but they are NOT the same. While studying this chapter you should be continually aware of which of these motions is being discussed.

Waves are sometimes classified as transverse or longitudinal, according to the direction the medium moves relative to the direction the wave moves. In a <u>longitudinal</u> wave, the medium is displaced in a direction that is _____ to the direction the wave travels, and in a <u>transverse</u> wave it is displaced in a direction that is _____ to the direction the wave travels. Many waves, water waves among them, are neither transverse nor longitudinal.

The <u>wave speed</u> is the speed with which the distortion moves. The graph below shows a distorted string at time $t = 0$. The string stretches along the x axis and its displacement is denoted by y. Suppose the distortion travels without change in shape to the right at 2.0 m/s. On the coordinates provided sketch the string at $t = 1$ s and $t = 2$ s.

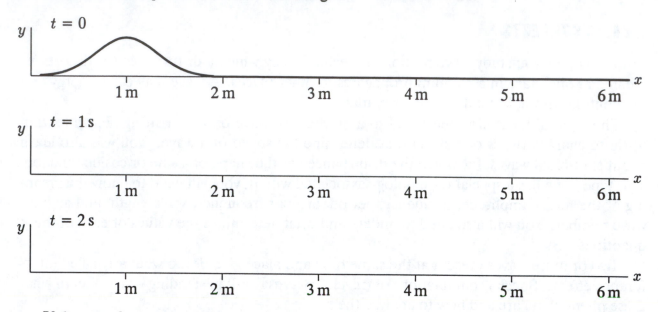

If the waveform moves with constant speed v and without change in shape, then any particular point on the waveform (the maximum displacement, for example) moves so its coordinate x at time t is given by $x - vt =$ constant for waveforms traveling in the *positive* x direction and by $x + vt =$ constant for waveforms traveling in the *negative* x direction. Carefully note the signs in these expressions.

For the waveform shown on the three graphs above take x to be the coordinate at the maximum string displacement and fill in the following table:

t (s)	x (m)	$x - vt$ (m)
0		
1		
2		

If you did not get the same answer for $x - vt$ each time, you probably did not draw the graphs correctly.

A wave on a string is described by giving the string displacement $y(x, t)$ as a function of position and time. For example, the graphs above show $y(x, t)$ for three different values of t. If you know the function $y(x, t)$, you can find the displacement of any point on the string at any time. Simply substitute the value of the coordinate x of the point and the value of the time t into the function.

If you are given the wave speed and the function $f(x)$ that describes the string displacement at $t = 0$, then you can find the displacement $y(x, t)$ for later times. If the waveform is moving in the positive x direction, you substitute _____ for x in the function $f(x)$. If the waveform is traveling in the negative x direction, you substitute _____ for x. All traveling waves are functions of $x - vt$ or $x + vt$. The variables x and t cannot enter in any way but these two combinations.

Sinusoidal waves. A sinusoidal wave is shown below for time t = 0.

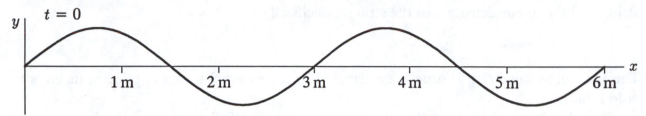

The function that describes the string displacement at $t = 0$ is $f(x) = y_m \sin(kx)$, where y_m and k are constants. y_m is called the _____ of the wave. Indicate it on the graph. k is called the _____ of the wave. It is related to the <u>wavelength</u> λ by $k =$ _____. Indicate the wavelength on the graph by marking an appropriate distance with the symbol λ. Any two points that are separated by a multiple of λ have identical displacements at every instant. The wavelength of the wave shown is about _____ m.

If the wave travels with speed v in the positive x direction, the displacement $y(x, t)$ at any time t is given by

$$y(x, t) = y_m \sin[k(x - vt)] = y_m \sin(kx - \omega t),$$

where $\omega = kv$.

Every point of the string moves in simple harmonic motion with <u>angular frequency</u> _____. In terms of the angular frequency the <u>frequency</u> of its motion is given by $f =$ _____ and the <u>period</u> of its motion is given by $T =$ _____. The period of an oscillation tells us _____

_____

and the frequency tells us _____

_____

The graph below shows the displacement of a point on the string as a function of time. On the time axis use the symbol T to label a time interval equal to a period.

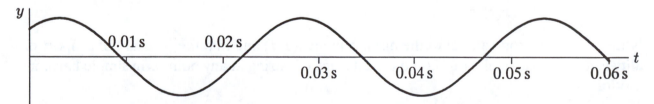

The period of the oscillation shown is about _____ s and the frequency is about _____ Hz.

When ω is written in terms of the period T and k is written in terms of the wavelength λ, the equation $\omega = kv$ becomes $\lambda = vT$. That is, the wave moves a distance equal to one wavelength in a time equal to _____. This result should be obvious to you. Consider a point on the string that has maximum displacement at $t = 0$. There is another point with the same displacement a distance λ away. By the time this part of the wave gets to the first point that point has moved through one complete cycle.

If the wave shown moves in the *negative* x direction, then

$$y(x,t) =$$

Wave speed. Carefully read Section 17–6 of the text. There, you are shown that Newton's second law leads directly to an expression for the wave speed in terms of the tension τ in the string and the linear density μ of the string. Specifically,

$$v =$$

The tension in the string is usually determined by the external forces applied at its ends to hold it taut.

You should carefully note that the wave speed does NOT depend on the frequency or wavelength. If the frequency is increased, the wavelength must decrease so that $v = \lambda f$ has the same value. Suppose, for example, the wave speed on a certain string is 5.0 m/s. Then, the wavelength of a 100 Hz sinusoidal wave is _____ m. If a 200 Hz sinusoidal wave travels on the same string with the same tension, its speed is _____ m/s and its wavelength is _____ m.

Particle velocity The transverse velocity $u(x, t)$ of the string at the position with coordinate x can be found as a function of time by differentiating _____ with respect to time. If the displacement is given by $y_m \sin(kx - \omega t)$, then

$$u(x, t) =$$

You should distinguish between wave and particle velocities. The wave velocity is associated with the motion of _____, while the particle velocity is associated with the motion of _____. Consider a transverse wave on a string and list some characteristics that are different for these two velocities: _____

_____

_____

The acceleration of a point on a string can be found by differentiating _____ with respect to time. For a sinusoidal wave the result is

$$a(x, t) =$$

Notice that it is proportional to the displacement $y(x, t)$ and that the constant of proportionality is negative, a result you should expect since the string at any point is in simple harmonic motion.

Energy and power. The mass in a segment of string of infinitesimal length dx is $dm =$ _____ if μ is the linear density. If u is the speed of the segment, then its kinetic energy is $dK =$ _____. This energy is transferred to a neighboring segment in time $dt = dx/v$, where v is the wave speed. Thus, the rate at which kinetic energy is carried by the string is given by $dK/dt =$ _____. For a sinusoidal wave, with $y(x, t) = y_m \sin(kx - \omega t)$,

$$\frac{dK}{dt} =$$

The average rate over a complete period is given by

$$\left(\frac{dK}{dt}\right)_{avg} =$$

since the average value of $\cos^2(kx - \omega t)$ is $1/2$.

The string has potential energy because _____

_____

The average rate with which potential energy is transported is exactly the same as the average rate with which kinetic energy is transported so, for a sinusoidal wave, the average rate of energy transport is given by

$$P_{avg} =$$

Notice that it depends on the square of the amplitude and on the square of the frequency.

Superposition and interference. To derive some results in this and the next chapter, you should know the trigonometric identity

$$\sin \alpha + \sin \beta = 2 \sin \left(\frac{\alpha + \beta}{2}\right) \cos \left(\frac{\alpha - \beta}{2}\right) ,$$

valid for any angles α and β. It is proved in the Mathematical Skills section.

When two waves, one with displacement $y_1(x, t)$ and the other with displacement $y_2(x, t)$, are simultaneously on the same string, the displacement of the string is given by their sum: $y(x, t) = y_1(x, t) + y_2(x, t)$, provided the amplitudes are small. The addition of waveforms is called <u>superposition</u>. Carefully note that displacements, not transmitted energies or powers, add. If the amplitudes are not small, the presence of one wave might change the shape of a second wave. This situation is not considered here.

The addition of waveforms leads to some interesting and important phenomena, called <u>interference</u> phenomena. Consider two sinusoidal waves with the same frequency, traveling in the same direction on the same string. Suppose the waves have the same amplitude but different phase constants and let $y_1(x, t) = y_m \sin(kx - \omega t)$ and $y_2(x, t) = y_m \sin(kx - \omega t + \phi)$. The waves are identical except that at every instant the second is shifted along the x axis from the first by an amount that depends on the value of the phase constant ϕ. Eq. 17–xx gives the equation for the resultant string displacement. Copy it here:

$$y(x, t) =$$

Now, use the trigonometric identity given above to prove the result:

Notice that the composite wave is sinusoidal with the same frequency and wavelength as either of the constituent waves.

Concentrate on the amplitude of the composite wave. It is a function of ϕ and, in particular, is given by _____. The maximum amplitude is given by _____ and the amplitude is a maximum when ϕ has any of the values _____. Maximum amplitude occurs when ϕ is adjusted so the crests of one of the constituent waves fall exactly on the _____ of the other. This condition is called _____ interference. The minimum amplitude is zero and the amplitude is a minimum when ϕ has any one of the values _____. Minimum amplitude occurs when ϕ is adjusted so the crests of one of the constituent waves fall exactly on the _____ of the other. This condition is called _____ interference. For other values of ϕ the interference is destructive but it is not complete.

Phasors. A phasor is a rotating vector that is used to represent a wave and to carry out calculations of interference phenomena. The upper diagram on the right shows a phasor associated with $y = y_m \sin(kx - \omega t)$. Its length (to some scale) is _____. Label the length with this symbol. Indicate the direction of rotation of the phasor. Identify the angle $kx - \omega t$ and label it on the diagram. The horizontal component of the phasor is _____ and the vertical component is _____. Both of these have the same mathematical form as a traveling wave.

The lower diagram on the right represents the addition of two waves with the same frequency and wavelength. The arrow closest to the origin represents $y_1 = y_{1m} \sin(kx - \omega t)$ and the other represents $y_2 = y_{2m} \sin(kx - \omega t + \phi)$. Label the lengths of the phasors and indicate the direction of rotation. Identify the phase angle ϕ and label it on the diagram. Draw the resultant phasor that represents the composite wave. Use the law of cosines to find an expression for the amplitude of the resultant wave in terms of y_{1m}, y_{2m}, and ϕ. Its square is

$$y_m^2 =$$

Standing waves and resonance. In a standing wave each part of the string oscillates back and forth but the waveform does not move, as it does in a traveling wave. No energy is transmitted from place to place.

A standing wave can be constructed as the superposition of two traveling waves with the same amplitude and frequency but moving in opposite directions. Let $y_1(x, t) = y_m \sin(kx - \omega t)$ and $y_2(x, t) = y_m \sin(kx + \omega t)$ represent the two traveling waves. The sum is given by Eq. 17–xx of the text. Copy it here:

$$y(x, t) =$$

Now, use the trigonometric identity given above to prove the result:

Note that x and t do NOT enter the expression for $y(x,t)$ in either of the forms $x - vt$ or $x + vt$. The wave is NOT a traveling wave. The shape changes as the string moves but it does not travel along the string. Points of maximum displacement remain at the same places, as do points of minimum displacement. Also note that each point on the string vibrates in simple harmonic motion with an amplitude that varies with position along the string. In fact, the amplitude of the oscillation of the point with coordinate x is given by _____.

At certain points, called <u>nodes</u>, the amplitude is zero. Since their displacements are always zero, these points on the string do not vibrate at all. For the standing wave given above, nodes occur at positions for which kx is a multiple of _____ or, what is the same thing, x in terms of the wavelength is a multiple of _____. Nodes are _____ wavelength apart.

At other points, called _____, the amplitude is a maximum. In terms of y_m it is _____. For the standing wave given above, these points occur at positions for which kx is an odd multiple of _____ or, what is the same thing, x is an odd multiple of _____.

The phase constant for both traveling waves was chosen to be zero but that is not a necessity. If the phase constants of the constituent traveling waves are different, the nodes and antinodes are still the same distances apart but they are shifted in position from those that occur when the phase constants are the same.

One way to generate a standing wave is simply to allow a sinusoidal wave to be reflected from an end of the string. The reflection, of course, travels in the opposite direction. If such a wave is reflected from a *fixed* end, the reflected and incident waves at that end have _____ signs and cancel each other there. The fixed end is a _____ of the standing wave pattern. On the other hand, if the same wave is reflected from a free end, the incident and reflected waves have _____ signs there and the free end of string is an _____ of the standing wave pattern.

For a string with both ends fixed, both ends are nodes of a standing wave pattern. This means that the traveling waves producing the pattern may have only certain wavelengths. Possible wavelengths are determined by the condition that the length of the string must be a multiple of _____, where λ is the wavelength. If v is the wave speed for traveling waves on the string, then the standing wave frequencies are given in terms of the string length L by $f =$ _____, where n is a positive integer, called the _____ number.

The lines below represent a string with fixed ends. Draw the amplitude as a function of position for the standing waves with the three lowest frequencies.

Standing wave frequencies are called the _____ frequencies of the string. If the string is driven at one of its standing wave frequencies by an applied sinusoidal force, the amplitude at the antinodes becomes large. The driving force is said to be in <u>resonance</u> with the string. Of course, the string can be driven at another frequency but then the amplitude remains small.

When an applied driving force is in resonance with a string, the energy supplied by the applied force is dissipated in internal friction. In the absence of friction the standing wave amplitude would grow without bound. This is to be contrasted with the situation in which the driving force is not in resonance with the string. Then, in addition to frictional dissipation, the string loses energy by doing work on _____.

II. PROBLEM SOLVING

Many of the problems involving sinusoidal waves on a string deal with the relationships $v = \lambda f = \lambda/T = \omega/k$, where v is the wave speed, λ is the wavelength, f is the frequency, T is the period, ω is the angular frequency, and k is the angular wave number. Typical problems might give you the wavelength and frequency, then ask for the wave speed, or might give you the wave speed and period, then ask for the wavelength or angular wave number.

Sometimes the quantities are given by describing the motion. For example, a problem might tell you that the string at one point takes a certain time to go from its equilibrium position to maximum displacement. This, of course, is one-fourth the period. In other problems the frequency of the source (a person's hand or a mechanical oscillator, for example) might be given. You must then recognize that the frequency of the wave is the same as the frequency of the source.

For some problems you must be able to write an expression for the displacement as a function of position and time for a sinusoidal wave. Since it has the form $y(x, t) = y_m \sin(kx \pm \omega t + \phi)$, you must be able to determine k, ω, and ϕ from data given in the problem statement. In many cases the initial conditions are not given. You may then select the time $t = 0$ so that $\phi = 0$ and write $y = y_m \sin(kx \pm \omega t)$. The sign in front of ωt is determined by the direction of travel. In other problems you may be given the displacement as a function of coordinate and time and asked to identify various quantities.

Some problems deal with the wave speed. For waves on a string, the fundamental equation is $v = \sqrt{\tau/\mu}$, where τ is the tension in the string and μ is the linear density of the string. The tension and linear density may not be given directly but, if the problem asks for the wave speed, sufficient information will be given to calculate these quantities.

The equation $P_{avg} = \frac{1}{2}\mu v \omega^2 y_m^2$ for the average power being transmitted at any time past any point is typically used in a straightforward manner to compute the P_{avg}, given the parameters of the wave. You may also be asked to find the value of one of the parameters (μ, ω, v, τ, y_m, or a related quantity) to achieve a given level of power transmission.

Interference problems may give the phase difference of two waves and ask for the resultant wave. Use $y_m \sin(kx - \omega t) + y_m \sin(kx - \omega t + \phi) = 2y_m \cos(\phi/2) \sin(kx - \omega t + \phi/2)$. In variations the resultant is given and you are asked for the phase difference. Solve for $\cos(\phi/2)$ and double the inverse cosine.

Some problems dealing with standing waves on a string give you the amplitude, frequency, and angular wave number (or related quantities) for the traveling waves and ask for the standing wave pattern. The inverse problem gives the standing wave in the form $y = A \sin(kx) \cos(\omega t)$ and asks for the component traveling waves. They have the form $y_m \sin(kx \pm \omega t)$ with $y_m = A/2$.

Instead of the wavelength or angular wave number of the traveling waves you might be told the distance between successive nodes or successive antinodes. Double it to find the wavelength. If you are told the distance between a node and a neighboring antinode, multiply it by 4 to find the wavelength.

If a standing wave is generated in a string with both ends fixed, the wave pattern must have a node at each end of the string. This means the length L of the string and the wavelength λ of the traveling waves must be related by $L = n\lambda/2$, where n is an integer. If one end is fixed and the other is free, the fixed end is a node and the free end is an antinode, so the length must be an odd multiple of $\lambda/4$.

III. MATHEMATICAL SKILLS

Partial derivatives. Partial derivatives are important for understanding much of this chapter. The displacement $y(x, t)$ of a string carrying a wave is a function of two variables, the coordinate x of a point on the string and the time t. You may differentiate with respect to either variable.

The notation $\partial y(x, t)/\partial x$ stands for the derivative of y with respect to x, with t treated as a constant. Similarly, $\partial y(x, t)/\partial t$ means the derivative of y with respect to t, with x treated as a constant. The result of either differentiation may again be a function of x and t. For example, $\partial \sin(kx - \omega t)/\partial x = k \cos(kx - \omega t)$ and $\partial \sin(kx - \omega t)/\partial t = -\omega \cos(kx - \omega t)$.

You should understand the physical significance of a partial derivative as well as be able to evaluate it, given the function. The partial derivative $\partial y(x, t)/\partial x$ gives the slope of the string at the point x and time t; here you are evaluating the rate at which the displacement changes with *distance* along the string at some instant of time. For this to be meaningful the definition must make use of displacements for slightly separated points on the string *at the same time*, in the limit as the separation tends to zero. That is why t is treated as a constant.

The partial derivative $\partial y(x, t)/\partial t$ gives the string velocity at the point x and time t; here you are evaluating the rate at which the displacement changes with *time* at a given point. Clearly this is associated with the displacement *of the same point* but at slightly different times, in the limit as the time interval approaches zero. That is why x is treated as a constant.

A trigonometric identity. The identity

$$\sin \alpha + \sin \beta = 2 \sin \left(\frac{\alpha + \beta}{2} \right) \cos \left(\frac{\alpha - \beta}{2} \right)$$

plays an important role in this chapter. You should be able to show its validity. Write

$$\sin \alpha + \sin \beta = \sin \left(\frac{\alpha + \beta}{2} + \frac{\alpha - \beta}{2} \right) + \sin \left(\frac{\alpha + \beta}{2} - \frac{\alpha - \beta}{2} \right)$$

and use the rules for expanding the sine of the sum and the sine of the difference of two angles: $\sin(\theta_1 + \theta_2) = \sin \theta_1 \cos \theta_2 + \cos \theta_1 \sin \theta_2$ and $\sin(\theta_1 - \theta_2) = \sin \theta_1 \cos \theta_2 - \cos \theta_1 \sin \theta_2$. These give

$$\sin \alpha + \sin \beta =$$
$$\sin \left(\frac{\alpha + \beta}{2} \right) \cos \left(\frac{\alpha - \beta}{2} \right) + \cos \left(\frac{\alpha + \beta}{2} \right) \sin \left(\frac{\alpha - \beta}{2} \right)$$
$$+ \sin \left(\frac{\alpha + \beta}{2} \right) \cos \left(\frac{\alpha - \beta}{2} \right) - \cos \left(\frac{\alpha + \beta}{2} \right) \sin \left(\frac{\alpha - \beta}{2} \right)$$
$$= 2 \sin \left(\frac{\alpha + \beta}{2} \right) \cos \left(\frac{\alpha - \beta}{2} \right) .$$

IV. NOTES

Chapter 18
WAVES — II

I. BASIC CONCEPTS

In this chapter the ideas of wave motion introduced in the last chapter are applied to sound waves. Pay particular attention to the dependence of the wave speed on the properties of the medium in which sound is propagating. Although the idea is the same, the properties are different from those that determine the speed of a wave on a taut string. Completely new concepts include beats and the Doppler shift. Be sure you understand what these phenomena are and how they originate.

This chapter makes use of the concepts of displacement, velocity, speed, and acceleration from Chapter 2, bulk modulus from Chapter 13, pressure and density from Chapter 15, amplitude, frequency, period, and angular frequency from Chapters 16 and 17, and wavelength, angular wave number, wave speed, harmonic, power transmitted by a wave, wave interference, standing wave, and resonance from Chapter 17. Be sure you understand them all.

General description. Sound waves are propagating distortions of a material medium. In fluids they are longitudinal: particles of the medium move back and forth along the line of _____. In solids they may be longitudinal, transverse, or neither.

Since particles at slightly different positions move different amounts the medium becomes compressed or rarefied as a sound wave passes. That is, we can consider a sound wave to be the propagation of a local increase or decrease in density. Since a change in pressure is associated with a change in density a sound wave may also be considered to be the propagation of a deviation in local pressure from the ambient pressure.

To understand this chapter, you must know the relationships between density, volume, and pressure discussed in Chapter 15. Consider a tube of fluid with cross-sectional area A, density ρ, and bulk modulus B, at pressure p. Suppose an element of fluid originally of length ℓ is uniformly elongated by $\Delta\ell$ so its new length is $\ell + \Delta\ell$ and its new volume is $A(\ell + \Delta\ell)$. The mass of the fluid in the element is given by $m = \rho A\ell$, so the density after elongation is $\rho' = m/A(\ell + \Delta\ell) = \rho\ell/(\ell + \Delta\ell)$. This expression for ρ' can be written $\rho' = \rho + \Delta\rho$, where $\Delta\rho$ is the change in density. In the space below show that if $\Delta\ell$ is much smaller than ℓ, then $\Delta\rho = -\rho\Delta\ell/\ell$:

Notice that $\Delta\rho$ is negative if $\Delta\ell$ is positive. The density decreases because the same mass occupies a larger volume after elongation. The change in pressure that accompanies the elongation is given by $\Delta p = -B\Delta V/V = -B\Delta\ell/\ell$. It is also proportional to $\Delta\ell$. The same expressions are also valid for a compression. Then, $\Delta\ell$ is negative.

The relationships $\Delta\rho = -\rho\Delta\ell/\ell$ and $\Delta p = -B\Delta\ell/\ell$ are used to derive an expression for the speed of sound in terms of properties of the medium and to relate displacement, density, and pressure waves to each other.

Speed of sound.   The speed of sound in a fluid is determined by the density ρ and bulk modulus B of the fluid. Specifically, it is given by

$$v = $$

To understand this expression, apply Newton's second law to an element of fluid in which a sound pulse is traveling and make use of the relations between density, volume, and pressure given above. The derivation follows closely the derivation of the expression for the speed of a wave on a string, given in Section 17–6. You may wish to review that derivation.

We view a compressional pulse in a fluid from a reference frame that moves with the speed of sound v, so the pulse is stationary and the fluid moves with speed v into the right end and out of the left end of the pulse, as shown in Fig. 18–5 of the text. An element of fluid with cross-sectional area A and length ℓ (when just outside the pulse) takes time $\Delta t = \ell/v$ to completely enter the pulse. In terms of v and Δt, the volume of the element when it is outside is _____ and if ρ is the fluid density outside, then the mass of the element is $m =$ _____ .

Now, consider the fluid element when it is partly inside and partly outside the pulse. If the pressure outside the pulse is p and the pressure inside the pulse is $p + \Delta p$, then the net force on the element is $F =$ _____ . Since the pressure inside the pulse is greater than the pressure outside, this force slows the element as it enters the pulse, so its velocity changes from v to $v + \Delta v$, where Δv is negative. The acceleration of the element is $a = \Delta v/\Delta t$. In $F = ma$ replace F with $-A\Delta p$, m with $\rho v A\Delta t$, and a with $\Delta v/\Delta t$, then solve for v:

Multiply the result by v to obtain $v^2 = -v\Delta p/\rho\Delta v$.

The key to the remainder of the derivation is to observe that the volume of the fluid element changes as it enters the pulse and that the fractional change in volume is the same as the fractional change in speed: that is, $\Delta V/V = \Delta v/v$. The leading edge of the element, inside the pulse, travels slower than the trailing edge, outside the pulse, by Δv. In time Δt it goes a distance $\Delta v\Delta t$ less, and this must be the amount by which the element is shortened. Thus, $\Delta V = A\Delta v\Delta t$. Now, $V = Av\Delta t$ is the volume V of the element when it is completely outside. Divide one of these results by the other to obtain $\Delta V/V = \Delta v/v$. Substitute $\Delta V/V$ for $\Delta v/v$ in the expression for v^2, then replace $-\Delta p/(\Delta V/V)$ with the bulk modulus B and solve for v:

You should have obtained $v = \sqrt{B/\rho}$. Thus, the expression for the speed of sound can be derived straightforwardly from Newton's second law and is a direct result of the forces neighboring portions of the fluid exert on each other. You should know that the speed of sound is about _____ m/s in air, about _____ m/s in water, and about _____ m/s in solids.

Pressure and density waves. To describe a traveling sound wave, you can use any of three quantities: the particle displacement, the deviation of the density from its ambient value, and the deviation of the pressure from its ambient value. All propagate as waves. They are related to each other and you should understand the relationships.

Suppose a sound wave is traveling along the x axis through a fluid and suppose further that the displacement of the fluid at coordinate x and time t is given by a known function $s(x,t)$. Consider an element of fluid that in the absence of the wave extends from x to $x + \Delta x$. Its length is Δx. In the presence of the wave the same fluid element extends from $x + s(x,t)$ to $x + \Delta x + s(x + \Delta x, t)$ so its change in length is $\Delta \ell = s(x + \Delta x, t) - s(x,t)$. In the limit as Δx becomes small $\Delta \ell$ becomes $[\partial s(x,t)/\partial x]\Delta x$. Thus, the density in the presence of a wave is $\rho'(x,t) = \rho + \Delta\rho(x,t)$, where $\Delta\rho(x,t) = -\rho\Delta\ell/\Delta x = -\rho\partial s(x,t)/\partial x$.

Similarly, the pressure in the presence of a wave is $p'(x,t) = p + \Delta p(x,t)$ where $\Delta p = -B\Delta\ell/\Delta x = -B\partial s(x,t)/\partial x$ and B is the bulk modulus.

Suppose the sound wave is sinusoidal and the fluid displacement is given by $s(x,t) = s_m \cos(kx - \omega t)$, where k is the wave number and ω is the angular frequency. In terms of these quantities, the deviation of the density from its ambient value is given by $\Delta\rho(x,t) = $ _____ and the deviation of the pressure from its ambient value is given by $\Delta p(x,t) = $ _____ .

These expressions can be written

$$\Delta\rho = \Delta\rho_m \sin(kx - \omega t)$$

and

$$\Delta p = \Delta p_m \sin(kx - \omega t),$$

where the density amplitude $\Delta\rho_m$ and the pressure amplitude Δp_m are given in terms of the displacement amplitude s_m, the wave speed v, and the wave number k by $\Delta\rho_m = $ _____ and $\Delta p_m = $ _____ .

Notice that the pressure and density waves are not in phase with the displacement wave. At points where the displacement is a maximum the deviation of the pressure from its ambient value is a _____ and the deviation of the density from its ambient value is a _____ . These results make physical sense because the fluid displacement is nearly uniform in the neighborhood of a displacement maximum. An element of fluid is neither compressed nor elongated there. A _____ value of Δp corresponds to a compression, a _____ value corresponds to an expansion.

At points where the displacement is zero the deviation of the pressure from its ambient value is a _____ and the deviation of the density from its ambient value is a _____ . Here the rate of change of the displacement with distance has its greatest magnitude. The elongation or compression of the fluid is greater here than anywhere else. Thus, we expect deviations of the pressure and density to be the greatest at these points.

Interference. To find the particle displacement when two or more sound waves are simultaneously present, add the _____ due to the individual waves. Let $s_1 = s_m \cos(kx - \omega t)$ represent one wave and $s_2 = s_m \cos(kx - \omega t + \phi)$ represent another. These waves interfere constructively if the phase difference ϕ has any of the values _____ rad, where n is an integer. The interference will be fully destructive if ϕ has any of the values _____ rad, where n is _____.

If the waves are generated by sources that are in phase but they travel to the detector along different paths, they may have different phases at the detector. If one wave travels a distance x to the detector and the other travels a distance $x + \Delta d$, then the phase difference at the detector is given by $\phi =$ _____, where k is the angular wave number. Since $k = 2\pi/\lambda$, where λ is the wavelength, the phase difference can be written $\phi = 2\pi\Delta d/\lambda$. Fully constructive interference occurs if Δx is a multiple of _____; fully destructive interference occurs if Δx is an odd multiple of _____.

Energy transport and intensity. If $s(x,t)$ is the particle displacement at coordinate x and time t, then $v_s =$ _____ is the particle velocity. The mass contained in an infinitesimal length dx of fluid is given by d$m =$ _____, where ρ is the density and A is the cross-sectional area. This kinetic energy is transported from the segment of fluid to a neighboring segment in time d$t =$ dx/v, where v is the speed of sound. Thus the rate at which kinetic energy flows in a sound wave is given by

$$\frac{\mathrm{d}K}{\mathrm{d}t} =$$

For a sinusoidal sound wave, for which $s = s_m \cos(kx - \omega t)$, this is

$$\frac{\mathrm{d}K}{\mathrm{d}t} =$$

The average over a cycle is

$$\left(\frac{\mathrm{d}K}{\mathrm{d}t}\right)_{\mathrm{avg}} =$$

since the average value of $\sin^2(kx - \omega t)$ is $1/2$. Potential energy is transported at the same rate, so the average rate of transport of mechanical energy is given by

$$P_{\mathrm{avg}} =$$

The <u>intensity</u> of a sound wave is the average rate of energy flow per unit _____ and, for the sinusoidal wave discussed above, is given by

$$I =$$

where s_m is the displacement amplitude. The SI units for intensity are _____.

<u>Sound level</u> is often used instead of intensity. The sound level β associated with intensity I is defined by $\beta =$ _____, where I_0 is the standard reference intensity (_____ W/m^2). Sound level is measured in units of _____, abbreviated _____.

Carefully note that the sound level is defined in terms of the *logarithm to the base* 10 of I/I_0. If $I = I_0$, the sound level is _____ db. If the intensity is increased by a factor of 10, the sound level increases by _____ db.

The standard reference intensity is roughly the threshold of human hearing. The intensity at the upper end of the human hearing range (called the threshold of pain) is about _____ W/m^2 and the sound level is about _____ db. Look at Table 18–2 of the text for some other sound level values.

Standing sound waves. Two sinusoidal sound waves with the same frequency and amplitude but traveling in opposite directions combine to form a standing wave. At a displacement node the displacement is always _____. At a displacement antinode the displacement oscillates between _____ and _____, where s_m is the displacement amplitude of either of the traveling waves. At other points the displacement oscillates with an amplitude that is given by _____, where k is the angular wave number of either of the traveling waves and x is the coordinate of the point.

If there are no losses, standing waves are created in pipes by the superposition of a sinusoidal wave and its reflection from the end of the pipe. A displacement _____ exists at a closed end of a pipe. The open end of a pipe is very nearly a displacement _____.

If both ends of a pipe are open, the wavelengths associated with possible standing waves are such that the pipe length is a multiple of _____. In terms of L, the standing wave wavelengths are given by

$$\lambda =$$

and the standing wave frequencies are given by

$$f =$$

where v is the speed of sound for the fluid that fills the pipe and n is _____.

For the lowest three standing wave frequencies use the coordinates below to plot the displacement amplitude A as a function of position along the pipe.

If one end of a pipe is open and the other is closed, the wavelengths associated with possible standing waves are such that the pipe length is an odd multiple of _____. In terms of L, the standing wave wavelengths are given by

$$\lambda =$$

and the standing wave frequencies are given by

$$f =$$

where v is the speed of sound for the fluid that fills the pipe and n is _____.

For the lowest three standing wave frequencies use the coordinates below to plot the displacement amplitude A as a function of position along the pipe.

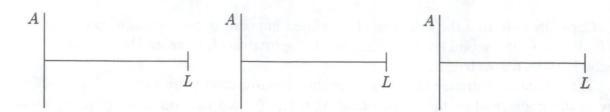

A string of a stringed instrument or the air in an organ pipe can vibrate with a combination of its standing wave (or natural) frequencies. Which sound is produced depends on how the instrument is excited. Usually the lowest frequency dominates but higher frequency sound is mixed in. The admixture of higher frequencies gives any instrument the quality of sound peculiar to that instrument and, for example, allows us to distinguish a violin from a piano.

The lowest frequency is called the _____ frequency or first harmonic, while higher frequencies are higher harmonics.

Beats. Beats are created when two sound waves with nearly the same _____ are simultaneously present. We can view the resultant wave as one with a frequency that is the average of the frequencies of the constituent waves and an amplitude that varies with time, but much more slowly than either of the constituent waves.

Since we are interested in the time dependence of the wave, we can study the displacement at a single point in space and ignore variations with position. Let $s_1 = s_m \cos(\omega_1 t)$ represent the displacement at the point due to one of the waves and $s_2 = s_m \cos(\omega_2 t)$ represent the displacement at the same point due to the other wave. The resultant displacement is the sum of the two and, once the trigonometric identity

$$\cos\alpha + \cos\beta = 2\cos\frac{\alpha - \beta}{2}\cos\frac{\alpha + \beta}{2}$$

is used, it can be written as Eq. 18–43 of the text:

$$s(t) =$$

Notice that there are two time dependent factors, both periodic. One has an angular frequency $\omega =$ _____, the average of the two constituent angular frequencies. This is the greater of the angular frequencies associated with the two factors and if the two constituent frequencies are nearly the same, it is essentially equal to either of them.

The angular frequency of the second time dependent factor is $\omega' =$ _____. Note that it depends on the difference in the two constituent frequencies. If ω_1 and ω_2 are nearly the same, the factor associated with this angular frequency is slowly varying. We may think of it as a slow variation in the amplitude of the faster vibration. The effect can be produced, for example, by blowing a note on a horn at the angular frequency ω, but modulating it so it is periodically loud and soft.

A <u>beat</u> is a maximum in the *intensity* and occurs each time $\cos(\omega' t)$ goes from +1 to _____. Thus, the beat angular frequency ω_{beat} is NOT the same as ω'. In fact, $\omega_{\text{beat}} =$ _____ω', or in terms of ω_1 and ω_2

$$\omega_{\text{beat}} =$$

If one constituent wave has a frequency of 1000 Hz and the other has a frequency of 1005 Hz, then the beat frequency is $f_{\text{beat}} =$ _____ Hz and the beat angular frequency is $\omega_{\text{beat}} =$ _____ rad/s.

Doppler effect. Suppose a sustained note with a well-defined frequency f is played by a stationary trumpeter. If you move rapidly *toward* the trumpeter, you will hear a note with a _____ frequency. If you move rapidly *away*, from the trumpeter you will hear a note with a _____ frequency. Similar effects occur if you are stationary and the trumpeter is moving. The note has a higher frequency if the trumpeter is moving _____ and a lower frequency if the trumpeter is moving _____. These are examples of the Doppler effect. The next time you hear a police or ambulance siren on the highway listen carefully as the vehicle approaches and then recedes from you. If it is going sufficiently fast, you should hear the Doppler shift in frequency.

To understand how the Doppler effect comes about, suppose a sound detector is moving with speed v_D toward a source of sound with frequency f and wavelength λ ($= v/f$, where v is the speed of sound). If the detector were at rest, it would receive _____ wave crests in time t. Because it is moving toward the source it receives _____ more crests in the same time. Thus, the number of crests it receives in time t is _____ and the frequency it detects is this number divided by t, or

$$f' =$$

Write down the equations for the other possibilities:

If the detector is moving with speed v_D *away from* a stationary source, then the frequency it detects is

$$f' =$$

If the detector is stationary and the source is moving with speed v_S *toward* it, then the frequency it detects is

$$f' =$$

If the detector is stationary and the source is moving *away from* it with speed v_S, then the frequency it detects is

$$f' =$$

Eq. 18–47 in the text covers all possibilities. Write it here:

$$f' =$$

You can easily determine which signs to use in any particular situation by remembering that motion of the source toward the observer or the observer toward the source results in hearing a higher frequency while motion of the source away from the observer or the observer away from the source results in hearing a lower frequency than would be heard if both were stationary.

You should understand that the velocities in the Doppler effect equation are measured relative to the medium in which the wave is propagating (the air, for example). What counts is not the motion of the source relative to the observer but the motions of both the source and observer relative to the medium of propagation. You should also understand that the Doppler effect equations given above are valid only if the motion is along the line joining the source and detector. For motion in other directions v_D and v_S must be interpreted as components of the velocities along that line.

To obtain an understanding of the speeds involved, estimate the speed with which you would have to move toward a stationary source of sound in still air to hear a 10% increase in frequency: _____ m/s. Can this speed be achieved by walking slowly, walking fast, riding a bicycle, driving a car at a moderate speed, or driving a car at high speed?

The Doppler effect also occurs for electromagnetic radiation. Light from stars that are moving at high speeds away from Earth is shifted toward the red and the extent of the shift is used to calculate the speeds of the stars. Doppler shifts of radar waves reflected from moving objects can be used to find their speeds. Police use the effect to detect speeders and TV technicians use it to find the speeds of thrown baseballs.

If a source of sound is moving through a medium faster than the speed of sound in the medium, a shock wave is produced. Then, a wavefront has the shape of a _____, with the source at its _____, as shown in Fig. 18–23 of the text. The half angle θ of the cone is given by

$$\sin \theta =$$

where v is the speed of _____ and v_S is the speed of _____. Note that no shock wave is produced if $v < v_S$.

Shock waves are responsible for sonic booms. A listener hears a sonic boom when

_____

_____.

II. PROBLEM SOLVING

Since many of the problems are similar to those of the last chapter, you might want to review Chapter 17 of this manual.

One difference is that the speed of sound in an isotropic medium is given by $\sqrt{B/\rho}$, where B is the bulk modulus and ρ is the density of the medium. If you are given the bulk modulus and density, you can calculate the speed of sound. Alternatively, you might use $v = \lambda f$ (or a related expression) to find the speed of sound, then use $v = \sqrt{B/\rho}$ to solve for B or ρ, given the other.

Interference problems are quite similar to those of the last chapter. For fully destructive interference the two waves are out of phase by an odd multiple of π rad; for fully constructive interference their phases differ by zero or a multiple of 2π rad. In some cases the phase difference comes about because the two waves travel different distances from their sources to the detector. This phase difference is given by $\phi = 2\pi \, \Delta d/\lambda$, where Δd is the difference in the distances traveled and λ is the wavelength. There may be an additional phase difference because the sources are not in phase with each other. You may be given the difference in the distances traveled and asked for the phase difference or the amplitude of the resultant wave at the detector. To calculate the resultant amplitude use $2s_m \cos(\phi/2)$, where s_m is the amplitude of either one of the waves.

Sound waves can also form standing wave patterns. The requirements are the same as for waves on a string: two traveling waves with the same amplitude and frequency but traveling in opposite directions form a standing wave.

To solve power, intensity, and sound level problems, use the equations $P_{\text{avg}} = \frac{1}{2}\rho A v \omega^2 s_m^2$ for the average power transmitted through area A, $I = \frac{1}{2}\rho v \omega^2 s_m^2$ for the intensity, and $\beta = (10\,\text{db})\log(I/I_0)$ for the sound level. Remember that the intensity is the average power per unit area. If a source emits isotropically (the same in all directions), then energy crossing every sphere centered at the source is the same. Since the surface area of a sphere is $4\pi r^2$, the intensity a distance r from the source is given by $P_{\text{avg}}/4\pi r^2$.

Many problems deal with the production of sound by pipes with either both ends open or one end open and one end closed. Remember that a displacement node and a pressure antinode occur at a closed end while a displacement antinode and a pressure node occur at an open end. Also remember that nodes are half a wavelength apart and antinodes occur halfway between adjacent nodes.

Some problems deal with the production of beats by two sinusoidal sound waves with nearly the same frequency. You may be given the frequency f_1 of one of the waves and the beat frequency f_{beat}, then asked for the frequency f_2 of the other wave. Since $f_{\text{beat}} = |f_1 - f_2|$, it is given by $f_2 = f_1 \pm f_{\text{beat}}$. You require more information to determine which sign to use in this equation. One way to give this information is to tell you what happens to the beat frequency if f_1 is increased (or decreased). If the beat frequency increases when f_1 increases, then f_1 must be greater than f_2 and $f_2 = f_1 - f_{\text{beat}}$. If the beat frequency decreases, then f_1 must be less than f_2 and $f_2 = f_1 + f_{\text{beat}}$.

Nearly all Doppler shift problems can be solved using

$$f' = f\left[\frac{v \pm v_D}{v \mp v_S}\right],$$

where v is speed of sound, v_S is the speed of the source, v_D is the speed of the observer, f is the frequency of the source, and f' is the frequency detected by the observer. The upper sign in the numerator refers to a situation in which the observer is moving toward the source; the lower sign refers to a situation in which the observer is moving away from the source. The upper sign in the denominator refers to a situation in which the source is moving toward the observer; the lower sign refers to a situation in which the source is moving away from the observer. Remember that all speeds are measured relative to the medium in which the sound is propagating. You might be given the velocities and one of the frequencies, then asked for the other frequency. In other situations you might be given the two frequencies and one of the velocities, then asked for the other velocity. In all cases, simple algebraic manipulation of the equation will produce the desired expression.

Solutions to most shock wave problems start with $\sin\theta = v/v_S$, where θ is the half angle of the shock envelope, v is the speed of sound, and v_S is the speed of the source. For an airplane flying horizontally at altitude h you may also need to make use of $\tan\theta = h/d$, where d is the horizontal distance from the intersection of the shock wave with the ground to the point on the ground under the plane. Some problems give the time interval t from when the plane is overhead to when the sonic boom is heard. Use $d = v_S t$.

III. NOTES

OVERVIEW II
THERMODYNAMICS

Thermodynamic phenomena are familiar to everyone. During a long trip, the tires of your car get hot and the air pressure in them increases; the car engine and coolant also warm, but the fan blows cooler outside air over them and keeps them at appropriate operating temperatures; perhaps the car has an air conditioner to cool the air inside and you notice that you use fuel at a slightly greater rate when it is on; the brake shoe and disks become hot when you brake but, after the car has stopped for sufficient time, they have the same temperature as the surrounding air.

Thermodynamics deals chiefly with the transfer of energy between large systems of particles and the changes that thereby occur in the systems. Simple gases are used to illustrate thermodynamic phenomena and, for them, the quantities considered are the volume, pressure, and temperature. The list is longer for other systems. Chapter 19 deals with the concept of temperature and, when you study that chapter, you will learn its precise definition and how it is measured.

An energy transfer can change both the internal energy of the system and the kinetic energy associated with the motion of the system as a whole. As you may recall from your study of Chapter 8, the internal energy is the sum of the kinetic energies of the particles, as measured in a reference frame that moves with the center of mass, and the potential energies of their interactions with each other. A kinetic energy is associated with the motion of the system as a whole but, for simplic-ity, we treat situations in which the center of mass remains at rest and only the internal energy changes. You will study ideal gases, for which the internal energy is simply the sum of the kinetic energies of the particles, including their rotational kinetic energies. These gases have no internal potential energy.

The study of energy transfers themselves form an important part of thermodynamics, apart from the changes they bring about in a system. You will learn that energy can be transferred to or from a system in two distinct ways, through work done on or by the system and through heat exchanged between the system and its environment. You are already familiar with work; heat is probably new to you. For processes involving a system with its center of mass at rest, work involves a change in volume. Heat, on the other hand, is energy that is transferred because the system and its environment are at different temperatures and does not involve a change in volume.

Work done by a system tends to decrease the internal energy and temperature of the system. Energy entering a system (as heat, for example) tends to increase those quantities. In fact, the principle of energy conservation, modified for a system with its center of mass at rest, tells us that the difference between the energy entering a system and the work done by the system is equal to the change in internal energy of the system. This principle, known as the first law of thermodynamics, is an important theme of Chapter 19. You will use it in many of your calculations involving changes in internal energy, heat energy, and work.

The pressure, volume, and temperature of a gas are related to each other: a change in one brings about changes in one or both of the others. The relationship is different for different systems but such a relationship, called an equation of state, exists for every system. Every equation of state contains at least three variables and so does not by itself determine the changes that accompany an exchange of energy. More information about the process must be known. You will study, for example, processes in which the volume, pressure, or temperature are held constant. Chapter 20 deals with the equation of state for an ideal gas.

To complete the study of changes in a system brought about by the transfer of energy, you must know how the internal energy depends on the other thermodynamic variables. The internal energy of an ideal gas depends only on the temperature and changes in these quantities are directly proportional to each other. This important result is obtained in Chapter 20.

Thermodynamics, although it deals with macroscopic systems, is firmly based on the mechanics of particles. In Chapter 20, you will learn how the pressure of a gas is related to the speeds of the particles. You will also learn how the total kinetic energy of the particles is related to the temperature.

There are certain natural limitations on thermodynamic processes. For example, if no work is done on or by a system, then energy flows out of the system as heat if its temperature is higher than that of its environment and flows into the system as heat if its temperature is lower. Only if work is done can energy flow as heat from the colder to the hotter body. Another natural limitation concerns processes in which energy enters a system as heat and the system does work. No process can convert all the heat energy into work; at some point in the process, heat energy must flow out of the system or else the internal energy must change. These limitations and others are codified in what is known as the second law of thermodynamics.

The second law is stated in terms of an important property of any system, called its entropy. In Chapter 21, you will learn to calculate entropy changes for various processes. You will also learn that the second law can be stated as: The total entropy of a system and its environment never decreases. You will see how this law limits thermodynamics processes.

The limitations described by the second law have important practical consequences for the design of refrigerators, air conditioners, engines, electrical power generators, and many other devices. More importantly for our purposes, they also give us a glimpse at the way nature operates.

Chapter 19
TEMPERATURE, HEAT,
AND THE FIRST LAW OF THERMODYNAMICS

I. BASIC CONCEPTS

You now begin the study of thermodynamics. Temperature is familiar to you but you have probably not thought about it in detail. Be sure you understand its definition and how it is measured. Pay particular attention to the zeroth law of thermodynamics, which codifies the characteristic of nature that allows temperature to be defined. You will also study the phenomenon of thermal expansion, the familiar change in the dimensions of an object that occurs when its temperature is changed. The first law of thermodynamics, derived from the conservation of energy principle, governs the exchange of energy between a system and its environment. Be sure to distinguish between the two mechanisms of energy transfer, work and heat. Learn how to calculate the energy absorbed or rejected as heat in terms of the heat capacity of the system and learn how to calculate the work done by a system, given the pressure as a function of volume.

Review the following concepts from previous chapters: work (Chapter 7), thermal energy (Chapter 8), and manometer and pressure (Chapter 15).

Temperature.   Temperature is intimately related to the idea of <u>thermal equilibrium</u>. If two objects that are not in thermal equilibrium with each other are allowed to exchange energy, they will do so and, as a result, some or all of their macroscopic properties will change. When the properties stop changing, there is no longer a net flow of energy and the objects are said to be in thermal equilibrium with each other. They are then at the same temperature.

Temperature is the property of an object that _____

_____.

Note carefully that, when two systems are in thermal equilibrium with each other, all macroscopic quantities except temperature might have different values for the two systems. Suppose the systems are gases. When in thermal equilibrium with each other, their pressures may be different, their volumes may be different, their particle numbers may be different, and their internal energies may be different. Their temperatures, however, are the same.

The law of nature that legitimizes the temperature as a property of a body in thermal equilibrium with another is the <u>zeroth law of thermodynamics</u>. State the law here:

_____

_____

_____

Suppose the law were not valid and suppose further that body A and body B are in thermal equilibrium and therefore at the same temperature. If body C is in equilibrium with A but not with B, then the temperature of C is not a well-defined quantity and cannot be considered a property of C.

Another important consequence of the zeroth law is that it allows us to select some object for use as a thermometer. Suppose that when the thermometer is in thermal equilibrium with body A it is also in thermal equilibrium with body B. Then, because the law is valid, we know that A and B have the same temperature. We do not need to place A and B in thermal contact to test if they are equilibrium with each other.

Measurement of temperature. In general, temperature is measured by measuring some property of a system. You are familiar with an ordinary mercury thermometer, for which the _____ of the mercury column is measured. The text describes a constant-volume gas thermometer, for which the _____ of the gas in a bulb is measured. The gas must always have the same volume, no matter what its temperature. Look at Fig. 19–5 and tell how the gas is kept at the same volume. At high temperatures the gas in the bulb expands and pushes the top of the left-hand mercury column downward. The reservoir is then _____ to bring the top of the column back to _____. At low temperatures the gas contracts and the top of the left-hand column moves closer to the bulb. The reservoir is then _____ to bring the top of the column back to _____.

The temperature T in kelvins is taken to be proportional to the pressure: $T = Cp$, where the constant of proportionality is chosen so $T =$ _____ at the _____ point of water. Thus, $T = T_3(p/p_3)$, where T_3 is _____ K and p_3 is _____. Actually the ratio p/p_3 must be evaluated in the limit as the _____ of gas in the bulb approaches _____. In this limit, the measured value of T does not depend on the type of gas used.

Three types of temperature scales are in common use: Celsius, Fahrenheit, and Kelvin (or absolute). A constant-volume gas thermometer gives the temperature on the Kelvin scale. The Celsius temperature T_C is defined in terms of the Kelvin-scale temperature T by $T_C =$ _____. The Fahrenheit temperature T_F is given in terms of the Celsius temperature by $T_F =$ _____. A Celsius degree is the same as a kelvin and is _____ as large as a Fahrenheit degree.

Thermal expansion. Most solids and liquids expand when the temperature is increased and contract when it is decreased. If the temperature is changed by ΔT, a rod originally of length L changes length by $\Delta L =$ _____, where α is the coefficient of linear expansion. Over small temperature ranges α may be considered to be independent of temperature. This is an approximation. For a situation in which the temperature is raised by ΔT, then lowered by the same amount, repeated application of the equation predicts the rod is shorter than its original length, whereas it actually returns to its original length.

Consult Table 19–2 of the text for values of some coefficients of linear expansion. You may need them to solve some problems. Carefully note that when values of α from the table are used ΔT must be in Celsius degrees or kelvins.

When the temperature of a homogeneous solid changes, the length of every line in the solid changes by the same *fractional* amount, given by $\Delta L/L = \alpha \Delta T$. The length of a scratch on the surface of a homogeneous solid also changes by the same fractional amount. The fractional change in the diameter of a round hole in an object, for example, is given by $\Delta D/D =$ _____, where α is the coefficient of linear expansion of the object.

When the temperature changes by a small amount ΔT, the area A of a face of a solid changes by $\Delta A =$ _____ and the volume V of the solid changes by $\Delta V =$ _____, where α is the coefficient of linear expansion. For a fluid the change in volume is given by the same expression. In the space below, prove the first result for yourself. Consider a rectangular face with sides L_1 and L_2. Its area before a temperature change is $A = L_1 L_2$ and its area after the change is $A + \Delta A = (L_1 + \Delta L_1)(L_2 + \Delta L_2)$. Substitute $\Delta L_1 = \alpha L_1 \Delta T$ and $\Delta L_2 = \alpha L_2 \Delta T$ for the changes in length and retain only terms that are proportional to ΔT when you multiply the two factors in parentheses. You should obtain $\Delta A = 2\alpha A \Delta T$. The expression for the change in volume can be obtained in a similar manner. The change in volume of a fluid is given by $\Delta V = \beta V \, \Delta T$, where β is the <u>coefficient</u> of <u>volume</u> <u>expansion</u>.

Water near 4°C and a few other substances decrease in volume when the temperature is increased. These materials have negative coefficients of thermal expansion in the temperature range for which such a contraction occurs.

Heat.   Recall from your study of Chapter 8 that the work done on an object by external forces may change its internal (thermal) energy. There is another way in which the internal energy may change: the object may absorb or reject energy as heat. Heat is the energy that flows between a system and its environment because _____

_____.

Heat and work are alternate ways of transferring energy. Energy flows as heat from the environment to the system when the temperature of the environment is _____ than the temperature of the system.

A sign convention is adopted for work and heat. Work W is taken to be positive when _____ and heat Q is taken to be positive when _____.

The SI units of heat are _____.   Other units in common use and their SI equivalents are 1 cal = _____ J, 1 Btu (British thermal unit) = _____ J, and 1 Cal = _____ J. The Calorie unit (with a capital C) is commonly used in nutrition.

Heat capacity and specific heat.   The heat capacity relates the energy transferred as heat into or out of the system to the change in temperature of the system. If during some process energy Q is absorbed as heat and the temperature increases by a small increment ΔT, then the heat capacity of the system for that process is given by $C =$ _____.   The heat capacity depends not only on the kind of material in the system but also the amount of material. A related property that depends only on the kind of material and not the amount is the specific heat, defined in terms of C by $c =$ _____, where m is the mass of the system. Another is the molar specific heat (or heat capacity per mole), defined by $C' =$ _____, where n is the number of moles in the system. A mole is _____ elementary units (molecules for a gas). This number is called _____ number and is denoted by N_A. In this chapter of this *Student's Companion*, C' is used to denote a molar specific heat.

The heat capacity depends on the process used to transfer energy as heat. It is different for constant volume processes and constant pressure processes, for example. Thus, there are many different heat capacities, one for each type process.

Suppose a certain system has molar specific heat C' for the process being considered. Then, the heat capacity of n moles is $C =$ _____ and the heat capacity of N molecules is $C =$ _____, where N_A is Avogadro's number. Suppose a certain system has specific heat c for a process and each of its molecules has mass m. Then, the heat capacity of N molecules is $C =$ _____ and the heat capacity of n moles is $C =$ _____. The molar specific heat for this system and process is $C' =$ _____.

Consider an object of mass m, composed of material with specific heat c for some process. Assume c is independent of temperature. If the temperature of the material is changed from T_i to T_f by the process, then the energy exchanged as heat is given by $Q =$ _____. Strictly speaking, the heat capacity and specific heat are temperature dependent, but they may usually be approximated by constants if the temperature change is small.

Carefully note that $Q = C(T_f - T_i)$ is valid regardless of which temperature is higher. If $T_f > T_i$, then Q is positive, indicating that energy enters the object. If $T_f < T_i$, then Q is negative, indicating that energy leaves the object.

When two objects, initially at different temperatures, are placed in thermal contact with each other and isolated from their surroundings, the hotter substance cools, and the cooler substance warms until they reach the same temperature. The heat capacities of the objects can be used to find the final common temperature. The principle used is that the energy leaving the hotter object as heat has the same magnitude as the energy entering the cooler object as heat. The algebraic relationship is $Q_A + Q_B = 0$. If object A has mass m_A, specific heat c_A, and initial temperature T_A, then $Q_A =$ _____, where T_f is the final temperature. If object B has mass m_B, specific heat c_B, and initial temperature T_B, then $Q_B =$ _____. Note that the energy corresponding to the hotter object is negative, indicating that energy is actually transferred *from* it. Since $Q_A + Q_B = 0$, the final temperature is given by $T_f =$ _____. You should use the specific heats (for constant volume, constant pressure, or some other process) appropriate to the actual process. If more than two objects are involved, use $\sum Q = 0$.

Both heat and work tend to change the internal energy of the system and thus tend to change the temperature. The change in temperature that accompanies an exchange of energy as heat may be greater or less than it would be if no work were done and the same energy were exchanged as heat. For example, if energy Q is absorbed by an object as heat and positive work W is done on it, then the heat capacity is _____ than it would be if the same amount of energy is absorbed as heat with no work being done.

Heats of transformation. A substance accepts or releases energy as heat when it changes phase (melts, freezes, vaporizes, or condenses, for example), even though the temperature remains constant during a phase change. In words, a <u>heat of transformation</u> is _____ _____.

If L is the heat of transformation for a certain phase change of a system with mass m, then the magnitude of the energy transferred as heat during a phase change is given by $|Q| =$ _____. Q is positive (energy absorbed) for melting and vaporization; it is negative (energy released) for freezing and condensing. The <u>heat</u> <u>of</u> <u>fusion</u> L_F is the energy per unit mass transferred during _____ and _____. The <u>heat</u> <u>of</u> <u>vaporization</u> L_V is the energy per unit mass transferred during _____ and _____. Some values are given in Table 19–4.

Work. The <u>thermodynamic</u> <u>state</u> of a gas can be described by giving the values of three thermodynamic variables: _____, _____, and _____. The state is changed by doing (positive or negative) work on the gas or by transferring energy as heat between the gas and its environment. Look at Fig. 19–xx of the text. Work is done on the gas by _____ and energy is transferred as heat between the gas and _____.

Thermodynamics deals with the work done *by* the system, rather than with the work done *on* the system. These are the negatives of each other. When the volume of a gas changes from V_i to V_f, the work that is done *by* the gas is given by the integral of the pressure p:

$$W =$$

Processes exist for which work is done and the volume does not change but these necessarily involve an acceleration of the center of mass or an angular acceleration about the center of mass and we do not consider them. Rather, we take into account only changes in the *internal* energy of a system.

The integral for work is valid only if the process that changes the volume is carried out so the gas is nearly in thermal equilibrium at all times. Only then is the pressure well defined throughout the process. If $V_f > V_i$, then the sign of the work done by the system is _____; if $V_i > V_f$, then the sign of the work done by the system is _____.

The process can be plotted as a curve on a graph with p as the vertical axis and V as the horizontal axis. Then, the work W is the _____ under the curve.

The work done on the system is different for different processes, represented as different functional dependencies of p on V and plotted as different curves on a p-V diagram. To calculate the work, you must know how the pressure varies as the volume changes.

The diagram on the right shows an initial state, labelled i, and a final state, labelled f. Draw two different paths from i to f, illustrating two different processes. Label the curves 1 and 2, then tell for which process the gas does the greater work: _____

The first law of thermodynamics. The first law of thermodynamics postulates the existence of an internal energy and states that its change as the system goes from any initial thermal equilibrium state to any final thermal equilibrium state is independent of how the change is brought about. It then equates the change in the internal energy to $Q - W$, evaluated for any path between the initial and final equilibrium states. For an infinitesimal change in state the mathematical form of the first law of thermodynamics is

$$dE_{\text{int}} =$$

Carefully note the signs here. Positive work done by the system and energy leaving the system as heat (dW positive and dQ negative) both tend to decrease the internal energy.

The internal energy is actually the sum of the kinetic energies of all the particles in the system and the potential energy of their interactions. We consider only systems for which the center of mass is at rest so the kinetic energy of translation of the system as a whole is zero.

Carefully contrast the internal energy with heat and work. When a system undergoes a change from one equilibrium state to another, the work done by the system and the energy absorbed as heat by the system depend on the actual process by which the state is changed and may be different for different processes. A change in internal energy, like a change in temperature, pressure, or volume, does NOT depend on the process but only on the initial and final states.

Because a change in the internal energy is a function of only the initial and final equilibrium states, the process used to change states is immaterial for its calculation. At intermediate stages of the process the system might not even be in thermal equilibrium.

You should understand some special cases. If a system undergoes a change of state at constant volume, then the work done by it is _____ and the change in internal energy is related to the energy absorbed as heat by $\Delta E_{int} = $ _____.

When energy Q is absorbed as heat at constant volume, the internal energy increases by exactly that amount. If the change of state is adiabatic then the energy absorbed as heat is _____ and the change in internal energy is related to the work by $\Delta E_{int} = $ _____. When work W is done by the system in an adiabatic process, the internal energy decreases by exactly that amount.

If the process is cyclic, so the initial and final states are the same, then the change in internal energy is _____ and the work done by the system is related to the energy absorbed as heat by $W = $ _____.

The free expansion of a gas is an important example of a process for which the system is not in thermal equilibrium during intermediate stages. Initially a partition confines the gas to one part of its container, which is thermally insulated. Then, the partition is removed and the gas expands into the other part. For this process, $W = $ _____ and $Q = $ _____. Thus, according to the first law, $\Delta E_{int} = $ _____.

Heat transfer. There are three mechanisms by which energy flows as heat from one place to another. Briefly describe each of them in words, being sure the descriptions distinguish among them:

conduction: _____

_____

convection: _____

_____

radiation: _____

_____

The important quantity is the *rate* of energy transfer, with symbol P_{cond}. If energy ΔQ is transferred as heat in a short time Δt, then the rate of transfer over that time is given by $P_{cond} = $ _____. The flow of energy is always from a _____ to a _____ region.

To discuss conduction, we consider a homogeneous bar of material with one end held at temperature T_H (hot) and the opposite end held at temperature T_C (cold). Let the x axis be along the bar, with $x = 0$ at the hot face and $x = L$ at the cold face. The temperature in the bar varies from point to point along its length and so is a function of x.

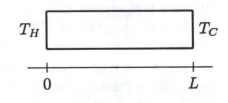

The rate of energy flow through the bar is proportional to the temperature difference of its ends and its cross-sectional area A. The constant of proportionality k is called the _____ _____ of the material. Mathematically the rate of energy flow through the bar is given by

$$P_{\text{cond}} = \underline{\qquad\qquad}$$

The thermal conductivity can be considered to be independent of the temperature over small temperature ranges but does, in fact, vary slightly with T. Table 19–6 gives some values.

Building materials are often characterized by their thermal resistances or R values rather than by their thermal conductivities. The R value is related to the thermal conductivity k by $R = \underline{\qquad}$.

The rate P_{rad} with which an object emits energy as radiation is given by

$$P_{\text{rad}} = \underline{\qquad\qquad}$$

where σ has the value _____ W/m²·K⁴ and is called the _____ constant. ϵ is the _____ of the surface and has a value between ___ and ___.

II. PROBLEM SOLVING

Some problems of this chapter deal with the definition of temperature. You will need to manipulate the relationship $T_1/T_2 = x_1/x_2$, where x_1 is the value at temperature T_1 of some property of a object that is used to measure temperature (the pressure of a gas, for example) and x_2 is the value of the same property at temperature T_2. You must know three of the quantities that appear in the equation or else the ratio of two of them and the value of a third, then you can solve for the fourth. In some cases, one of the temperatures is the triple point of water, 273.16 K. In some cases, the pressure of a gas is used and you must take the limit as the amount of gas becomes vanishingly small.

You should know how to convert from one common temperature scale to another. The relevant equations are $T_C = T - 273.15$ (Kelvin to Celsius) and $T_F = (9/5)T_C + 32°$ (Celsius to Fahrenheit). The most straightforward problems give the temperature on one scale and ask for the temperature on another.

Another set of problems is concerned with thermal expansion. The important relationship is $\Delta L = \alpha L \Delta T$. You might be given the original length L, the coefficient of linear expansion α, and the temperature change ΔT, then asked for the change in length ΔL or else the new length $L + \Delta L$. In other problems, you might be asked for the temperature change required to achieve a given change in length or a given final length. Sometimes the lengths for two different temperatures are given and you must calculate the coefficient, assumed to be independent of temperature. The fractional change $\Delta L/L$ might be given instead of the initial and final lengths. In some cases, you are asked to compare the expansions of two objects.

Some thermal expansion problems deal with areas and volumes. These are essentially the same as the problems dealing with linear dimensions but the coefficient of expansion is different. If α is the coefficient of linear expansion, then 2α is the coefficient of area expansion and 3α is the coefficient of volume expansion. Some problems deal with two objects in contact. Each individually obeys the law of thermal expansion and the sum of the two expansions is the total expansion of the composite system.

The heat of transformation is the energy per unit mass transferred as heat during a phase change. It is a property of the material that is changing phase. In the equation $Q = mL$, m is the mass of material that undergoes a phase change and is not necessarily the mass of all the material present.

If there is no phase change, the energy Q absorbed or released as heat by a substance is related to the change in temperature ΔT by $Q = C\,\Delta T$, where C is the heat capacity, or by $Q = mc\,\Delta T$, where m is the mass and c is the specific heat.

You should know how to calculate the final temperature of two or more objects, initially at different temperatures, when they have been in contact long enough to achieve thermal equilibrium. Use the condition that the energy rejected as heat by one object is absorbed as heat by the others. If the final temperature is known, this condition can be used to calculate the specific heat or mass of one of the objects.

Some problems test your understanding of the first law of thermodynamics: $\Delta E_{\text{int}} = Q - W$. Be sure you understand the sign convention you must use with this equation.

Thermal conductivity problems all involve $P_{\text{cond}} = kA(T_H - T_C)/L$. Usually all but one of the quantities that appear in this equation are given and you are asked for the other one. More complicated problems involve several rods, for example, welded either end to end or side by side. In the first case, the rate of energy conduction must be the same in all rods, while in the second it is the sum of the rates for the individual rods. In other problems, the energy transfer accomplishes a purpose, such as the melting of ice at one end of a rod.

III. NOTES

Chapter 20
THE KINETIC THEORY OF GASES

I. BASIC CONCEPTS

Kinetic theory is used to relate macroscopic quantities like pressure, temperature, and internal energy to the energies and momenta of the particles that comprise a gas. Here you will gain an understanding of the relationship between the microscopic and macroscopic descriptions of a gas. The example used throughout this chapter and the next is known as an ideal gas. Pay careful attention to its properties.

The following concepts, which were discussed in previous chapters, are used in this chapter: speed from Chapter 2, kinetic energy from Chapter 7, thermal energy from Chapter 8, linear momentum from Chapter 9, elastic collisions from Chapter 10, pressure from Chapter 15, and temperature, heat, and work done by a gas from Chapter 19. Be sure you understand them.

Ideal gas law. Since you will be dealing with ideal gases you should know what one is. Describe the properties that distinguish an ideal gas from a real gas: _____

_____

_____

The quantity of matter in the system of interest is important for most of the discussions of this chapter. It might be given as the number of particles or as the number of moles. The number of molecules in a mole of molecules is _____ . This is Avogadro's number, denoted by N_A. If a gas contains N molecules, then it contains _____ moles of molecules. If the mass of one mole is M, then the mass of one molecule is $m = $ _____ . If m is the molecular mass and M is the molar mass, then a gas containing N molecules has a total mass of _____ and a gas containing n moles has a total mass of _____ .

To understand the derivations of this chapter, you must first understand the <u>ideal gas law</u> (or ideal gas equation of state), which expresses a relationship between the temperature, pressure, volume, and number of particles for an ideal gas. In terms of the number of moles n of gas and the universal gas constant R, it is

$$pV = $$

Here p is the _____ , V is the _____ , and T is the _____ of the gas. It is important to recognize that for this equation to be valid, the Kelvin or ideal gas temperature scale must be used for the temperature. The SI value of R is $R = $ _____ J/mol·K.

The ideal gas law can also be written in terms of the number of molecules N and the Boltzmann constant k. In terms of the gas constant R and Avogadro's number N_A, $k = $ _____ J/K. The ideal gas law is then

$$pV = $$

Real gases obey the ideal gas law only approximately; the approximation is better when

_____ .

This is chiefly because the ideal gas law assumes the gas molecules are point particles that do not interact with each other except for elastic collisions of extremely short duration. In a real gas, molecules have extent in space and interact continuously with each other. Molecular sizes and interactions play less of a role as the number of molecules per unit volume decreases.

Work. The work done by the system as its volume changes from V_i to V_f is given by the integral $W = $ _____ . Its value depends on the function $p(V)$ for the pressure. Different dependencies of p on V are represented by different paths on a p-V graph. You should know how to compute the work for the three special processes described below. The device depicted in Fig. 20–10a of the text allows the macroscopic properties of a gas to be varied individually. You should also understand how to manipulate the device to carry out each of the special processes.

For each of the processes described below, all carried out on n moles of an ideal gas, write out the function $p(V)$ when requested, carry out the integration for the work done by the gas, and write the final expression for the work. Give expressions for the changes requested. On the p-V axes to the right, draw a curve that represents the process. Finally, describe how you would manipulate the device of Fig. 20–10a of the text to carry out the work.

1. Change in pressure from p_i to p_f at constant volume V (isochoric process)
   $p(V)$ is immaterial
   $W = $

   Change in temperature: $\Delta T = $
   How would you manipulate the device?

   _____

   _____

2. Change in volume from V_i to V_f at constant pressure p (isobaric process)
   $p(V) = $ constant
   $W = $

   Change in temperature: $\Delta T = $
   How would you manipulate the device?

   _____

   _____

3. Change in volume from V_i to V_f at constant temperature T (isothermal process)

$p(V) =$

$W =$

Change in pressure: $\Delta p =$

How would you manipulate the device?

_____

_____

Pressure. The pressure in a gas is intimately related to the average of the squares of the speeds of the molecules. The relationship is

$$p =$$

where n is the number of moles, M is the molar mass, V is the volume, and v_{avg}^2 is the average value of the squares of the molecular speeds.

Pressure can be computed as the average force per unit area exerted by the molecules of a gas on the container walls. We assume that at each collision between a molecule and a wall the normal component of the molecule's _____ is reversed. By computing the change per unit time we can find the force exerted by the molecule on the wall. When this is divided by the area of the wall, the result is the contribution of the molecule to the pressure.

Consider an ideal gas in a cubical container with edge L. If m is the mass of a molecule, v_x is the component of its velocity normal to the wall, and Δt is the time from one collision to the next with the same wall, then the average force exerted by the molecule on the wall is given by $F_{avg} =$ _____. In terms of L and v_x, the time between collisions is $\Delta t =$ _____. Thus, the average force of the molecule is $F_{avg} =$ _____ and its contribution to the pressure is _____. This is now summed over all molecules. Replace the sum of v_x^2 with the product of its average value and the number of molecules nN_A. You obtain $p =$ _____. Finally, replace the average of v_x^2 with one-third the average of v^2, mN_A with M, and L^3 with V. These substitutions complete the derivation. You obtain $p =$ _____, which should be the expression you wrote above.

Explain why the average of v_x^2 is one-third the average of v^2:

_____

_____

_____

The <u>root-mean-square</u> <u>speed</u> of the molecules in a gas is not quite the same as the average speed but it gives us a rough idea of the average. Its definition, in terms of the average of the squares of the molecular speeds, is $v_{rms} =$ _____ and, in terms of the pressure and density, it is given by

$$v_{rms} =$$

You can use this equation in conjunction with the ideal gas law to estimate the root-mean-square speed of a given gas with a known molar mass M, at a given temperature. Substitute $p = nRT/V$ into the expression for v_{rms} to obtain

$$v_{rms} = $$

Carefully note that v_{rms} depends only on the temperature and the molar mass. Thus, v_{rms} remains the same if the temperature does not change as the pressure and volume change; that is, v_{rms} does not change if the product pV does not change. v_{rms}^2 is directly proportional to the temperature and inversely proportional to the molar mass.

Suppose molecules in an ideal gas have a root-mean-square speed v_{rms}. If the temperature is doubled, the root-mean-square speed becomes _____.

Suppose two ideal gases are in thermal equilibrium with each other. Molecules in the first gas have mass M and root-mean-square speed v_{rms}. If molecules in the second gas have mass $2M$, then their root-mean-square speed is _____.

Some values of v_{rms} are listed in Table 20–1. For gases at room temperature they run from a low of _____ m/s (for _____) to a high of _____ m/s (for _____). You should know that the root-mean-square speed is roughly the same as the speed of sound for any given gas.

Kinetic energy and temperature.   For an ideal gas the average translational kinetic energy per molecule is proportional to the temperature. In terms of the universal gas constant R and Avogadro's number N_A, the exact relationship is

$$K_{avg} = \frac{1}{2}mv_{rms}^2 = $$

where m is the molecular mass. In terms of the Boltzmann constant k and the temperature T,

$$K_{avg} = $$

Carefully note that the molecules of two ideal gases, in thermal equilibrium with each other have exactly the same average translational kinetic energy. The root-mean-square speeds, however, are different if the molecular masses are different. Also note that the average translational kinetic energy depends on the temperature but not separately on the pressure or volume of the gas.

Mean free path.   The mean free path of a molecule in a gas is the average distance it travels between _____ with other molecules. It can be estimated theoretically by considering the number of molecules that are encountered by any given molecule as it moves through the gas. Assume the molecules each have a diameter d. Concentrate on one that is moving with speed v and assume all others are stationary. The moving molecule has a collision with another molecule if the distance between their centers is less than _____. In terms of the diameter d and speed v, the number of collisions it has in time t equals the number of molecules in the volume of a cylinder with cross-sectional area $A = $ _____ and length

$\ell =$ _____. Since N/V is the number of molecules per unit volume in the gas, the number of collisions in time t is given by _____. The mean free path λ is the distance vt divided by the number of collisions in time t or, in terms of d and N/V, $\lambda =$ _____.

Because each molecule collides with moving rather than stationary molecules the mean free path is actually somewhat less. It is, in fact, given by

$$\lambda =$$

The mean free path of air molecules at sea level is about _____; the mean free path is about _____ at an altitude of 100 km and about _____ at an altitude of 300 km.

The distribution of molecular speeds. Not all molecules in a gas have the same speed. The distribution of speeds is described by the <u>Maxwell</u> <u>speed</u> <u>distribution</u> $P(v)$, defined so that $P(v)\,dv$ gives the fraction of molecules with speeds in the range from v to $v + dv$. If a gas at temperature T has molar mass M, then the Maxwell speed distribution is given by

$$P(v) =$$

Sketch the function $P(v)$ on the axes below:

To find the number of molecules with speed between v_1 and v_2, evaluate the definite integral

$$N(v_1, v_2) = \int_{v_1}^{v_2} P(v)\,dv.$$

If the interval $v_2 - v_1$ is small, you can approximate the integral by $(v_2 - v_1)P(v_1)$. See Sample Problem 20–5 of the text.

The Maxwell speed distribution can be used to calculate the most probable speed, the average speed, and the root-mean-square speed. The <u>most</u> <u>probable</u> <u>speed</u> v_p is the one for which $P(v)$ is _____. In terms of the molar mass and the temperature, it is given by

$$v_p =$$

The <u>average</u> <u>speed</u> is given by

$$v_{\text{avg}} = \int_0^\infty v P(v)\,dv =$$

where the explicit expression for the Maxwell distribution was substituted for $P(v)$ and the integral was evaluated. The mean-square speed is given by

$$v_{\text{avg}}^2 = \int_0^\infty v^2 P(v)\, dv =$$

where the explicit expression for the Maxwell distribution was again substituted for $P(v)$ and the integral was evaluated. The root-mean-square speed v_{rms} is the square root of this. When the square root is taken, the result is

$$v_{\text{rms}} =$$

Note that all three speeds (v_p, v_{avg}, and v_{rms}) depend on the temperature as well as on the molar mass. When the temperature increases, they all _____. For a given gas at a given temperature the greatest of the three characteristic speeds is _____ and the least is _____.

Internal energy and specific heats. For a *monatomic* ideal gas the internal energy is the total kinetic energy of the molecules. Since the average kinetic energy of a molecule is $K_{\text{avg}} = \frac{3}{2}kT$, the internal energy of n moles of an ideal gas is given in terms of R and T by $E_{\text{int}} =$ _____.

Note that the internal energy of an ideal gas depends only on the temperature and the amount of gas. This information will prove useful for solving ideal gas problems. For other systems the internal energy may also depend on the pressure or other macroscopic quantities.

Suppose now that the temperature is changed by ΔT while the volume is held constant. Then the internal energy changes by $\Delta E_{\text{int}} =$ _____. Since no work was done, the entire increase or decrease in internal energy is due to energy absorbed or rejected as heat. The energy is $Q =$ _____. The heat capacity at constant volume is $Q/\Delta T =$ _____ and the molar specific heat at constant volume, in terms of R, is $C_V =$ _____. Its value is _____ J/mol·K. Notice that it does not depend on the temperature. Carefully note that in this chapter and the next, C is used to represent a *molar* specific heat, not a heat capacity.

You should understand that $\Delta E_{\text{int}} = nC_V\Delta T$ for any process, no matter if it involves energy as heat or not, and no matter if the volume changes or not. However, $Q = nC_V\Delta T$ only for a constant volume process, during which no work is done.

The molar specific heat at constant pressure C_p is related to the molar specific heat at constant volume by $C_p =$ _____. More energy is absorbed as heat at constant pressure than at constant volume for a given rise in temperature. Since the internal energy depends only on the temperature, its change is the same. What happens to the extra energy absorbed at constant pressure? _____

The equipartition theorem tells us that, in thermal equilibrium, the internal energy is equally divided among the degrees of freedom of a system, with the energy associated with each degree of freedom being _____ kT per molecule or _____ RT per mole. One degree of freedom is associated with each independent energy term. For example, _____ degrees of freedom are associated with the translational kinetic energy of a molecule since it can move in any of three independent directions. Other degrees of freedom are associated with rotation of the molecules.

Consider three gases. The first consists of monatomic molecules, the second of diatomic molecules, and the third of polyatomic molecules. Fill in the following table giving the translational, rotational, and total internal energies per mole in terms of RT.

	MONATOMIC	DIATOMIC	POLYATOMIC
Number of translational degrees of freedom	_____	_____	_____
Translational energy	_____	_____	_____
Number of rotational degrees of freedom	_____	_____	_____
Total internal energy	_____	_____	_____

Fill in the following table for the molar specific heats for constant volume processes and for constant pressure processes.

	MONATOMIC	DIATOMIC	POLYATOMIC
C_V	_____	_____	_____
C_p	_____	_____	_____

Knowing how the internal energy depends on the temperature allows us to compute the change in temperature when work is done on the gas or energy is absorbed as heat by the gas. If, for example, work W is done on n moles of a monatomic ideal gas in thermal isolation (no heat transfer), then the increase in internal energy is ΔE_{int} = _____ and the increase in temperature is ΔT = _____ .

Adiabatic processes. As an ideal gas undergoes an adiabatic process the quantity pV^γ is _____ . Here γ is the ratio _____ . If p_i and V_i are the pressure and volume of the initial state and p_f and V_f are the pressure and volume of the final state, then

$$p_i V_i^\gamma =$$

Use $pV = nRT$ to eliminate p from this expression. You should find that if T_i is the temperature of the initial state and T_f is the temperature of the final state, then

$$T_i V_i^{\gamma - 1} =$$

If two of the three quantities p, V, and T are given for one state and one of them is given for another state, connected to the first by an adiabatic process, then you can use the equations given above and $pV = nRT$ to find values for the other thermodynamic variables. Notice that p, V, and T are not constant during an adiabatic process.

II. PROBLEM SOLVING

You should know the relationships between the mass of a molecule, the mass of a mole of molecules, and the total mass of a collection of molecules. If a gas contains N molecules, each of mass m, the total mass is Nm. Since there are N_A molecules in a mole, the gas contains N/N_A moles. The molar mass M is the mass of N_A molecules: $M = N_A m$. If the gas contains n moles, its total mass is nM. Note that this is exactly the same as Nm.

Many problems require you to use the ideal gas law $pV = nRT$ to compute one of the quantities that appear in it. You should also know how to compute changes in one of the quantities for various processes. For a change in volume (and temperature) at constant pressure, $p\Delta V = nR\Delta T$; for a change in pressure (and temperature) at constant volume, $V\Delta p = nR\Delta T$; and for a change in volume (and pressure) at constant temperature, $p\Delta V = -V\Delta p$. In each case, the amount of gas was assumed to remain constant.

You should understand how to compute the work done by a gas during a given process. Use $W = \int p\,dV$. First, you will need to find the functional form of the dependence of the pressure p on the volume V for the process. The ideal gas law is often helpful here. Use $p = nRT/V$. If the temperature is constant, the integral can be evaluated immediately. If the temperature is not constant you need to know how it changes.

You should know how the root-mean-square speed of molecules in a gas depends on the temperature: $v_{\mathrm{rms}} = \sqrt{3RT/M}$, where M is the molar mass. You should also know how the average translational kinetic energy depends on the temperature: $K_{\mathrm{avg}} = \frac{3}{2}RT/N_A$. The translational kinetic energy per mole is $\frac{3}{2}RT$, the energy of n moles is $\frac{3}{2}nRT$, and the translational kinetic energy for N molecules is $\frac{3}{2}NRT/N_A = \frac{3}{2}NkT$. For monatomic molecules the last two expressions give the internal energy. For diatomic and polyatomic molecules other terms, giving the rotational energy, must be added. Other problems deal with the mean free path.

You should be able to solve problems involving the first law of thermodynamics: $\Delta E_{\mathrm{int}} = Q - W$. For some of these you need to know the specific heat or the molar specific heat. Remember that the change in the internal energy is given by $\Delta E_{\mathrm{int}} = nC_V\Delta T$ and the energy input as heat is given by $Q = nC\Delta T$, where C is the molar specific heat for the process.

Some problems deal with ideal gases as they undergo adiabatic processes. Remember that the combination pV^γ remains constant for such processes. Here γ is the ratio of the specific heats: $\gamma = C_p/C_V$. The ideal gas law is still valid and for some problems $pV = nRT$ must be solved simultaneously with $pV^\gamma = \text{constant}$.

III. MATHEMATICAL SKILLS

Average and root-mean-square values. This chapter introduces some of the ideas of statistics, the most important of which are the average and root-mean-square of a collection of numbers. To find the average, add the numbers and divide the result by the number of terms in the sum. To find the root-mean-square, add the squares of the numbers, divide the sum by the number of terms, and take the square root of the result. The root-mean-square is the square root of the average of the squares of the numbers.

For practice, suppose the speeds of five molecules are 350 m/s, 225 m/s, 432 m/s, 375 m/s, and 450 m/s. Find their average and root-mean-square speeds. The following table might help.

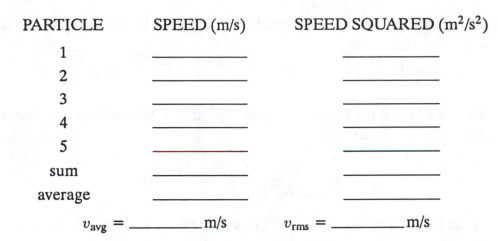

PARTICLE	SPEED (m/s)	SPEED SQUARED (m²/s²)
1	_____	_____
2	_____	_____
3	_____	_____
4	_____	_____
5	_____	_____
sum	_____	_____
average	_____	_____

$$v_{\text{avg}} = \underline{\qquad} \text{ m/s} \qquad v_{\text{rms}} = \underline{\qquad} \text{ m/s}$$

Your answers should be $v_{\text{avg}} = 366$ m/s and $v_{\text{rms}} = 375$ m/s. Notice that the average and root-mean-square values are different. In fact, the root-mean-square speed is greater than the average speed.

Integrals of the Maxwell distribution function. This chapter contains several integrals that are difficult to evaluate using only a knowledge of introductory calculus. They are associated with the average and mean-square speed for molecules with a Maxwellian speed distribution. The average speed is given by

$$v_{\text{avg}} = 4\pi \left[\frac{M}{2\pi RT} \right]^{3/2} \int_0^\infty v^3 e^{-Mv^2/2RT} \, dv \,,$$

and the mean-square speed is given by

$$v_{\text{avg}}^2 = 4\pi \left[\frac{M}{2\pi RT} \right]^{3/2} \int_0^\infty v^4 e^{-Mv^2/2RT} \, dv \,.$$

Most standard integral tables list the following integrals:

$$\int_0^\infty x^{2a} e^{-px^2} \, dx = \frac{(2a-1)!!}{2(2p)^a} \sqrt{\frac{\pi}{p}}$$

and

$$\int_0^\infty x^{2a+1} e^{-px^2} \, dx = \frac{a!}{2p^{a+1}} \,.$$

Here p is any positive number and a is any integer. $a!$ is the factorial of a; that is, $a! = 1 \cdot 2 \cdot 3 \cdot 4 \ldots a$. $(2a-1)!!$ is the product of all odd integers from 1 to $2a-1$; that is, $(2a-1)!! = 1 \cdot 3 \cdot 5 \cdot 7 \ldots (2a-1)$. We use the first integral when x to an *even* power multiplies the exponential in the integrand and the second when x to an *odd* power multiplies the exponential.

To evaluate the integral for the average speed of a Maxwellian distribution, substitute $v = x$ and $M/2RT = p$ into the equation for v_{avg}. You should obtain

$$v_{\text{avg}} = 4\pi \left[\frac{M}{2\pi RT} \right]^{3/2} \int_0^\infty x^3 e^{-px^2} \, dx \,.$$

The integral is the same as the second integral taken from the tables, with $a = 1$. So

$$v_{\text{avg}} = 4\pi \left[\frac{M}{2\pi RT}\right]^{3/2} \frac{1}{2p^2} = 4\pi \left[\frac{M}{2\pi RT}\right]^{3/2} \frac{2R^2T^2}{M^2} = \sqrt{\frac{8RT}{\pi M}}.$$

To evaluate the integral for the mean-square speed of a Maxwellian distribution, make the same substitutions to obtain

$$v_{\text{avg}}^2 = 4\pi \left[\frac{M}{2\pi RT}\right]^{3/2} \int_0^\infty x^4 e^{-px^2}\, \mathrm{d}x.$$

The integral is the same as the first one taken from the tables, with $a = 2$. So

$$v_{\text{avg}}^2 = 4\pi \left[\frac{M}{2\pi RT}\right]^{3/2} \frac{3}{2(2p)^2} \sqrt{\frac{\pi}{p}} = 4\pi \left[\frac{M}{2\pi RT}\right]^{3/2} \frac{3R^2T^2}{2M^2} \sqrt{\frac{2\pi RT}{M}} = \frac{3RT}{M}.$$

Thus, the root-mean-square speed is $\sqrt{3RT/M}$.

IV. NOTES

Chapter 21
ENTROPY AND THE SECOND LAW
OF THERMODYNAMICS

I. BASIC CONCEPTS

Here you study the second law of thermodynamics and a closely associated property of any system, its entropy. Understand the two definitions: in terms of heat and temperature and in terms of microstates of the system. Pay careful attention to calculations of entropy changes during various processes. Learn to distinguish between reversible and irreversible processes and understand how entropy behaves for each case. The second law has many far-reaching consequences; we can imagine many phenomena that do not violate any other laws of physics, including the conservation laws, but, nevertheless, do not occur. The second law gives the reason.

You should consider reviewing the following topics that were discussed in previous chapters: internal energy from Chapter 8, pressure from Chapter 15, and temperature, heat, and work done by a gas from Chapter 19.

Reversible processes.   The concepts of <u>reversible</u> and <u>irreversible</u> processes are important for understanding entropy and the second law of thermodynamics. Explain what each is, being sure to distinguish between them.

reversible: _____
_____

irreversible: _____
_____

Reversible processes must be carried out <u>quasi-statically</u>; that is, extremely slowly. After a reversible process is carried out on a system, carrying out the steps in the reverse order brings the system back to its original state. This means, for example, that friction cannot be present.

If two different processes are used to take a system from the same initial equilibrium state to the same final equilibrium state, one reversible and one irreversible, the work and heat may be different but changes in the _____, _____, _____, and _____ will be the same. These quantities are called <u>state</u> <u>functions</u>. Their changes during any process that begins and ends at an equilibrium state depends only on the initial and final states and not on the process. Over a cycle, the net change in any state function is _____ .

Entropy.   If energy dQ is received as heat by the system when its state changes from an initial equilibrium state to a nearby equilibrium state, then the entropy of the system changes by

$$d S =$$

where T is the temperature on the Kelvin scale. This defines entropy. If the state changes reversibly from some initial equilibrium state i to some final equilibrium state f, the change in entropy can be computed by evaluating the integral

$$\Delta S = S_f - S_i =$$

Recognize that the process connecting the initial and final states must be reversible for this equation to be valid. Notice that the integral $\int_i^f dQ$ depends on the process but the integral $\int_i^f dQ/T$ does not.

Entropy is a state function. This means that whenever a system is in the same thermodynamic state it has the same entropy, just as it has the same temperature, pressure, volume, and internal energy.

To evaluate the integral for ΔS, you must pick some reversible process that connects the initial and final states but, because entropy is a state function, it need not be the process actually carried out on the system. To evaluate ΔS for an irreversible process, pick a *reversible* process with the same initial and final states and use that process to find the value of the integral. Sample Problems 21–xx and 21–xx of the text show how to calculate the entropy changes for several situations. Here are some more examples.

1. Change in phase.   Suppose mass m of a substance with heat of fusion L_F melts at temperature T. For this process T is constant and $\Delta S = Q/T$. Since $Q = mL_F$, $\Delta S =$ _____. The change in entropy on freezing is $\Delta S =$ _____. Note that the entropy of the substance increases on melting and decreases on freezing.

2. Constant volume process.   Suppose the molar specific heat at constant volume C_V for a certain gas is independent of T, V, and p. If an n-mole sample of this gas undergoes an infinitesimal reversible change of state at constant volume, so its temperature changes by dT, then the energy exchanged as heat is $dQ =$ _____ and the change in entropy is $dS = dQ/T =$ _____. For a finite change in temperature, from T_1 to T_2, say, the change in entropy is the integral of this expression. In the space below, carry out the integration to show that $\Delta S = nC_V \ln(T_f/T_i)$:

3. Constant pressure process.   Suppose the molar specific heat at constant pressure C_p for a certain gas is independent of T, V, and p. If an n-mole sample of this gas undergoes an infinitesimal reversible change of state at constant pressure, so its temperature changes by dT, then the energy absorbed as heat is $dQ =$ _____ dT and the change in entropy is $dS = dQ/T =$ _____. For a finite change in temperature, from T_i to T_f, the change in entropy is the integral of this expression. In the space below, carry out the integration to

show that $\Delta S = nC_p \ln(T_f/T_i)$:

4. **Isothermal process.** If an ideal gas undergoes an infinitesimal reversible change of state at constant temperature, then $dE_{int} = nC_V dT = 0$ and $dQ = dW = p\,dV$. Thus, $dS = (p/T)\,dV$ and $\Delta S = \int (p/T)\,dV$, where the limits of integration are the initial volume V_i and the final volume V_f. To evaluate this integral, p must be known as a function of volume and temperature. The equation of state gives this information. For n moles of an ideal gas, $p/T = $ _____. In the space below, show that $\Delta S = nR \ln(V_f/V_i)$:

As an example of an irreversible process consider the adiabatic free expansion of an ideal gas from volume V_i to volume V_f. The net energy transferred as heat is $Q = $ _____, the net work done is $W = $ _____, the change in the internal energy is $\Delta E_{int} = $ _____, and the change in the temperature is $\Delta T = $ _____. The change in entropy is calculated using a reversible isothermal expansion from V_i to V_f. Explain why a reversible process is used: ___

Explain why an isothermal process is used: _____

The change in entropy is

$$\Delta S =$$

Notice that no energy is transferred as heat but the entropy increases.

The second law of thermodynamics. The second law of thermodynamics is expressed in terms of entropy changes. In words, it is: _____
_____
_____.

Notice you must consider the *total* entropy of the system *and* its environment. Many processes lower the entropy of a system but, according to the second law, they must then _____ the entropy of the environment.

If a change of state is reversible, then the change in the total entropy of the system and its environment is _____. If the entropy of the system decreases in a reversible process, then the entropy of the environment must _____ by exactly the same amount. If the system changes state through an irreversible process, then the change in the total entropy of the system and its environment is _____. All real processes are irreversible to some extent, so we expect the total entropy of every real system and its environment to _____ in any process.

Suppose two identical systems each go from the same initial state to the same final state. For system 1, the process is carried out reversibly while for system 2, it is carried out irreversibly. Suppose the change in the entropy for system 1 is ΔS_1. Then, the change in the entropy of the environment of system 1 is _____. The change in entropy of system 2 is _____. The change in the magnitude of the entropy of the environment of system 2 is _____ than $|\Delta S_1|$.

Heat engines and refrigerators. In terms of work and heat, describe what a heat engine does: _____

_____

A heat engine makes use of a working substance whose thermodynamic state varies during the process but that periodically returns to the same state. The same sequence of steps is performed on it during every cycle. If in each cycle a heat engine receives energy of magnitude $|Q_H|$ as heat from a high-temperature thermal reservoir and rejects energy of magnitude $|Q_C|$ as heat to a low-temperature reservoir, then according to the first law of thermodynamics, it does work $W =$ _____ over a cycle. Note that because the engine works in cycles the change in the internal energy is _____.

The thermal efficiency of a heat engine is defined in terms of $|Q_H|$ and $|W|$ by

$$\varepsilon =$$

In terms of $|Q_H|$ and $|Q_C|$, it is

$$\varepsilon =$$

In a sense, Q_H is the energy cost of getting work W. $|Q_C|$ is the energy that is lost.

A *perfect* heat engine takes in energy as heat and does an identical amount of work. No energy is rejected as heat. Since $|W| = |Q_H|$ for a perfect heat engine, its efficiency is _____. Such an engine does not violate the first law of thermodynamics but it does violate the second law.

Consider a reversible heat engine (called ideal) and let T_H be the temperature of the hot reservoir and T_C be the temperature of the cold reservoir. Over a cycle, the change in entropy of the hot reservoir is $\Delta S_H =$ _____, the change in entropy of the cold reservoir is $\Delta S_C =$ _____, and the change in entropy of the working substance is $\Delta S_w =$ _____. In terms of Q_H, Q_C, T_H, and T_C, the change in the total entropy of the engine and reservoirs is $\Delta S =$ _____. According to the second law, this must be _____. Thus,

$$\frac{|Q_C|}{|Q_H|} =$$

and the efficiency is given in terms of T_H and T_C by

$$\varepsilon =$$

Observe that the efficiency cannot be 1 unless $T_C = 0$, an impossibility. The first law tells us that in a cycle, energy input as heat is turned to work and energy output as heat. The second law tells us it cannot be turned completely to work; some must be output as heat.

The derivation of the expression for ε shows that all reversible engines operating between the same two temperatures have the same efficiency. Irreversible engines have a _____ efficiency.

A refrigerator provides another application of the second law. In terms of work and heat, tell what a refrigerator does: _____

_____

Suppose in each cycle a refrigerator takes in energy of magnitude $|Q_C|$ as heat at a low-temperature reservoir and rejects energy of magnitude $|Q_H|$ as heat at a high-temperature reservoir. It does (negative) work W in each cycle. Its coefficient of performance is defined in terms of $|Q_C|$ and $|W|$ by

$$K = $$

In terms of $|Q_C|$ and $|Q_H|$, the coefficient is

$$K = $$

For a *perfect* refrigerator $W = $ _____ and $K = $ _____. If such a refrigerator existed, work would not be required to run it. A perfect refrigerator does not violate the first law of thermodynamics but it does violate the second law and cannot be built.

Consider a reversible refrigerator (called ideal) and let T_H be the temperature of the hot reservoir and T_C be the temperature of the cold reservoir. Over a cycle the change in entropy of the hot reservoir is $\Delta S_H = $ _____, the change in entropy of the cold reservoir is $\Delta S_C = $ _____, and the change in entropy of the working substance is $\Delta S_w = $ _____. In terms of Q_H, Q_C, T_H, and T_C, the change in the total entropy of the engine and reservoirs is $\Delta S = $ _____. According to the second law, this must be _____. Thus,

$$\frac{|Q_C|}{|Q_H|} = $$

and the coefficient of performance is given in terms of T_H and T_C by

$$K = $$

Because $T_H > T_C$, $|Q_H|$ must be greater than $|Q_C|$. A refrigerator necessarily delivers more energy as heat to the high-temperature reservoir than it receives from the low-temperature reservoir. This means work must be done on the refrigerator. The derivation of the expression for K shows that all reversible refrigerators operating between the same two temperatures have the same coefficient of performance. Irreversible refrigerators have a _____ coefficient.

Stirling cycles. A Stirling engine provides a concrete example. The cycle consists of two isothermal processes at different temperatures, linked by two constant volume processes. Here are the steps when the cycle is run as a heat engine. Tell how the device of Fig. 21–xx of the text can be manipulated to carry out each step.

1. Isothermal expansion at high temperature: _____
   _____

2. Constant volume change in temperature to the lower temperature: _____
   _____

3. Isothermal compression at low temperature: _____
   _____

4. Constant volume change in temperature to the higher temperature: _____
   _____

Diagram a Stirling cycle on the p-V axes to the right. Identify and label each of the steps. Assume the cycle is run as a heat engine and mark the portion of the cycle during which energy flows as heat from a reservoir to the working substance and the portion during which energy flows as heat from the working substance to a reservoir. If the gas is ideal, the energy taken in as heat during the increase in temperature at constant volume is the same in magnitude as the energy released as heat during the decrease in temperature. Why?

_____

_____

Thus we need to consider only the energy Q_H taken in as heat from the high-temperature reservoir and the energy Q_C released as heat to the low-temperature reservoir.

Suppose the working substance is a monatomic ideal gas and assume the engine is operated reversibly. Suppose the volume is expanded from V_i to V_f at temperature T_H and contracted from V_f to V_i at temperature T_C. Then, during a single cycle, the energy received as heat from the high-temperature reservoir is given by $Q_H =$ _____, the energy rejected as heat to the low-temperature reservoir is $Q_C =$ _____, and the total work done is $W =$ _____. In terms of the temperatures of the reservoirs, the efficiency of the engine is $\varepsilon =$ _____. This should match the equation you wrote earlier for an ideal engine.

Entropy and statistics. There are many different arrangements of atoms that lead to the same thermodynamic properties. All arrangements with the same properties are said to have the same configuration. Each arrangement is called a <u>microstate</u>. A different number of microstates may be associated with different configurations.

The fundamental hypothesis of statistical mechanics is that every microstate, consistent with energy conservation, is equally probable. Thus, a system with a large number of particles will, with overwhelming probability, be in the configuration with the greatest number of microstates.

The text considers the free expansion of an ideal gas from its original confinement in half of a box into the entire box. Each molecule is labeled with its position, either in the left (L) or right (R) portion of the box. Explain how two microstates differ: _____

_____

_____

Explain how one configuration differs from another: _____

_____

_____

The gas is most likely to have the configuration with the most microstates. Describe this configuration: _____

_____

The entropy of a system in a given configuration is related to the number of microstates associated with that configuration. If W is the number of microstates, then the relationship is

$$S =$$

where k is _____.

For N identical molecules, with n_L in the left portion of the box and n_R in the right portion,

$$W =$$

If all N molecules are initially in the right portion, then the initial entropy is given by

$$S_i =$$

and if the final configuration has half the molecules in each portion, then the final entropy is given by

$$S_f =$$

The change in entropy on expansion is

$$S_f - S_i =$$

Use Stirling's approximation, $\ln N! \approx N(\ln N) - N$, valid for large N, to show that ΔS agrees with the calculation you carried out earlier:

II. PROBLEM SOLVING

You should know how to calculate changes in entropy for various processes. Remember that $\Delta S = \int (1/T)\,dQ$ for a reversible process between the initial and final state. To evaluate the integral you need to substitute for dQ and what you substitute depends on the process involved. In most cases you will use a heat capacity or specific heat to determine dQ in terms of the temperature change dT. In some cases you will need the first law of thermodynamics to determine dQ in terms of the work dW and the change dE_{int} in the internal energy. The change in the internal energy can always be written in terms of the heat capacity or specific heat at constant volume and the change in temperature.

Many problems deal with heat engines and refrigerators. In each case, energy is absorbed as heat over part of the cycle, energy is rejected as heat over another part, and work is done on or by the working substance. Given any one of these quantities and the efficiency or coefficient of performance, you should be able to calculate the others. The relevant equations are the first law of thermodynamics and the definition of either the efficiency or the coefficient of performance. Other problems give you two of the energy terms and ask for the efficiency or coefficient of performance.

Some problems ask you to test data for heat engines or refrigerators to see if they violate the first or second laws of thermodynamics. Remember that the second law places upper limits on the efficiency of an engine and the coefficient of performance of a refrigerator.

You should also practice calculating the heat and work for various processes (isothermal, adiabatic, constant volume, and constant pressure). Use the heat capacity or specific heat to compute the heat, use $\int p\,dV$ to compute the work.

Some problems deal with the microscopic view of entropy. You must calculate the number W of microstates for a given configuration and use $S = -k \ln W$.

III. NOTES

OVERVIEW III
ELECTRICITY AND MAGNETISM

Electric and magnetic phenomena pervade our lives. We use electric power to heat and light our homes, schools, and places of work. We use it to carry out a vast number of industrial and manufacturing tasks. We use it, via our radios, television sets, and movie projectors, to obtain information and entertainment. Magnetic forces are used to run all electric motors and, in every electric generating plant, a coil of wire is rotated near a large magnet to produce electric power.

As important as all these uses are, electricity is even more vital to us in another area: electric forces bind particles together to form atoms and atoms together to form all materials. We literally could not exist without electrical forces! Nor could any solid or liquid, or, for that matter, any atom.

Visible light, microwaves, x rays, and radio waves are all electromagnetic in nature, and differ only in their frequencies. Optics, x-ray imaging, radio transmission and reception, and microwave production and detection are all logically part of the field of electricity and magnetism.

Particles such as electrons and protons exert electric and magnetic forces on each other; other particles do not. The property of a particle that gives rise to these forces is called charge. Charged particles at rest interact only via electric forces; charged particles in motion exert both electric and magnetic forces on each other.

Electric forces between charged particles at rest are described in Chapter 22. You will learn there are two types of charge, called positive and negative, and that like charges repel each other while unlike charges attract each other. In each case, the force is along the line that joins the charges and is proportional to the reciprocal of the square of the distance between them. Electric forces, of course, are vectors and, when two or more charges act on another, the resultant force is the vector sum of the individual forces.

We may think of a charge creating electric and magnetic fields in all of space; the fields then exert forces on other charges. The problem of finding the force exerted by a collection of charges is then broken down into two somewhat simpler problems: *what fields are produced by the collection?* and *what force do the fields exert on the other charge?* Electric and magnetic interactions can be described much more easily in terms of fields than in terms of forces. When charges are moving or produce electromagnetic radiation, the field viewpoint is almost a necessity.

You start your study of electric fields, in Chapter 23, by learning about the force exerted by a field on a charge. If no counterbalancing forces act, the charge accelerates. You will apply Newton's second law and your knowledge of kinematics to examine the motions of charges in electric fields.

The relationship between a collection of charges and the electric field it produces is discussed in Chapters 23 and 24. It can be formulated in two equivalent, but quite different, ways. In the first chapter, it is given in terms of the charges and their positions while in the second, it is given in terms of an integral of the field over a surface. You should understand both formulations. The first is useful when you are given the charges and their positions, then asked to find the electric field; the second, known as Gauss' law, is useful when you know the field and want to find the charge distribution. It can also be used to find the field in certain situations of high symmetry.

Electrostatic forces, those of charges at rest, are conservative, so a potential energy can be associated them. Chapter 25 deals with electric potential energy and a closely related quantity, electric potential. You also will learn to calculate the work that must be done to assemble a collection of charges.

In Chapter 26, the ideas of electric field and potential, as well as Gauss' law, are put to practical use in a study of capacitance. Capacitors are extremely useful devices for the storage of electrical energy, which can then be recovered in short intense bursts. Small capacitors are used in electronic flash units for cameras. Large capacitors are used to power the extremely intense lasers used in an attempt to produce controlled nuclear fusion. In a more fundamental vein, you will learn here that you may think of the potential energy of a collection of charges as being stored in the electric field and you will learn how to calculate the energy per unit volume stored in any field.

An electric current is simply a collection of moving charges. Currents are studied for two reasons. First, electrical circuits carrying current have many important practical applications. Think of the miniature circuits that operate computers and the house and industrial circuits that operate lights, appliances, and machinery. Secondly, currents produce magnetic fields. Chapters 27 and 28, in which currents are studied, form a bridge to later chapters on the magnetic field.

In Chapter 27, you will learn that the current in any non-superconducting material is determined by the electric field. You will also learn about resistivity, the property of materials that determines the current for a given electric field. It is closely related to the transfer of energy from moving charges to the material through which they move. Chapter 28 deals with batteries and other devices used to produce currents. Here you will apply the ideas of this and the previous chapter to the analysis of simple electrical circuits carrying time-independent currents. Energy balance in electrical circuits, important because current is often used to carry energy from one place to another, is also studied in this chapter.

In Chapter 29, you start the study of magnetic fields by learning about the force exerted by a magnetic field on a moving charge. This is extended to a discussion of the force and torque exerted on a current-carrying circuit.

The relationship between a current and the magnetic field it produces is discussed in Chapter 30. Just as for electric fields, two equivalent views are presented. In the first, known as the Biot-Savart law, the field produced by an infinitesimal element of current is postulated and this expression is integrated over the entire current. In the second, known as Ampere's law, the integral of the magnetic field around a closed loop is related to the current passing through the loop. The first viewpoint is of value when the current circuit is known and the field is sought; the second is of value when the field is known and the current is sought. Ampere's law can also be used to find the magnetic fields of highly symmetric current distributions.

A time-varying current produces an electric as well as a magnetic field, even if the net charge is zero in every region of space.

The relationship between the fields is given by Faraday's law and is discussed in Chapter 31. The law also deals with situations in which an object moves in a constant magnetic field. The result is exactly as if an infinitesimal battery were placed at each point in the object. Current is induced if the object forms part of an electrical circuit. Faraday's law is also applied to electrical circuits in this chapter. You will learn about inductors: circuit elements that are used to control the rate of change of current in a circuit and that store energy in magnetic fields just as capacitors store energy in electric fields.

Electromagnetic theory is completed by two additional relationships, both described in Chapter 32. The first, Gauss' law for magnetism, is equivalent to the statement that magnetic monopoles do not exist. The second, called Maxwell's law, tells us that a magnetic field is always associated with a varying electric field, just as an electric field is always associated with a varying magnetic field. You may think of a varying magnetic field as the source of an electric field and a varying electric field as the source of an magnetic field.

Gauss' law for electricity, Gauss' law for magnetism, Faraday's law, and the Ampere-Maxwell law provide a complete description of the electric and magnetic fields produced by any given charge and current distribution. They are known as Maxwell's equations and, taken with the equations for the electric and magnetic forces on a charge, they provide a complete classical description of electromagnetic phenomena. They are collected and reviewed in Chapter 32.

Chapter 32 also contains a discussion of the magnetic properties of materials. When an externally generated magnetic field is applied to a substance, the substance produces a field of its own. You may think of currents circulating within the material. For param-

agnetic materials, the induced field is in the same direction as the applied field and the total field is greater than the applied field. For diamagnetic materials, the induced field is in the opposite direction and the total field is smaller than the applied field. Certain materials, such as iron, are ferromagnetic and can be permanently magnetized. They produce a magnetic field even in the absence of an applied field.

In an electrical circuit that contains both an inductor and a capacitor, the current oscillates. Such circuits have great practical use in the tuning of radios and television sets. In Chapter 33, they are used to demonstrate the storage and flow of energy in a circuit. Most electrical power is transmitted by means of alternating currents, for which the current direction changes periodically. You will study this type circuit in Chapter 33.

The force that one charge exerts on another, whether electric or magnetic, depends on the relative positions of the charges. If one charge moves, the force it exerts on the other charge does not change immediately. On the contrary, the electric and magnetic fields first change in the near vicinity of the moving charge and these changes propagate as waves. The force on the second charge is not altered until the wave reaches it. Electromagnetic waves travel with the speed of light and, in fact, visible light is such a wave with frequency in a certain range. In Chapter 34, you will learn that the wave-like propagation of electric and magnetic fields is predicted by Maxwell's equations. Energy transport by electromagnetic radiation is also discussed in this chapter.

If the electric field of an electromagnetic wave is always parallel to the same line, the wave is said to be linearly polarized. By way of contrast, the field in other waves, said to be unpolarized, jumps randomly. Some materials, such as are used in many sunglasses, transmit only light that is linearly polarized in a certain direction. Polarization is studied in Chapter 34.

In Chapter 34, you also study reflection and refraction. Two new waves are generated when an electromagnetic wave impinges on the boundary between two regions with different optical properties. The reflected wave propagates back into the region of the incident wave and the refracted wave continues into the other region. The direction of travel of the refracted wave is different from that of the incident wave and depends on the optical properties of both regions.

Once you understand the nature of electromagnetic waves, you are ready for Chapters 35 through 37, the optics chapters. In Chapter 35, you will learn how mirrors and lenses form images. You will be able to find the positions and sizes of the images. You then apply the ideas to microscopes and telescopes.

Electromagnetic waves obey a superposition principle like other waves: if two or more waves are simultaneously present the total electric field, for example, is the vector sum of the electric fields associated with the individual waves. Superposition gives rise to interference effects, the topic of Chapter 36. If two waves are in phase, then they interfere constructively and the resultant amplitude is larger than either constituent amplitude. If their phases differ by 180°, they interfere destructively and tend to cancel each other. If the difference in phase comes about because the waves travel different distances, the effect is wavelength dependent; destructive interference of red light, for example, occurs at different places than the destructive interference of blue light. Interference is responsible for the colors on soap bubbles and oil films.

Waves passing by the edge of an object travel into the shadow region and produce interference-like fringes in that region: alternating dark and bright bands appear if only a single wavelength is present. Diffraction effects such as this are studied in Chapter 37. Diffraction gratings, which consist of closely spaced, extremely narrow, transparent lines ruled on glass or plastic, are used to analyze the spectrum of light from materials. X rays are diffracted from crystals and the pattern they produce can be used to find the atomic structure of the crystal. Diffraction by gratings and crystals is also discussed in Chapter 37.

Chapter 22
ELECTRIC CHARGE

I. BASIC CONCEPTS

You start your study of electromagnetism with Coulomb's law, which describes the force that one stationary charged particle exerts on another. Pay attention to both the magnitude and direction of the force and learn how to calculate the total force when more than one charge acts.

Be sure you understand the concept of force. You might review the relevant parts of Chapter 5.

Electric charge. Electric charge is defined in terms of electric current, which in turn is defined in terms of the magnetic force exerted by two current-carrying wires on each other. You must delay a precise understanding of charge until you study magnetism. For now you should know that charge is a property possessed by some particles and that two charged particles exert electric forces on each other. The SI unit for charge is _____ (abbreviated _____).

There are two types of charge, called <u>positive</u> and <u>negative</u>. In many of the equations of electromagnetism, a positive number is substituted for the value of a positive charge and a negative number is substituted for the value of a negative charge.

The charge on an electron is _____ C; the charge on a proton is _____ C. Charge is quantized: the charge on all particles detected so far is a positive or negative multiple of a fundamental unit of charge e. The value of e is _____ C.

All macroscopic materials contain enormous numbers of electrons and protons but if the number of electrons in an object equals the number of protons, then the net charge is zero and the object is said to be <u>neutral</u> (or uncharged). If there are more electrons than protons, the object has a net negative charge; if there are more protons than electrons, it has a net positive charge. In either case, it said to be <u>charged</u>. Notice that the net charge is computed by algebraically adding the charges of all particles in the object, taking their signs into account. Normally, all macroscopic bodies are neutral.

Objects may be given net charges by rubbing them together. When a glass rod is rubbed with silk, the rod becomes _____ charged. When a plastic rod is rubbed with fur, the rod becomes _____ charged.

Charge is conserved. The net charge of a closed system (with no particles entering or leaving) is always the same. After rubbing a glass rod, silk becomes _____ charged and, if no particles leak on or off, the magnitude of its charge is the same as the magnitude of the charge on the rod. Carefully note that it is the algebraic sum of the charges that remains the same. Before rubbing, the net charge was zero. After rubbing, it is still zero, the sum of the negative charge on one object and the positive charge on the other.

The force that one point charge exerts on another is along the line that joins the charges. If the charges have the same sign, then the force is one of repulsion; if the charges have different signs, then the force is one of attraction. The diagram below shows three possible pairings. At each charge draw a vector that indicates the force exerted on it by the other charge in the pair.

$\oplus$ $\oplus$ $\ominus$ $\ominus$ $\oplus$ $\ominus$

Conductors, insulators, and semiconductors. The *electrons* in materials move and are transferred to or from other objects by rubbing; the atomic nuclei (containing protons) are nearly immobile. Materials are often classified according to the freedom with which electrons can move. Electrons in an insulator are not free to move far; any charge placed on an insulator remains where it is placed. On the other hand, conductors contain many electrons that are free to move throughout the conductor. When you touch a charged conductor to an uncharged conductor, charge is transferred and, as a result, both conductors are charged. When you touch a charged insulator to a neutral conductor, very little, if any, charge is transferred. Scraping or rubbing is required to transfer charge to or from an insulator.

Give some examples of conductors and insulators.

conductors: _____

insulators: _____

Your body is a moderately good conductor, particularly if there is moisture on it. Although you can charge a conductor by rubbing it, you cannot touch it while you do or else the charge will move to your body. Conductors in the laboratory are usually equipped with insulating handles.

Electrical force. Suppose two point particles, one with charge q_1 and the other with charge q_2, are a distance r apart. The magnitude of the force exerted by either of the charges on the other is given by Coulomb's law:

$$F =$$

You should write this equation with the factor $1/4\pi\epsilon_0$. The constant of nature ϵ_0 is called _____ and has the value $\epsilon_0 =$ _____ $C^2/N\cdot m^2$ (to three significant figures). The factor $1/4\pi\epsilon_0$ has the value _____ $N\cdot m^2/C^2$ (to three significant figures). When you use this force equation, you must remember to enter the magnitudes of the charges. If you also enter their signs, you might get a negative result, which is not correct for the magnitude of a vector.

Notice that the magnitude of the force is proportional to the inverse square of the distance between the charges and is also proportional to the product of the magnitudes of the charges. If the distance between the charges is doubled, the force is _____ as large; if one of the charges is doubled, the force is _____ times as large; if both charges are doubled, the force is _____ times as large.

Force is a vector. To find the direction of the electric force one charge exerts on another, draw the line that joins the charges, then decide if the force is attractive (opposite sign charges) or repulsive (like sign charges). If it is attractive, the force on either charge is along the line and toward the other charge. If it is repulsive, it is along the line and away from the other charge. Electrostatic forces between two charges obey Newton's third law: they are equal in _____ and opposite in _____.

When more than two charges are present, the force on any one of them is the *vector* sum of the forces due to the others. Each force is computed using Coulomb's law. This is the principle of <u>superposition</u> for electric forces.

To illustrate the principle of superposition, consider four identical charges q at the corners of the square with edge a, as shown. Develop an expression for the total force on the charge labelled 1 in the diagram. Fill in the following table, giving the magnitude of each force, the x component, and the y component, all in terms of q and a.

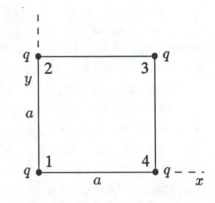

	magnitude	x component	y component
force of 2 on 1			
force of 3 on 1			
force of 4 on 1			

The x component of the total force on 1 is given by

$$F_x =$$

and the y component is given by

$$F_y =$$

Shell theorems. These are two theorems about electrical forces between a point charge and a special distribution of charge, a uniform spherical shell. You will find them extremely useful.

Describe what is meant by a uniform spherical shell of charge: _____

_____

_____

If a point charge q is outside a uniform shell of charge, a distance r from its center, the force exerted by the shell on the point charge has a magnitude that is given by $F =$ _____, where Q is the charge on the shell. The force is along the line joining the center of the shell and the point charge; it is away from the shell if q and Q have the same sign and toward the shell if they have opposite signs.

If a point charge q is anywhere inside the shell, the force exerted on it by the shell is

_____.

Forces on neutral objects. Charged objects attract neutral macroscopic objects. A neutral conductor is attracted to a charged rod because the electrons within the inductor are redistributed, making some regions positively charged and others negatively charged. The forces on the negative and positive regions are, of course, in opposite directions but the force is greater on the region nearer to the charged object.

The diagram on the right shows a positively charged insulating rod held near a neutral conductor. Indicate the distribution of charge on the conductor and draw arrows of different length to indicate the forces on the positive and negative regions of the conductor. The net force should be toward the rod.

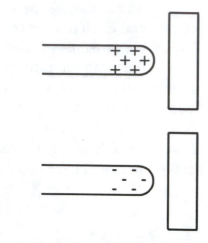

The diagram on the right shows a negatively charged insulating rod held near a neutral conductor. Indicate the distribution of charge on the conductor and draw arrows of different length to indicate the forces on the positive and negative regions of the conductor. The net force should be toward the rod.

Suppose you touch the conductor with your hand while the charged insulating rod is nearby. Electrons flow from you to the conductor if the insulating rod is positive or from the conductor to you if the rod is negative. Remove your finger, then remove the rod. The conductor is now charged. It is _____ if the insulating rod was positive and _____ if the rod was negative. This process is called <u>charging by induction</u>.

Charged objects also attract neutral insulators but the force is much smaller than the force of attraction for a conductor. Under the influence of the external charge, electrons in an insulator move slightly from their normal orbits so the center of the negative charge is slightly displaced from the center of the positive charge. This results in a net force. A charged rod can be used to pick up small pieces of paper and a rubbed balloon "sticks" to a wall.

II. PROBLEM SOLVING

The fundamental problems of this chapter deal with Coulomb's law. Given two charges and their separation you should be able to calculate the force (both magnitude and direction) each exerts on the other. You should know how to use vector addition to find the resultant force when more than one charge interacts with the charge of interest. Variations include problems that ask you to find one of the charges or the position of a charge, given the force.

To find the value of one or more of the charges in some problems, a preliminary calculation must be performed prior to using Coulomb's law. For example, you may need to calculate the total positive charge in an object. The number of atoms in an object can be found from the mass of an atom and the mass of the object; the number of protons in an atom is given by the atomic number. Sometimes the molar mass is given (or can be found in the periodic table of the chemical elements) and you will need to divide that quantity by the Avogadro constant (6.022×10^{23} mol^{-1}) to calculate the mass of an atom.

Other problems deal with conducting spheres and you will need to apply the shell theorems. In addition, you should know that when two identical conducting spheres touch each other, the total charge is shared equally between them.

Finally, you should know the relationship between charge and current: current is the charge that moves into or out of a region per unit time.

III. NOTES

Chapter 23
ELECTRIC FIELDS

I. BASIC CONCEPTS

Here the idea of an electric field is introduced and used to describe electrical interactions between charges. It is fundamental to our understanding of electromagnetic phenomena and is used extensively throughout the rest of this course. Build a firm foundation by learning well the material of this chapter.

Some of the concepts you should review are: force (Chapter 5), electric charge (Chapter 22), and permittivity constant (Chapter 22).

The electric field. A <u>field</u> is an important concept in physics. It gives some property of a region as a function of position. An example of a scalar field is _____ _____ ; an example of a vector field is _____ .

To describe the electric force that one charge exerts on another, we associate an <u>electric field</u> with the first charge and say that this field exerts a force on the second charge. Consider two charges Q and q, for example. Charge Q creates an electric field and that field exerts a force on q. Charge q also creates an electric field and *that* field exerts a force on Q. The electric field of a charge exists throughout all space, so anywhere a second charge is placed it will experience a force.

The electric field at any point in space is defined as the electric force per unit test charge on a stationary positive test charge placed at that point, in the limit as the test charge becomes vanishingly small. If the test charge is q_0 and $\vec{F}$ is the electric force on it, then the electric field at the position of the test charge is

$$\vec{E} =$$

You should be aware that this field is *not* the field created by q_0; it is the total field created by all *other* charges. The test charge does not exert a force on itself. An electric field is associated with the charges that create it and exists irrespective of whether a test charge is present. The SI units of an electric field are _____ .

Explain why the limit as q_0 becomes vanishingly small must be included in the definition:

_____

_____

Suppose a collection of stationary charges creates an electric field $\vec{E}$ and a charge q is placed in this field. Then the force exerted on q is given by

$$\vec{F} =$$

If $\vec{E}$ varies from place to place, as it usually does, then you use the value at the position of q.

If an electric force is the only force acting on a charge, then Newton's second law takes the form $q\vec{E} = m\vec{a}$ and gives the acceleration of the charge. If you know the initial position and velocity of the charge, then you can, in principle, predict its subsequent motion.

Electric field lines.   Electric fields are often represented in diagrams by <u>electric field lines</u> (also called lines of force). These lines graphically depict both the direction and magnitude of the electric field throughout a region of space. Tell how the direction of the electric field at a point is related to the field line through that point: _____

_____

Arrows are placed on field lines to indicate the direction of the field but electric field lines themselves are *not* vectors.

Tell how the magnitude of the electric field at a point is related to the field lines in the vicinity of the point: _____

_____

Electric field lines emanate from _____ charge and terminate on _____ charge.

To see the electric field lines for some charge distributions, look carefully at Figs. 23–2, 3, 4, and 5 of the text. Be aware that the configuration of lines is different for different charge distributions: the electric field of a point charge, for example, is *not* the same as the electric field of a dipole. For each figure, notice the directions of the lines, where they are close together, and where they are far apart. You should recognize that the number of field lines is proportional to the charge. If all charge that creates the field is doubled, then the number of lines doubles in every region.

The diagram to the right shows some field lines. Indicate on it a region where the magnitude of the electric field is relatively large and a region where it is relatively small. Suppose an electron is at the point marked •. Draw an arrow to indicate the direction of the force on it and label the arrow $\vec{F}$.

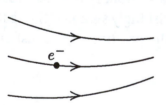

Point charges.   The magnitude of the electric field at a point a distance r from a single point charge q is given by

$$E =$$

The field is parallel to the line that joins the charge q and the point. If q is positive, it points away from q; if q is negative, it points toward q.

If more than one point charge is responsible for the electric field, you calculate the total field by summing the fields due to the individual charges. Don't forget that electric fields are vectors and the sum is a vector sum. You can sum the magnitudes only if the fields are in the same direction.

In any given situation, you must distinguish between the total electric field and the field that acts on a given charge. Suppose, for example, there are three charges: q_1, q_2, and q_3. The total field at any given point is the vector sum of the fields due to the three charges. The field that acts on q_1, however, is the vector sum of the fields due to q_2 and q_3 only.

Electric fields of continuous charge distributions. A continuous charge distribution is characterized by a <u>charge density</u>. If the charge is on a line (straight or curved), the appropriate charge density is the _____ charge density, denoted by _____ and defined so that the charge dq in an infinitesimal segment ds of the line is given by dq = _____. If the charge is distributed on a surface, the appropriate charge density is the _____ charge density, denoted by _____ and defined so that the charge dq in an infinitesimal element of area dA is given by dq = _____. A uniform charge distribution means that the appropriate charge density is the same everywhere along the line or on the area.

Carefully study the derivations of expressions for the electric field on the axis of a uniform ring of charge and on the axis of a uniform disk of charge. You should understand why the field is parallel to the axis in each case.

The diagram on the right shows an infinitesimal segment ds of the ring and the field it produces at point P on the axis, a distance z from the ring center. On the diagram, show the location of another infinitesimal segment such that the sum of its field and the field shown is along the axis. Draw the electric field vector at P for this segment. Since every infinitesimal segment of the ring can be paired in this way with another segment, the total field must be along the axis.

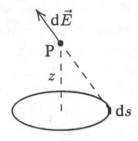

If the ring has radius R and total charge q, then the linear charge density is λ = _____, the charge in an infinitesimal segment ds is dq = _____, the magnitude of the field produced at P by the segment is

$$dE = $$

and the component along the axis is

$$dE_\parallel = $$

You must now integrate around the ring. The integration, of course, sums the components along the axis of the fields due to the ring segments. The result for the total field is

$$E = $$

To calculate the field produced by a uniform disk of charge, the disk is divided into rings, each having infinitesimal width dr. The area of a ring of radius r is dA = _____ and if σ is the area charge density, then dq = _____ is the charge in a such a ring. The magnitude of the field it produces at a point on the axis a distance z from the ring center is

$$dE =$$

To find an expression for the total field, this expression is integrated from $r = 0$ to $r = R$. The result is:

$$E =$$

By taking R to be much larger than z, you can find an expression for the field produced by a uniform plane of charge. Notice that if $R \gg z$, then $z/\sqrt{z^2 + R^2}$ is much smaller than 1 and

$$E =$$

You will find this expression extremely useful in your study of later chapters. It is a good approximation to the field of a finite plane of charge at points that are close to the plane and far from the edges. Notice that the field does not depend on distance from the plane.

Note that at points near a ring, disk, or plane, superposition produces total fields that are quite different from each other and quite different from the field of a point charge. At points far from a charge distribution, however, the electric field tends to become like that of a point particle with charge equal to the net charge in the distribution. For a ring of charge with uniform linear density λ, the net charge is $q =$ _____ and the field far away ($z \gg R$) is given by $E =$ _____. For a disk with uniform area density σ, the net charge is $q =$ _____ and the field far away ($z \gg R$) is given by $E =$ _____.

Electric dipoles.   The geometry is shown in Fig. 23–8 of the text. An electric dipole consists of a _____ charge and a _____ charge, of equal _____. It is characterized by a dipole moment $\vec{p}$, a vector with magnitude given by $p =$ _____, where q is _____ _____ and d is _____.
The direction of $\vec{p}$ is _____.
Look carefully at Fig. 23–5 of the text to see the lines of force produced by a dipole.

Go over the derivation of Eq. 23–6 of the text for the electric field at point P in Fig. 23–8. Be sure you understand why the field there is in the positive z direction. In terms of q, z, and d, the magnitude of the field produced at P by the positive charge is given by $E_+ =$ _____ and the magnitude of the field produced there by the negative charge is given by $E_- =$ _____. The results are summed to produce an expression for the total field. It is

$$E =$$

You will need to know about atomic dipoles, for which the separation of the charges is less than a nanometer and much less than z. Use the binomial theorem to expand $(z - d/2)^2$ and $(z + d/2)^2$. Retain only those terms that are proportional to d. See the Mathematical Skills section. In terms of the dipole moment, the field is

$$E =$$

Notice that the field is inversely proportional to z^3, not z^2.

Consider a dipole in a uniform electric field $\vec{E}$ (created by other charges). The net force on the dipole is _____ because _____.

The dipole does, however, experience a torque. If $\vec{p}$ is the dipole moment and $\vec{E}$ is the electric field, then the torque is given by

$$\tau =$$

This torque tends to rotate the dipole so that $\vec{p}$ _____.

When an electric dipole rotates in an electric field, the field does work on it. For a rotation during which the angle between the dipole and field changes from θ_0 to θ, the work done by the field is $W =$ _____. The sign of the work is _____ if the angle increases and _____ if the angle decreases. During a rotation, the potential energy of the field and dipole changes, the change being given by $\Delta U =$ _____. If we take the potential energy to be zero when the dipole moment is perpendicular to the field, the potential energy for any orientation can be written as the scalar product $U =$ _____. The potential energy is a maximum when _____ and is a minimum when _____.

Suppose a dipole is initially oriented with its moment perpendicular to an electric field. It is then rotated through 90° so it is in the same direction as the field. During this process the sign of the work done by the field is _____ and the sign of the change in the potential energy is _____. Suppose, instead, that the dipole is rotated so its direction becomes opposite to that of the field. During this process, the sign of the work done by the field is _____ and the sign of the change in the potential energy is _____.

II. PROBLEM SOLVING

You should know how to compute the electric field of a collection of point charges. Remember that the sum of the individual fields is a vector sum. Add the x components to get the x component of the sum; add the y components to get the y component of the sum. The magnitude of the electric field produced by a single point charge q, at a point a distance r away, is given by $E = |q|/4\pi\epsilon_0 r^2$. This must be multiplied by the sine or cosine of an angle to obtain a component. You must also determine the sign of the component.

Some problems deal with electric field lines. You should know that the electric field at any point is tangent to the line through that point, that the number of lines per unit area passing through an area perpendicular to the lines is proportional to the magnitude of the field, and that lines emanate from positive charge and end at negative charge.

You will also need to know that the force on a charge in an electric field is given by the product of the charge and the field. For a positive charge, it is in the direction of the field while for a negative charge it is directed opposite to the field.

You should know how to calculate the electric field of a continuous distribution of charge. Use the linear or area charge density to write an expression for the charge in an infinitesimal region, then carry out an integration over the entire charge distribution. Don't forget that the field is a vector and you must evaluate an integral for each component or, in special cases, use symmetry to show a field component vanishes.

Some problems deal with the trajectories of charges in electric fields. If the field is uniform, the acceleration of the charge is constant and you may use the kinematic equations of constant motion. The problems are quite similar to projectile motion problems; however, the acceleration is now due to the force of electric field rather than to gravity.

You should know how to calculate the torque of a uniform electric field on an electric dipole and the potential energy of a dipole in an electric field. In some cases, you are asked for the work that is done by the field on a dipole when the dipole turns. This is the negative of the change in the potential energy.

III. MATHEMATICAL SKILLS

Binomial theorem.   You will need to know the binomial theorem to derive an expression for the electric field of an electric dipole, far from the dipole. The first few terms of $(A + B)^n$ are

$$(A + B)^n = A^n + nA^{n-1}B + \frac{n(n-1)}{2}A^{n-2}B^2 + \frac{n(n-1)(n-2)}{2 \cdot 3}A^{n-3}B^3 + \dots$$

and, in general, term r in the sum is

$$\frac{n!}{r!(n-r)!}A^{n-r}B^r,$$

where $r!$ is the factorial of r: $r! = 2 \cdot 3 \cdot 4 \dots r$.

The exponent n may be any number, positive or negative. To estimate the square root of $A + B$, take $n = 1/2$ and to estimate the reciprocal of the square root take $n = -1/2$.

The series has a finite number of terms if n is a positive integer, the last term being B^n. Otherwise the series never ends. Notice that in successive terms of the series the exponent of A decreases while the exponent of B increases. Since you will want a sum in which successive terms are smaller than previous terms, take A to be the larger of the two quantities and B the smaller.

IV. NOTES

Chapter 24
GAUSS' LAW

I. BASIC CONCEPTS

Gauss' law is one of the four fundamental laws of electromagnetism. It relates an electric field to the charges that create it and is, therefore, closely akin to Coulomb's law. Unlike Coulomb's law, however, it is valid when the charges are moving, even at relativistic speeds. The central concept for an understanding of Gauss' law is that of electric flux, a quantity that is proportional to the number of field lines penetrating a given surface. You should pay careful attention to the definition of flux and learn how to compute it for various fields and surfaces. You should then learn how to use Gauss' law to compute the charge in any region if the electric field is known on the boundary and also how to use the law to compute the electric field in certain highly symmetric situations.

Here are some concepts you might review in connection with your study of this chapter: volume flow rate and fluid velocity from Chapter 15, charge, Coulomb's law, conductor, insulator, and permittivity constant from Chapter 22, and electric field from Chapter 23.

Electric flux. Electric flux is associated with any surface (real or imaginary) through which electric field lines pass. It is defined by the integral

$$\Phi =$$

where $\mathbf{E}$ is the electric field on the surface and d$\mathbf{A}$ is an infinitesimal element of surface area. This integral tells us to evaluate the scalar product $\mathbf{E} \cdot \mathbf{dA}$ for each element of the surface and sum the results for all elements. Notice that the infinitesimal area element is written as a vector. What is its direction? _____
For an open surface, such as the top of a table, d$\mathbf{A}$ may be in either of the two directions that are perpendicular to the surface (up or down for a horizontal surface). Which one you pick determines the sign but not the magnitude of the flux. For a closed surface, one that completely surrounds a volume, d$\mathbf{A}$ is always chosen to be _____.

The scalar product that appears as the integrand may be interpreted as the product of the infinitesimal area and the component of the field normal to the surface. Thus, only the field that pierces the surface contributes to the flux. Consider two identical surfaces and suppose a uniform electric field exists at each of them. If the fields have the same normal components, then the flux is the same even if they have different tangential components. A field that is parallel to a surface does not contribute to the flux through that surface.

The flux through a surface is proportional to the number of field lines that penetrate the surface. Recall that the number of lines per unit area through an infinitesimal area dA perpendicular to the field is proportional to the magnitude of the field. Thus, if dA is in the same direction as the field, the number of lines that penetrate it is proportional to $E\,\mathrm{d}A$. If the area is rotated so dA makes the angle ϕ with the field, the number of lines that penetrate the area decreases to $E\,dA\cos\phi$. Carefully study Fig. 24–2 of the text.

When studying this and subsequent chapters, you may think of the electric flux through a surface as giving the net number of lines crossing the surface. There are, however, two important distinctions you should make. First, Φ is not necessarily an integer. Second, Φ is negative if the field makes an angle of more than 90° with dA or, what is the same thing, the field lines cross the surface in a direction roughly opposite to dA. Field lines that cross a closed surface from inside to outside make a _____ contribution to Φ while lines that cross the surface from outside to inside make a _____ contribution. Since some parts of a closed surface may make positive contributions to the flux while other parts make negative contributions, the total flux through a surface may vanish even though an electric field exists at every point on the surface.

Gauss' law.   Gauss' law states that for any closed surface, the total flux through the surface is proportional to the net charge enclosed by the surface. The constant of proportionality involves ϵ_0, the permittivity constant. Copy Eq. 24–7 of the text here:

When you use the law, be sure to evaluate the integral for a *closed* surface. The small circle on the integral sign is a reminder. The surface is known as a *Gaussian surface*. When you use Gauss' law to solve a problem, you must identify the Gaussian surface you are using. It might be the physical surface of an object or it might be a purely imaginary construction. Also be very careful that the charge you use in the Gauss' law equation is the net charge *enclosed* by the Gaussian surface. Charge outside does not contribute to the total flux through a closed surface, although it does contribute to the electric field at the surface.

The law should not surprise you since the magnitude of the electric field and the number of field lines associated with any charge are both proportional to the charge. All lines from a single positive charge within a closed surface penetrate the surface from inside to outside, so the total flux is positive and proportional to the charge. All lines associated with a single negative charge within a closed surface penetrate from outside to inside so the total flux is negative and again proportional to the charge. Some lines associated with a charge outside a surface penetrate the surface but those that do, penetrate twice, once from outside to inside and once from inside to outside, so this charge does not contribute to the total flux through the surface. When more than one charge is enclosed, add the individual flux contributions, with their appropriate signs.

You should be aware that the electric field at points on a closed surface containing a net charge of zero is not necessarily zero. Each charge, whether in the interior or exterior, creates a field that does not vanish on the surface. Nevertheless, a net charge of zero inside means a total flux of zero. Zero total flux simply means there are just as many field lines entering the volume enclosed by the surface as there are leaving.

Gauss' law can be used to find an expression for the electric field of a point charge. Imagine a sphere of radius r with a positive point charge q at its center. Since the electric field is radially outward from the charge, the normal component of the field at any point on the surface of the sphere is the same as the magnitude E of the field. In terms of E and r, the total flux through the sphere is $\Phi =$ _____ . Equate this to q/ϵ_0 and solve for E. The result is: $E =$ _____ , in agreement with Coulomb's law.

Gauss's law can be written in differential form and used to solve for the electric field of any distribution of charge. In its integral form, it is useful in several ways. If the electric field is known at all points on a surface, the law can be used to calculate the net charge enclosed by the surface. If the charge is known, the law can be used to find the total electric flux through the surface. If the charge distribution is highly symmetric, so a symmetry argument can be used to show that $E \cos\theta$ has the same value at all points on the surface, then Gauss' law can be used to solve for the electric field at points on the surface.

Gauss' law and conductors. The electric field vanishes at all points in the interior of a conductor in electrostatic equilibrium (all charge stationary). What would happen if the field did not vanish? _____

_____

_____

Remember that $\mathbf{E} = 0$ inside a conductor even when excess charge is placed on it or when an external field is applied to it.

At electrostatic equilibrium, the electric field at all points just outside a conductor is perpendicular to the surface. It has only a normal, not a tangential component. The charge on the surface can bring about a change in the normal component of the field so it is zero inside and non-zero outside but it cannot change the tangential component. Since the tangential component must be zero inside, it must also be zero just outside.

The condition that the electric field is zero leads, via Gauss' law, to an interesting property of conductors: in an electrostatic situation, any excess charge on a conductor must reside on its surface; there can be no net charge in its interior. Imagine a Gaussian surface that is completely within the conductor. The electric field is _____ at every point on the Gaussian surface, so the total flux through the surface is _____ and, according to Gauss' law, the net charge enclosed by the surface is _____ . Any net charge in the conductor must lie outside the Gaussian surface. Since this result is true for *every* Gaussian surface that can be drawn completely within the conductor, no matter how close to its surface, we conclude that any excess charge must be on the surface of the conductor. Be careful! If the object is not a conductor, excess charge may be distributed throughout its volume.

You should understand conductors with cavities. The diagram on the right shows the cross section of a conductor with a cavity containing a point charge q_1. The net charge within the Gaussian surface indicated by the dotted line is _____ , so if q_2 is the charge on the inner surface of the conductor, $q_1 + q_2 =$ _____ and $q_2 =$ _____ . Furthermore, if Q is the total excess charge on the conductor and q_3 is the charge on its outer surface, then $q_2 + q_3 =$ _____ and $q_3 =$ _____ .

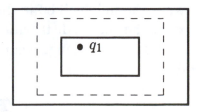

If the net charge on the conductor is zero, the charge on its outer surface is _____ . If

$Q =$ _____ , there is no net charge on the outer surface of the conductor. Notice that Gauss' law cannot tell us the distribution of charge on a surface, only the net charge there.

You should realize that the area charge density may not be the same at every point on the surface of a conductor. Regardless, the magnitude of the electric field at any point just outside is directly proportional to the area charge density at the corresponding point on the surface. The exact relationship is $E =$ _____ , where σ is the area charge density.

Gauss' law can be used to prove this result. Consider a small portion of the surface of a conductor and suppose its area is A and its area charge density is σ. The electric field is normal to the surface. Describe the Gaussian surface you will use: _____

_____

_____

One end of the Gaussian surface should be inside the conductor. The flux through this end is _____ . Another end should be outside the conductor and perpendicular to the field. If the magnitude of the field is E, the flux through this end is _____ . The sides are parallel to the field so the flux through them is _____ . In terms of E and A, the total flux through the Gaussian surface is _____ .

In terms of σ and A, the net charge enclosed by the surface is _____ . Gauss' law yields _____ = _____ . $E = \sigma/\epsilon_0$ follows immediately. You should realize that this field is produced by the charge on *all* parts of the surface and by external sources if they are present.

Calculating the electric field. Successful use of Gauss' law to solve for the electric field depends greatly on choosing the right Gaussian surface. First, it must pass through the point where you want the value of the field. Second, either the normal component of the electric field must have constant magnitude over the entire surface or else it must have constant magnitude over part of the surface and the flux through the other parts must be zero. A symmetry argument should be made to justify the use of the Gaussian surface you have chosen.

To understand how such an argument is made, consider a long straight wire with positive charge distributed uniformly on it. Take the wire to be infinitely long and use symmetry to show that the electric field at any point is radially outward from the wire. The outline of the argument is: Pretend that the field is *not* radially outward and show that this leads to a contradiction of the principle that identical charge distributions produce identical electric fields.

Assume the field at point P is in the direction shown in the diagram on the left below, a non-radial direction. Now, turn the wire over end for end, through a 180° rotation about the dotted line through P. If the direction chosen for the field is correct, then after turning the wire, the field will be in the direction shown in the diagram on the right. This is impossible since the charge distribution is exactly the same in the two cases. A radial field is the only one that does not change when the wire is turned.

Note that if the wire is not infinite, the argument works only if P is opposite the midpoint of the wire. Similarly, the argument does not work (and the field is not radial) if the charge is not distributed uniformly on the wire. In those cases, turning the wire changes the distribution of charge.

For an infinite wire with a uniform charge distribution, the electric field must have the same magnitude at all points that are the same distance from the wire. The charge distribution looks exactly the same from point P as it does from a point 2 m (or 2 km) farther along the wire, the same distance from the wire.

The points where the magnitude of the field are the same as at P form the rounded portion of the surface of a cylinder, so take a cylinder of length L and radius r, with P on the surface, for your Gaussian surface. In the space to the right, draw a diagram of the wire and the Gaussian surface. Label dA and $\mathbf{E}$ at point P.

The completion of the problem is now straightforward. First, evaluate the integral $\int \mathbf{E} \cdot \mathbf{dA}$ over the rounded portion of the cylinder. Since $\mathbf{E}$ and dA are parallel at all points on the surface, $\mathbf{E} \cdot \mathbf{dA} = $ _____ and since E is the same for all points on the surface, $\int \mathbf{E} \cdot \mathbf{dA} = E \int dA = $ _____, in terms of r and L. Since a Gaussian surface must be closed, you must also consider the circles that form the ends of the cylinder. No flux passes through either end because _____ .

Solve the Gauss' law equation. If q is the net charge enclosed by the cylinder, then the magnitude of the electric field is given by $E = $ _____. Usually the linear charge density λ of the wire is given. Since this is q/L, $E = $ _____ .

Now, consider some of the other charge distributions discussed in the text. Think of these derivations in two ways. First, they will give you a chance to practice solving problems using Gauss' law. Second, in some cases the results are useful for later work. You should particularly remember the expression for the electric field of a large plane sheet of charge. You will use it several times when you study capacitors.

Consider an infinite sheet with uniform area charge density σ. In the space to the right, sketch a portion of the sheet and the Gaussian surface used to find the electric field at a point away from the sheet. The field is perpendicular to the sheet and points away from it if it is positive. Draw a vector indicating the electric field at a point on the Gaussian surface. Also label the parts of the surface through which the flux is zero.

In terms of the magnitude E of the field and area A of one end of the Gaussian surface, the total flux through the surface is $\Phi = $ _____. In terms of σ and A, the total charge enclosed by the surface is _____. Gauss' law thus gives the following expression for the electric field in terms of the area charge density: $E = $ _____ .

You might think this result is inconsistent with the expression for the magnitude of the electric field just outside a conductor, $E = \sigma/\epsilon_0$. It is not. Consider an infinite plane conducting sheet with a uniform charge density σ on one surface. This charge produces an electric field with magnitude $\sigma/2\epsilon_0$, just like any other large uniform sheet of charge. The field exists on both sides of the sheet, in the interior of the conductor as well as in the exterior, and on each side, it points away from the surface if σ is positive.

To obtain an electric field of zero within the conductor, another field must be present, produced perhaps by charge on another portion of the conductor's surface or by external charge. In the diagram, E_1 is the magnitude of the field due to the charge layer and E_2 is the magnitude of the second field. In the interior, the second field exactly cancels the field due to the charge layer to produce a total field of zero and in the exterior, it augments the field due to the charge layer to produce a total field with magnitude σ/ϵ_0.

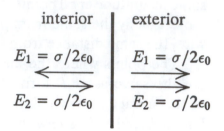

Another important charge configuration is a uniform spherical shell carrying total charge Q. Symmetry leads to us to conclude that the electric field must be radial so we use a Gaussian surface in the form of a sphere, concentric with the shell. If the radius of the Gaussian surface is r and the electric field has magnitude E at points on it, then the flux through the surface is $\Phi = $ _____. If the Gaussian surface is inside the shell, the charge enclosed is _____, so the electric field is $E = $ _____ at points inside the shell. If the Gaussian surface is outside the shell, the charge enclosed is _____, so the electric field is _____ at points outside the shell. Carefully study Section 24–9 to see how this result is applied to find the electric field inside a uniform spherical distribution of charge.

II. PROBLEM SOLVING

You should know how to calculate the electric flux through a given surface. A useful fact to remember is: if the field is uniform, then the total flux through a closed surface is zero. Sometimes you can easily calculate the flux of a uniform field through part of a closed surface but you have been asked for the flux through the remaining part. Since the two contributions to the total flux sum to zero, they are the negatives of each other.

Some problems deal with Gauss' law in the form $\epsilon_0 \Phi = q_{enc}$, where Φ is the electric flux through a closed surface and q is the charge enclosed by the surface. You might be given the charge and asked for the flux or you might be given the flux (or the information to calculate it) and asked for the enclosed charge.

When a problem deals with a conductor, you should remember that the electric field inside the conductor is 0, that the field just outside the conductor is perpendicular to the surface and has magnitude σ/ϵ_0, where σ is the area charge density, and that the total charge enclosed by any Gaussian surface completely inside the conductor is 0.

Many problems ask you to use Gauss' law to solve for the electric field in given situations. Become adept in finding an expression for the total flux through a Gaussian surface for the three special cases discussed in the text: cylindrical, spherical, and planar symmetry. For a cylindrical Gaussian surface, with a radial field that is uniform over the curved portion of the surface, $\int \mathbf{E} \cdot d\mathbf{A} = 2\pi r L E$, where r is the radius of the Gaussian cylinder and L is its length. For a spherical Gaussian surface with a radial field that is uniform over the surface, $\int \mathbf{E} \cdot d\mathbf{A} = 4\pi r^2 E$, where r is the radius of the Gaussian sphere. Also remember that the electric field due to a large plane with uniform area charge density σ is $\sigma / 2\epsilon_0$.

III. MATHEMATICAL SKILLS

Electric flux calculations.   To calculate the electric flux, you should be familiar with the evaluation of simple area integrals. If the region of integration is a rectangle with sides a and b, you will probably want to position the coordinate system so that the rectangle is in the x, y plane with two of its sides along the axes. The infinitesimal element of area can be taken to be a rectangle with sides dx and dy. The flux integral then becomes

$$\Phi = \int_{x=0}^{a} \int_{y=0}^{b} E_z(x, y) \, dx \, dy \, .$$

Integrations in x and y are carried out independently. If, for example, the z component of the electric field is given by $E_z = 9x^2 y + 2y$, then

$$\Phi = \int_{x=0}^{a} \int_{y=0}^{b} (9x^2 y^2 + 2y) \, dx \, dy = \int_{y=0}^{b} (3x^3 y^2 + 2xy) \Big|_{x=0}^{a} \, dy$$

$$= \int_{y=0}^{b} (3a^3 y^2 + 2ay) \, dy = (a^3 y^3 + ay^2) \Big|_{y=0}^{b} = a^3 b^3 + ab^2 \, ,$$

where the integration over x was carried out first, followed by the integration over y.

For many problems of this chapter a Gaussian surface can be chosen so that the normal component of the electric field has the same value at all points on a portion of it and is zero on the other portions. The integral for the flux then reduces to $\Phi = EA$, where A is the area of the region over which the normal component of the field is not zero and E is the magnitude of the normal component.

To evaluate the flux, you must know how to calculate the areas of various surfaces. The area of a rectangle with sides of length a and b is ab; the area of a circle with radius R is πR^2; the surface area of a sphere with radius R is $4\pi R^2$; and the surface area of the rounded portion of a cylinder with radius R and length L is $2\pi R L$.

If you have trouble remembering that $2\pi r L$ gives the area of a cylindrical surface, imagine a paper towel roll wrapped exactly once around with a towel. The surface area of the roll is the same as the area of the towel. Since the towel is a rectangle with sides of length $2\pi r$ and L, its area is $2\pi r L$.

Calculations of charge.   You must also be able to calculate the charge enclosed by a Gaussian surface when the volume, area, or linear charge density is given. If an object has a uniform volume charge density ρ, for example, the charge enclosed is given by ρV, where V is the volume of the part of the object that lies within the Gaussian surface. If the Gaussian surface is completely within the object and the object does not have any cavities, then the region enclosed by the Gaussian surface is completely filled with charge and V is the volume enclosed by the Gaussian surface. If the object has a cavity that is wholly within the Gaussian surface, then V is the volume enclosed by the Gaussian surface minus the volume of the cavity.

Suppose, for example, a sphere of radius R has a uniform volume charge density ρ and the Gaussian surface is a concentric sphere of radius r. If $r < R$, then the Gaussian sphere is filled with charge and the charge enclosed is $4\pi\rho r^3/3$. If, on the other hand, $R < r$, then the Gaussian surface is only partially filled with charge. The charge enclosed is the total charge, or $4\pi\rho R^3/3$.

If a spherical object has a spherical cavity with radius R_c and the Gaussian surface is within the object but outside the cavity, then the charge enclosed is $\rho[(4\pi r^3/3) - (4\pi R_c^3/3)]$. The first term in the brackets is the volume enclosed by the Gaussian surface and the second is the volume of the cavity. If $R < r$, the charge enclosed is $\rho[(4\pi R^3/3) - (4\pi R_c^3/3)]$.

Now, consider a solid cylinder with radius R and length L, having a uniform volume charge density ρ. Suppose the Gaussian surface is a concentric cylinder with radius r and the same length as the cylinder of charge. If $r < R$, the charge enclosed is $2\pi\rho rL$ and if $R < r$, it is $2\pi\rho RL$. If the cylinder has a cavity that is inside the Gaussian surface, its volume must be subtracted from the volumes in these expressions.

If the volume charge density varies from point to point in the object, you must evaluate the integral $\int \rho \, dV$ over the volume of that part of the object lying within the Gaussian surface. The most common example is a sphere with a charge density that varies only with distance r from the center. Carry out the integration by dividing the sphere into spherical shells with infinitesimal thickness dr. A typical shell extends from r to $r + dr$ and has a volume of $4\pi r^2 \, dr$. Notice that this is the product of the surface area of the shell and its thickness. The charge in the shell is $4\pi\rho(r) r^2 \, dr$. If the Gaussian surface is a sphere of radius r, concentric with the sphere of charge and entirely within it, the charge enclosed is $\int_0^r 4\pi\rho(r) r^2 \, dr$. If the Gaussian surface is entirely outside the charge distribution, the charge enclosed is $\int_0^R 4\pi\rho(r) r^2 \, dr$. Notice the upper limits of integration are different for these two cases.

For a cylinder with a charge density that depends only on the distance r from the axis, divide the cylinder into concentric cylindrical shells with thickness dr. The volume of a shell is $2\pi r L \, dr$, where L is the length of the cylinder. If the Gaussian surface is a concentric cylinder with radius r and is inside the charge distribution, then the charge enclosed is $\int_0^r 2\pi r L\rho(r) \, dr$. If the Gaussian surface is outside the distribution, the charge enclosed is $\int_0^R 2\pi r L\rho(r) \, dr$.

IV. NOTES

Chapter 25
ELECTRIC POTENTIAL

I. BASIC CONCEPTS

Because charges exert electrical forces on each other, work (perhaps negative) must be done to assemble a collection of charges, and because electrical forces are conservative, a potential energy is associated with the assembled collection. If the charges are released, potential energy is converted to kinetic energy as each charge moves in response to the forces of the other charges. In this chapter, you will learn to calculate and use the potential energy of a system of charges.

Electric potential is closely related to potential energy and plays a vital role in most succeeding discussions of electricity. Pay careful attention to its definition and learn how to compute it for collections of point charges and for continuous charge distributions. Remember that electric potential is a property of a collection of charges and does not depend on the test charge used to measure it.

Some of the ideas from previous chapters that you might want to review in connection with your study of this chapter: work from Chapter 7, potential energy from Chapter 8, electric field, electric field of a point charge, electric dipole, and electric field lines from Chapter 23.

Electric potential energy. Since the electric force is conservative, a potential energy function can be defined. Consider a collection of interacting charged particles. If one of them, with charge q, moves through an infinitesimal displacement ds, the electric field $\mathbf{E}$ acting on it does work

$$\mathrm{d}W =$$

and the potential energy of the collection changes by

$$\Delta U = -\mathrm{d}W =$$

If the displacement of q is not infinitesimal, but instead is from point i to point f, the work done by the field is given by the integral

$$W_{if} =$$

and the change in the potential energy of the collection is given by

$$U_f - U_i = -W_{if} =$$

Pay attention to the sign of the change in potential energy. If, for example, a positive charge moves in the direction of the electric field acting on it, then the sign of the work done by the field is _____ and the sign of the change in the potential energy of the system is _____ . If a positive charge moves in the direction opposite to the field acting on it, then the sign of the work done by the field is _____ and the sign of the change in the potential energy is _____ . These signs are reversed if the charge is negative.

Coulomb's law can be used to write the work and potential energy in terms of the charges and their separation. The simplest case is that of two point charges. Suppose charges q_1 and q_2 start a distance r_i apart and move so they end a distance r_f apart. Then, the work done by the electric field is

$$W_{if} =$$

and the change in the potential energy of the two-charge system is

$$\Delta U =$$

The result is the same if either charge remains stationary while the other moves or if both move. All that matters is their initial and final separations.

If the total charge in the collection is finite, the potential energy is usually taken to be zero for infinite charge separation. In the expression you wrote above for the change in the potential energy of two point charges, let the initial separation r_i become large without bound, replace r_f with r, and write the potential energy as a function of the final separation r:

$$U(r) =$$

This expression is valid no matter what the signs of the charges. If they have the same sign, the potential energy is positive; it decreases if their separation becomes greater and increases if their separation becomes less. Note that the electrical force is repulsive, so the field does positive work in the first case and negative work in the second. If the two charges have opposite signs, the potential energy is negative; it becomes less negative if their separation becomes greater and becomes more negative if their separation becomes less. The field now does negative work in the first case and positive work in the second.

To calculate the potential energy of a collection of point charges, sum the potential energies of all *pairs* of charges. If the system consists of four charges, for example, add the potential energies of q_1 and q_2, q_1 and q_3, q_1 and q_4, q_2 and q_3, q_2 and q_4, and q_3 and q_4. If r_{12} is the separation of q_1 and q_2, r_{13} is the separation of q_1 and q_3, etc., then the potential energy of this system is

$$U =$$

If one or more of the charges move, then some or all of the separations change and a change in the potential energy results. To find the change, calculate the initial and final potential energies, then subtract the initial potential energy from the final.

You may view the electric potential energy of the system as the work that must be done by an external agent to assemble the collection, bringing the charges from infinite separation to their final positions. The charges are at rest at the beginning and end of the process, so there is no change in kinetic energy. This work may be positive or negative.

You should recall how the principle of energy conservation is used. If the charges are released from some initial configuration, their mutual attractions and repulsions might cause them to move. The potential energy of the system will decrease as potential energy is converted to kinetic energy. You write $\Delta U + \Delta K = 0$, where K is the total kinetic energy of the charges. Thus, for example, if you know the initial and final configurations of the charge, you can compute the change in the total kinetic energy.

Electric potential. Electric potential and electric potential energy are closely related but they are not the same. Be careful to distinguish the two concepts. To find the electric potential of a collection of point charges, a reference point is chosen and the electric potential at that point is set equal to zero. Then, a positive *test* charge, not one of the charges in the collection, is moved from the reference point to any point P and the work done by the electric field on the test charge is calculated. The electric potential at P is the negative of this work divided by the test charge. Note that this is also the change in the electric potential energy per unit test charge of the system consisting of the original collection of charges and the test charge. The charges of the collection must remain in fixed positions as the test charge is moved. If the total charge is finite, the reference point is usually selected to be infinitely far removed from the charge collection. Note that a value for the electric potential is associated with each point in space and exists regardless of whether a test charge is present. To find the value for a particular point, the test charge is moved from the reference point to that point.

Since electric potential is an energy divided by a charge, its SI unit is J/C. This unit is called a _____ (abbreviation: _____). Since an electric potential is the product of an electric field and a distance, the SI unit of an electric field may be taken to V/m.

Equipotential surfaces. An equipotential surface is a surface (imaginary or real) such that the potential _____ at all points on it. An equipotential surface can be drawn by connecting neighboring points at which the potential has the same value and the surface can be labelled by giving the value of the potential on it. Equipotential surfaces do not cross each other. One and only one of them goes through any point in space. Figure 25–3 of the text shows some equipotential surfaces.

The electric field line through any point is _____ to the equipotential surface through that point. If the field has a non-vanishing component tangent to a surface, then a potential difference must exist between points on the surface and the surface cannot be an equipotential surface.

The equipotential surfaces of an isolated point charge are _____ that are centered on the charge. The equipotential surfaces of a uniform field are _____ that are perpendicular to the field. Equipotential surfaces associated with other charge distributions are more complicated but if the distribution has a net charge, they are _____ far from the distribution. Look at Fig. 25–3c of the text to see the surfaces associated with a dipole, for which the net charge is zero. All conductors are equipotential volumes and their boundaries are equipotential surfaces.

You can approximate the magnitude of the electric field in any region by calculating $\Delta V/\Delta s$, where Δs is the perpendicular distance between two equipotential surfaces that differ in potential by ΔV. Equipotential surfaces are close together in regions of high field and far apart in regions of low field.

If the equipotential surfaces for some charge distribution are known, they can be used to calculate changes in the potential energy and the work done by the electric field or an external agent as an additional charge is moved from one place to another. The diagram to the right shows a family of equipotential surfaces where they cut the plane of the page. Several points, labelled a, b, c, d, and e, are also shown.

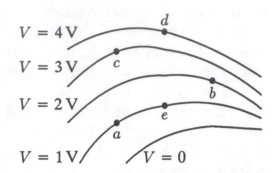

Calculate the change in potential energy as a particle is carried from one of these points to another, then calculate the work done by the electric field and the work done by the agent carrying the particle. Use the following table to record your answers. (P. E. stands for potential energy.)

PROCESS	CHANGE IN P. E.	WORK DONE BY FIELD	WORK DONE BY AGENT
electron from a to b	_____	_____	_____
electron from a to c	_____	_____	_____
electron from a to e	_____	_____	_____
proton from a to d	_____	_____	_____
proton from a to e	_____	_____	_____

The electric potential from the field. An expression for the potential difference between two points can be written as an integral involving the electric field at points along any path that connects the points. If $\mathbf{E}$ is the electric field, then the potential difference between the points a and b is given by an integral:

$$V_b - V_a =$$

You should understand that this integral is a line integral: the path is divided into infinitesimal segments, the integrand is evaluated for each segment, and the results are summed. Because the electric field is conservative, every path will give the same value for the potential difference of the given points.

You should also understand that the expression you wrote follows immediately from the definition of the potential difference as the negative of the work per unit test charge done by the field on a test charge as the test charge moves from a to b. The force of the electric field on the test charge q_0 is _____ and the work done by the field is _____ .

The integral expression for $V_b - V_a$ in terms of the electric field is valid for any situation, whether the field is produced by a single point charge or by a collection of charges. Specialize the expression for a situation in which the field is uniform, as it is outside a large sheet with uniform charge density, for example. If $\mathbf{E}$ is the electric field and $\Delta\mathbf{r}$ is the displacement from point a to point b, then $V_b - V_a =$ _____ .

For some situations, you must use different expressions for the electric field as the test charge passes through different regions. For example, the electric field of a uniform spherical distribution of charge is given by $E = Q/4\pi\epsilon_0 r^2$ for r outside the distribution and by $Qr/4\pi\epsilon_0 R^3$ for r inside. Here Q is the total charge and R is the radius of the distribution. The integral for the electric potential at a point inside must be broken into two parts. Use the first expression for the field as you integrate from far away to the surface of the charge distribution and the second expression as you integrate from the surface to a point inside.

Roughly speaking, if the potential varies from place to place, being high in one region and lower in a neighboring region, the electric field points from the region where the potential is _____ toward the region where the potential is _____. A positive charge is accelerated toward the _____ potential region; a negative charge is accelerated toward the _____ potential region.

You have been considering the problem of finding the potential when the field is given. The inverse problem can also be solved: if the electric potential is a known function of position in a region of space, then, to find the component of **E** in any direction, you calculate the rate of change of V with position along a line in that direction. Mathematically, if ds is the magnitude of an infinitesimal displacement in a given direction, then the component of **E** in that direction is given by the derivative

$$E_s =$$

Specialize this relationship to each of the three coordinate directions. Suppose the potential is given as some function $V(x, y, z)$. Then, the components of the electric field at the point with coordinates x, y, z are given by

$$E_x = \qquad\qquad E_y = \qquad\qquad E_z =$$

When you differentiate with respect to x to find E_x, you treat y and z as constants. Similar statements hold for E_y and E_z. The partial derivative symbol ∂ is used to remind you. See the Mathematical Skills section in Chapter 17 of this manual. When you are finished differentiating, but not before, you evaluate the result for the coordinates of interest. As an example, suppose the potential in volts is given by $V(x, y, z) = 3x^2 y - 5xy^2$, with the coordinates in meters. Then, $E_x = -\partial V/\partial x = -6xy + 5y^2$ and $E_y = -\partial V/\partial y = -3x^2 + 10xy$. At $x = 2$ m, $y = 3$ m, $E_x = -6 \times 2 \times 3 + 5 \times 3^2 = 9$ V/m and $E_y = -3 \times 2^2 + 10 \times 2 \times 3 = 48$ V/m.

Electric potentials of point charges. Suppose you wish to find the electric potential a distance r from a single isolated point charge q. Move a test charge q_0 from infinitely far away to a point a distance r from q. As you do this, the work done by the electric field of q is

$$W =$$

and the potential energy of the system consisting of q and q_0 changes from zero to

$$U =$$

The electric potential at the final position of the test charge is given by

$$V =$$

This expression is also valid if q is negative. The potential then has a negative value. Be sure to note that the point charge q does not have a potential energy (only collections of more than one charge have electric potential energies) but it does create an electric potential in the space around it.

The electric potential of a collection of point charges is the sum of the individual potentials of the charges in the collection. Since the electric potential is a scalar, the sum is algebraic. Suppose the system consists of charges q_1, q_2, and q_3. If you want to compute the potential at some point P, you must know the distance from each charge to P. Suppose q_1 is a distance r_1 from P, q_2 is a distance r_2, and q_3 is a distance r_3. Then, the potential at P is given by

$$V =$$

When you substitute values for the charges be sure you include the appropriate signs.

An electric dipole is another important example. It consists of a positive charge and a negative charge of the same magnitude, separated by a distance d. The magnitude of the dipole moment is given by $p = qd$, where q is the magnitude of either charge. The dipole moment is a vector, from the _____ charge toward the _____ charge. The potential at any point can be found by summing the potentials of the two charges. Let $\mathbf{r}$ be the vector from the midpoint of the line joining the charges to the point. The point is usually specified by giving the distance r and the angle θ between $\mathbf{r}$ and $\mathbf{p}$. If the charges are close together, compared to r, the potential is given by

$$V =$$

The electric potential of a collection of charges can be used to compute changes in the potential energy when another charge, not belonging to the collection, is moved from one point to another. If charge Q is placed at point a, where the electric potential due to charges in the collection is V_a, the potential energy of the system consisting of the collection and Q is $U =$ _____ and if Q is moved from that point to a point where the potential due to the collection is V_b, the change in the potential energy of the system is $\Delta U =$ _____ .

Electric potentials of continuous charge distributions. Charge may be distributed continuously along a line, on a surface, or throughout a volume. If it is, you divide the region into infinitesimal elements, each containing charge dq, and sum (integrate) the potential due to the regions. If an infinitesimal region is a distance r from point P, then the contribution of that region to the potential at P is

$$dV =$$

and the potential due to the whole distribution is the integral

$$V =$$

In practice, dq is replaced by $\lambda\,ds$ for a line distribution and by $\sigma\,dA$ for a surface distribution. Here λ is a linear charge density and σ is an area charge density. To carry out the integration, you will normally write the line element ds or area element dA in terms of variables that are appropriate to the geometry of the problem. Don't forget to write the distances in terms of these quantities also.

To see how the integration is done, study the calculations in the text of the potentials due to charge distributed uniformly along a straight line and uniformly over a disk.

Suppose you wish to find the potential due to a uniform line of charge, as shown in Fig. 25–13. If the linear charge density is λ, then the charge in an infinitesimal segment dx is $dq = $ _____. If the coordinate of the segment is x and you are calculating the potential at a point that is a distance d from the line at $x = 0$, then the distance from the segment to the point is $r = $ _____. The potential at the point due to the segment is given by

$$dV = $$

This expression can be integrated. Suppose the line extends from $x = 0$ to $x = L$. Then,

$$V = $$

Now calculate the potential produced by a uniform disk of charge at a point on the axis a distance z above the center of the disk. See Fig. 25–14. First, consider a ring with radius R' and width dR'. The area of the ring is $dA = $ _____ and if σ is the area charge density, the charge contained in the ring is $dq = $ _____. All charge on the ring is the same distance from the point; in particular, in terms of z and R', $r = $ _____. Thus, the potential produced by the ring at z is $dV = $ _____ and the total potential is given by the integral _____. Since $\int (R'^2 + z^2)^{-1/2} R'\,dR' = (R'^2 + z^2)^{1/2}$, the integral can be evaluated easily. Carefully note the limits of integration. The result is

$$V = $$

Isolated conductors. All points in the interior and on the surface of an isolated conductor have the same electric potential, once equilibrium is established. This follows immediately from the definition of the potential difference between two points and the condition $\mathbf{E} = $ _____ inside a conductor in equilibrium.

Consider a conducting sphere of radius R, with charge Q uniformly distributed on its surface. If r is the distance from the sphere center to a point outside the sphere, the electric potential there is given by

$$V = $$

If r is the distance from the sphere center to a point inside the sphere, the potential there is given by

$$V = $$

Notice that the potential inside does not depend on r and has the same value as the potential at the surface.

Suppose one conducting sphere has radius R_1 and charge Q_1 while another has radius R_2 and charge Q_2. They are far apart, which means the charge on one does not influence the distribution of charge on the other. Both have uniform distributions on their surfaces. The potential at the surface of the first sphere is

$$V_1 =$$

and the potential at the surface of the second is

$$V_2 =$$

If the spheres are then connected by a long metal wire, these potentials must be the same. This condition can be used to solve for the ratio Q_1/Q_2. If this was not the ratio of the charges when the wire was attached, charge flows through the wire until this ratio is obtained.

II. PROBLEM SOLVING

To solve the problems of this chapter, you should know how to calculate the electric potential of point charges and continuous distributions of charge. Use $V = q/4\pi\epsilon_0 r$ for single point charge and use the sum of the potentials due to individual charges for a collection of point charges. Use $V = (1/4\pi\epsilon_0) \int (1/r)\, dq$ for a continuous distribution of charge. Replace dq with $\lambda\, ds$ for a line of charge or with $\sigma\, dA$ for a surface of charge. You should know how to compute the work done by the electric field ($W = -q\Delta V$) and the work ($-W$) done by an external agent when a charge q is moved from one place to another.

You should know how to calculate the potential energy of a collection of charges: $U = q_1 q_2/4\pi\epsilon_0 r$ for two charges. You should also know that the potential energy of a collection of charges is the sum of the potential energies of all the pairs of charges in the collection. You should know how to compute the change in the potential energy when a charge moves. You should know how to interpret changes in potential energy in terms of the work done by the electric field or external agent. You should also know how to use the principle of conservation of energy to compute changes in the kinetic energy of a moving charge.

Some problems ask you to sketch equipotential surfaces. It is usually fairly easy to draw electric field lines and then sketch surfaces that are perpendicular to the lines.

III. NOTES

Chapter 26
CAPACITANCE

I. BASIC CONCEPTS

Capacitors are electrical devices that are used to store charge and electrical energy and, as you will see in a later chapter, are important for the generation of electromagnetic oscillations. This chapter is an excellent review of the principles you learned in previous chapters: you will make extensive use of Gauss' law and the concepts of electric potential and electrical potential energy.

Here are some topics that were covered in previous chapters and are used in this chapter: charge from Chapter (22), electric field and electric field lines from Chapter (23), Gauss' law from Chapter (24), electric potential, work done by an electric field, electric potential energy, and equipotential surfaces from Chapter (24), and electric dipole from both Chapters (23) and (25).

Capacitors and capacitance.   A capacitor is simply two _____ , isolated from each other. Each is called a <u>plate</u>. The symbol used to represent a capacitor in an electrical circuit is: _____ .

In normal use, one plate holds positive charge and the other holds negative charge of the same magnitude. The charge creates an electric field, pointing roughly from the positive plate toward the negative plate, and because an electric field exists in the region between the plates, the plates are at different electric potentials. The positive plate, of course, is at a higher potential than the negative.

The potential difference of the plates and the charge on either one are proportional to each other. If the magnitude of the potential difference is V and the magnitude of the charge on either plate is q, then

$$q =$$

where _____ is the <u>capacitance</u> of the capacitor. Because q and V are proportional, the capacitance is independent of both q and V. If q is doubled, _____ doubles, but _____ remains the same. A capacitor can be charged by transferring charge from one plate to the other. The capacitance is a measure of how much is transferred for a given _____ .

A battery connected to a capacitor transfers electrons from the plate at its positive terminal, through the battery, to the plate at its negative terminal. When fully charged, the potential difference across the plates is the same as the potential difference across the terminals of the battery. That is, for example, a 9 V battery will transfer charge until the potential difference of the plates is 9 V. Remember, however, that a potential difference exists between the plates of a charged capacitor if a battery is connected or not.

The SI unit for capacitance is a coulomb/volt. This unit is called a _____ and is abbreviated _____. Microfarad (abbreviated _____) and picofarad (abbreviated _____) capacitors are commonly used in electronic circuits. A 1 F parallel-plate capacitor must have a huge plate area or else an extremely small plate separation. Until recently, the technology was not available to construct such a capacitor in a reasonable volume.

Calculation of capacitance. To calculate capacitance, first imagine charge q is placed on one plate and charge $-q$ is placed on the other. Use Gauss' law to find an expression for _____ in the region between the plates. You will find that the field at every point is proportional to q. Once an expression for the electric field is known, use _____ to compute the magnitude V of the potential difference of the plates. V is also proportional to q. Finally, use _____ to find an expression for the capacitance. It will not depend on q or V, but in general terms it will depend on: _____
_____.

Carefully study the examples given in the text. They clearly show that the electric field and potential difference used in the calculation are due to the charge on the plates and, in each case, they explicitly give the geometric quantities on which the capacitance depends.

Suppose a capacitor consists of two parallel metal plates, each of area A and the two separated by a distance d. If charge q is paced on one and charge $-q$ is placed on the other, the electric field between the plates is given by

$$E =$$

and the potential difference of the plates is given by

$$V =$$

Thus the capacitance is given by

$$C =$$

Carefully note that the capacitance depends only on the permittivity constant ϵ_0, the plate area A, and the plate separation d, not on q or V.

Remember the expression for the capacitance of a parallel-plate capacitor. It will be used many times. Also remember that C is proportional to A and inversely proportional to d. If the plate separation is halved without changing the plate area, the capacitance is _____.
If the plate area is halved without changing the separation, the capacitance is _____.

Suppose a capacitor consists of two coaxial cylinders of length L, the inner one with radius a and the outer one with radius b. Charge q is placed on the inner cylinder and charge $-q$ is placed on the outer cylinder. The electric field between the cylinders is given by

$$E =$$

the potential difference of the cylinders is given by

$$V =$$

and the capacitance is given by

$$C =$$

Note that C depends only on ϵ_0, a, b, and L.

Suppose a capacitor consists of two concentric spherical shells, with radii a and b. Charge q is placed on the inner shell and charge $-q$ is placed on the outer shell. The electric field between the shells is given by

$$E =$$

the potential difference of the shells is given by

$$V =$$

and the capacitance is given by

$$C =$$

Capacitance is also defined for a single conductor. The other plate is assumed to be infinitely far away. To find an expression for the capacitance, imagine charge q is on the conductor, find an expression for the electric field outside the conductor, calculate the potential V of the conductor relative to the potential at infinity, and use $q = CV$ to find the capacitance. For example, the capacitance of a spherical conductor of radius R is given by

$$C =$$

Capacitors in series and parallel. Be sure you can distinguish between <u>parallel</u> and <u>series</u> combinations of two or more capacitors. For a _____ combination, the potential difference is the same for all the capacitors and for a _____ combination, the charge is the same. If neither the potential difference nor the charge is the same, then the capacitors do not form either a parallel or series combination.

In the space below draw two capacitors connected in parallel and two connected in series. Assume a potential difference V is applied to each combination and indicate on the diagrams where it is applied. Label the charge on each plate to show which plates have the same charge and which have different charges.

PARALLEL   SERIES

In each case, the capacitors can be replaced by a single capacitor with capacitance C_{eq} such that the charge transferred is the same as the total charge transferred for the original combination when the potential difference is the same as the potential difference across the original combination. If the capacitances are C_1 and C_2, then the value of C_{eq} for a parallel combination is given by

$$C_{eq} = $$

and the value of C_{eq} for a series combination is given by

$$C_{eq} = $$

For a parallel combination, the equivalent capacitance is greater than the greatest capacitance in the combination; for a series combination, the equivalent capacitance is less than the smallest capacitance in the combination.

Energy storage. As a capacitor is being charged, energy must be supplied by an external source, such as a battery, and energy is stored by the capacitor. You may think of the stored energy in either one of two ways: as the potential energy of the charge on the plates or as an energy associated with the electric field produced by the charges.

Suppose that, at one stage in the charging process, the positive plate has charge q' and the potential difference between the plates is V'. If an additional infinitesimal charge dq' is taken from the negative plate and placed on the positive plate, the energy is increased by $dU = $ _____ or, what is the same, by $dU = (q'/C)\,dq'$, where C is the capacitance. This expression is integrated from 0 to the final charge q. In terms of the final charge, the total energy required to charge an originally uncharged capacitor is

$$U = $$

In terms of the final potential difference V, it is

$$U = $$

Instead of associating the energy of a charged capacitor with the mutual interactions of the charges on its plates, it may be associated with the electric field associated with those charges. The text shows that the energy density (or energy per unit volume) in a parallel-plate capacitor is given by

$$u = $$

where E is the magnitude of the electric field between the plates.

This expression is quite generally valid for any electric field, in a capacitor or not. Most fields are functions of position, so a volume integral must be evaluated to calculate the energy required to produce the field: $U = \frac{1}{2}\epsilon_0 \int E^2\,dV$. Part c of Sample Problem 26–6 in the text shows how to evaluate this integral for a charged conducting sphere.

If, after charging a capacitor, the plates are connected by a conducting wire, then electrons will flow from the negative to the positive plate until both plates are neutral and the electric field vanishes. The stored energy is converted to kinetic energy of motion and eventually to internal energy in resistive elements of the circuit.

Dielectrics. If the space between the plates of a capacitor is filled with insulating material, then the capacitance is greater than if the space is a vacuum. In fact, if the space is completely filled, the capacitance is given by $C =$ _____, where C_0 is the capacitance of the unfilled capacitor and κ is the _____ of the insulator. This last quantity is a property of the insulator, is unitless, and is always greater than 1. See Table 26–1 of the text for some values.

Consider two capacitors that are identical except capacitor A has a dielectric between its plates while capacitor B does not. If the capacitors have the same charge on their plates, the potential difference across capacitor _____ is greater than the potential difference across the other capacitor. If the capacitors have the same potential difference, the charge on the positive plate of capacitor _____ is greater than the charge on the positive plate of the other capacitor.

Polarization of the dielectric by the electric field brings about an increase in capacitance. Suppose the dielectric is composed of polar molecules and describe what happens when it is polarized: _____

_____

Now, suppose the dielectric is composed of non-polar molecules and describe what happens:

_____

_____

The electric dipoles in a polarized dielectric produce an electric field that is directed opposite to the field produced by the charge on the conducting plates. Thus, the total field is weaker than the field produced by charge on the plates alone. It is, in fact, weaker by the factor κ. That is, if $\mathbf{E}_0$ is the field produced by the charge on the plates, then the total field is given by $\mathbf{E} =$ _____. If the electric field produced by charge on the plates is uniform, as it essentially is between the plates of a parallel-plate capacitor, then the dipole field and the total field are also uniform.

Since the electric field for a given charge on the plates is weaker if the capacitor is filled with a dielectric than if it is not, the potential difference is _____ when the dielectric is present. Since $q = CV$, this means the capacitance is larger when the dielectric is present.

For a parallel-plate capacitor the, effect of the dielectric is exactly the same as a uniform distribution of positive charge q' on the surface nearest the negative plate and a uniform distribution of negative charge $-q'$ on the surface nearest the positive plate. If the dielectric constant of the dielectric is κ and the charge on the positive plate is q, then

$$q' =$$

If the plates have area A, then the electric field due to the charge on the plates is _____, the field due to the dipoles is _____, and the total field is _____. In terms of q, A, d, and κ, the potential difference is given by

$$V =$$

The energy stored in a capacitor is still given by $U = \frac{1}{2}q^2/C = \frac{1}{2}CV^2$, where q is the magnitude of the charge on either plate and V is the potential difference. As a dielectric is inserted between the plates of a capacitor with the potential difference held constant (by a battery, say), the charge on the positive plate _____ and the stored energy _____. For the potential difference to remain the same, the battery must transfer charge as the dielectric is inserted. As a dielectric is inserted with the capacitor in isolation, so the charge cannot change, the potential difference _____ and the stored energy _____. If the dielectric is inserted by an external agent, the difference in energy is associated with _____. If no external agent acts, the difference in energy is associated with _____.

Although a capacitor filled with a dielectric holds more charge than an identical one at the same potential difference, but without a dielectric, the dielectric places an upper limit on the charge and potential difference. The <u>dielectric strength</u> of an insulator is _____ _____.

If a parallel-plate capacitor has capacitance C and plate separation d and the insulator filling it has dielectric strength E_{max}, then the maximum potential difference it can sustain is $V_{max} =$ _____ and the maximum charge that can be placed on the positive plate is $q_{max} =$ _____.

II. PROBLEM SOLVING

To solve many problems of this chapter, you should know the basic relationship between the magnitude q of the charge on either plate of a capacitor and the potential V difference across the plates: $q = CV$. You should also know how to calculate the capacitance of a parallel-plate capacitor, a cylindrical capacitor, and a spherical capacitor, all in terms of the geometry of the capacitor. You should also know that the capacitance of a capacitor that is filled with a dielectric of dielectric constant κ is $C = \kappa C_0$, where C_0 is the capacitance of the capacitor without the dielectric.

Some problems deal with capacitors in series or parallel. In each case, you should know how to compute the equivalent capacitance, the charge on each capacitor, and the potential difference across each capacitor, given the potential difference across the combination.

You should know how to compute the energy stored in a capacitor, given either its charge or potential difference. You should also know how to compute the energy density at points between the plates or, in general, at any point within a given electric field.

III. NOTES

Chapter 27
CURRENT AND RESISTANCE

I. BASIC CONCEPTS

Here you begin the study of electric current, the flow of charge. Pay close attention to the definitions of current and current density and understand how they depend on the concentration and speed of the charged particles. For most materials, a current is established and maintained only when an electric field is present. You will learn about the properties of materials that determine the magnitude and direction of the current for any given field.

Topics from previous chapters that you might want to review are: acceleration, velocity, and speed (Chapter 4), mass and Newton's second law of motion (Chapter 5), power (Chapters 7 and 8), charge (Chapter 22), electric field (Chapter 23), electric flux (Chapter 24), and electric potential and electric potential energy (Chapter 25).

Electric current and current density. As precisely as you can, tell in words what it means for a conducting wire to contain an electric current: _____

_____

The current i through any area is given by

$$i = \frac{dq}{dt},$$

where dq is _____ that passes through _____ in the time interval _____. Although current is a scalar, not a vector, it is assigned a direction. Positive charge moving to the right and negative charge moving to the left, for example, are both currents in the same direction, namely to the _____. Positive charge moving to the right and negative charge moving in the same direction are currents in _____ directions: the current associated with the positive charge is to the _____ while the current associated with the negative charge is to the _____.

Explain why the flow of neutral atoms or molecules is not an electric current: _____

_____

Explain why electric current is not a vector: _____

_____

For the steady flow of charge through a conductor, the current is the same for all cross sections along the conductor. Explain why: _____

_____

The SI unit for current is coulomb/second. This unit is called _____ and is abbreviated _____.

Current is associated with an area, like the cross-sectional area of a conductor. On the other hand, <u>current density</u> **J** is a related quantity that can be associated with each point in a conductor. At any point, it is the current per unit area through an area at the point, oriented perpendicularly to the charge flow, in the limit as the area shrinks to the point. Current density is a vector. If positive charge is flowing, **J** is in the direction of the velocity; if negative charge is flowing, **J** is in the direction opposite the velocity.

Consider any surface within a current-carrying conductor and let d**A** be an infinitesimal vector area, with direction perpendicular to the surface. If **J** is the current density, then the current i through the surface is given by the integral over the surface:

$$i =$$

The scalar product indicates that only the component of **J** normal to the surface contributes to the current. This is consistent with the statement that charge must flow through the surface for there to be a current through the surface. If the surface is a plane with area A and is perpendicular to the particle velocity and if the current density is uniform over the surface, then

$$i =$$

For a collection of particles with uniform concentration n, each with positive charge e and each moving with velocity $\mathbf{v}_d$, the current density is given in terms of e, n, and $\mathbf{v}_d$ by the vector relationship

$$\mathbf{J} =$$

This expression can be derived by considering the number of particles per unit time that pass through an area perpendicular to their velocity. Note that **J** and $\mathbf{v}_d$ are in the same direction for positively charged particles and in opposite directions for negatively charged particles. As Sample Problem 27–3 shows, the expression above can be used to compute the drift speed v_d for a given current in a given conductor.

All electrons in a current-carrying conductor do not actually move with the same velocity, but each of their velocities can be considered to be the sum of two velocities, one of which changes direction often as the electron collides with atoms. When all electrons are taken into account, this component averages to zero: it does not contribute to the current. The second component, the drift velocity, is much smaller in magnitude than the first and is the same for all electrons if the current density is uniform. This is the velocity that enters the expression for the current density, not the total velocity. Be careful to distinguish between drift and total velocities.

In an ordinary conductor, an electric field is required to produce a net drift and hence an electric current. Explain what a field does: _____

_____

For electrons, how is the direction of the drift velocity related to the direction of the electric field? _____ This means the current is in the same direction as the field and it is directed from a region of high electric potential toward a region of lower electric potential.

That an electric field can be maintained in a conductor does not contradict the statement proved earlier that the electrostatic field in a conductor is zero. Describe how the situation considered in this chapter differs from that considered in Chapter 24: _____

_____

Resistance and resistivity. The current in any material (except a superconductor) is zero unless an electric field exists in the material. Resistance is a measure of the current generated in a given conductor by a given potential difference. To determine resistance, a potential difference V is applied between two points in the material and the current i is measured. The resistance for that material and those points of application is defined by

$$R =$$

Remember that the resistance depends on where the current leads are attached, as well as on the material and its geometry. Resistance has an SI unit of volt/ampere, a unit that is called _____ and abbreviated _____ . In drawing an electrical circuit, an element whose function is to provide resistance, called a resistor, is indicated by the symbol _____ .

Resistance is intimately related to a property of the material called its resistivity. If at some point in the material the electric field is $\mathbf{E}$ and the current density is $\mathbf{J}$, then the resistivity ρ at that point is given by

$$\mathbf{E} =$$

We consider materials that are homogeneous and isotropic; the resistivity is the same at every point in the material and is the same for every orientation of the electric field.

The SI unit for resistivity is _____ . Values for some materials are given in Table 27–1 of the text. Note that typical semiconductors have resistivities that are greater than those of metals by factors of 10^5 to 10^{11} and insulators have resistivities that are greater than those of metals by factors of 10^{18} to 10^{24}. The conductivity σ of any substance is related to the resistivity by $\sigma =$ _____ .

Given the resistivity of the material and the points of application of a potential difference, the resistance can, in principle, be calculated. The simplest case is that of a homogeneous wire with uniform cross section. Let L be the length of the wire, A be its cross-sectional area, and ρ be its resistivity. If one end of the wire is held at potential 0 and the other is held at potential V, the electric field in the wire is given by $E =$ _____ . If the current is i then, since the current density is uniform, it is given by $J =$ _____ . Substitute $E = V/L$ and $J = i/A$ into $E = \rho J$ and solve for V/i $(= R)$. The result is

$$R =$$

Note that the resistance depends on a property of the material (ρ) and on the geometry of the sample (L and A). Resistivity, on the other hand, is a property of the material alone.

Temperature dependence of the resistivity. The resistivity of a metal increases with increasing temperature, chiefly because electrons suffer more collisions per unit time at high temperatures than at low temperatures. The temperature dependence is characterized by a quantity α, called the temperature coefficient of resistivity, which is a measure of the deviation of the resistivity from its value at a reference temperature. Let ρ_0 be the resistivity at the reference temperature T_0 and let ρ be the resistivity at a nearby temperature T. Then, in terms of α,

$$\rho - \rho_0 =$$

Temperatures are given in K or °C.

Table 27–1 lists values for some materials. Notice that α is negative for semiconductors. For them, the resistivity decreases as the temperature increases because the concentration of nearly free electrons increases rapidly with temperature. For metals, on the other hand, n is essentially independent of temperature.

Ohm's law. Ohm's law describes an important characteristic of the resistance of certain samples, called ohmic samples (or devices). It is: _____

_____

_____

Carefully note that $V = iR$ defines the resistance R and so holds for every sample. For an ohmic sample, V is a linear function of i or, what is the same, R does not depend on V or i. In addition, for ohmic samples, a reversal of the potential difference simply reverses the direction of the current without changing its magnitude. For some non-ohmic samples, such as a pn junction diode, a reversal of the potential difference not only changes the direction of the current but also its magnitude.

In the space below, sketch graphs of the current as a function of potential difference for an ohmic and for a non-ohmic sample. Include both positive and negative values of V.

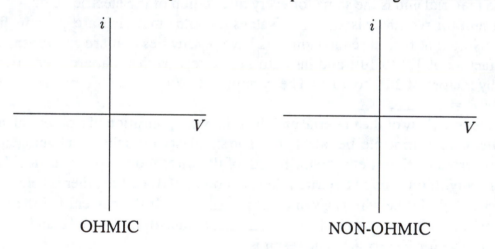

OHMIC   NON-OHMIC

For an ohmic substance, the resistivity is independent of the electric field and the current density is therefore proportional to the electric field. This implies that the drift velocity is also proportional to the field. Since an electric field accelerates charges, you should be surprised.

To show why Ohm's law is valid for some materials, the text considers the collection of so-called free electrons in a conducting sample. These electrons are accelerated by the electric field applied to the sample and suffer collisions with atoms of the material. We suppose that the effect of a collision is to stop the drift of an electron so that, as far as drift is concerned, an electron starts from rest at a collision and is accelerated by the field until the next collision. The collision stops it and the process is repeated. The time from the last collision, averaged over all electrons, is designated by τ. Because the collisions occur at random times, τ is independent of the time; its value is the same no matter when the averaging is done. In addition, τ gives the average time between collisions and is, in fact, called the <u>mean time between collisions</u> or the <u>mean free time</u>.

Since the magnitude of the acceleration is $a =$ _____, where E is the magnitude of the electric field and m is the mass of an electron, the average electron drift speed is given by $v_d =$ _____. Although each electron accelerates between collisions, v_d does not vary with time if the field is constant. Also note that v_d is proportional to E if τ does not depend on E. Explain why you expect τ to be essentially independent of E: _____

_____

In terms of E, m, e, and τ, the current density is given by $J = env_d =$ _____. This expression immediately gives $\rho =$ _____ for the resistivity. Because τ is independent of E, the resistivity is also independent of E and the material obeys Ohm's law.

Give a qualitative argument to convince yourself that a long mean free time leads to a small resistivity. In particular, explain why a long mean free time leads to a larger current than a short mean free time if the electric field is the same: _____

_____

_____

Because the mean time between collisions is determined to a large extent by thermal vibrations of the atoms, it is temperature dependent. As the temperature increases, we expect an electron to suffer more collisions per unit time, so τ _____ as the temperature increases.

Energy transfers. If current i passes through a potential difference V, from high to low potential, the moving charges lose potential energy at a rate given by

$$P = dU/dt =$$

For a resistor, the energy is transferred to atoms of the material in collisions. The result is an increase in the vibrational energy of the atoms and is usually accompanied by an increase in the temperature of the resistor. The phenomenon finds practical application in toasters and electrical heaters.

Since $V = iR$ for a resistor, the expression for the rate of energy loss can be written in terms of the current i and resistance R as $P =$ _____ or in terms of the potential difference V and resistance R as $P =$ _____.

Suppose the resistance of resistor A is greater than that of resistor B. If the same potential difference is applied to them, then the internal energy of resistor _____ will increase at the greater rate. If they have the same current, then the internal energy of resistor _____ will increase at the greater rate.

II. PROBLEM SOLVING

You should know the definitions of current and current density, their relationship to each other, and the relationship between current density and drift velocity. If net charge Δq passes through an area in time Δt then the current through the area is $i = \Delta q/\Delta t$. If area A is perpendicular to the charge flow then the magnitude of the current density is $J = i/A$, where i is the current through the area. If positive charge is flowing the current density is in the direction of its velocity; if negative charge is flowing the current density is opposite its velocity.

The magnitude of the current density is related to the drift speed v_d by $J = nev_d$, where n is the number of charge carriers per unit volume in the current.

You should know the definition of resistance and be able to compute the resistance R of a wire, given its resistivity ρ, length L, and cross-sectional area A. The relationship is $R = \rho L/A$.

You should understand that resistivity is determined to a large extent by the mean time between collisions and that this quantity is temperature dependent because the average velocity of the electrons is temperature dependent. The relationship between the resistivity ρ and mean time τ between collision is $\rho = m/e^2 n\tau$, where m is the mass of a charge carrier and n is the number of charge carriers per unit volume in the current.

Charge carriers lose kinetic energy in collisions with the atoms of a resistor. You should know how to compute the rate of energy dissipation, given the resistance and either the current in the resistor or the potential difference along the length of the resistor. Use $P = i^2 R$ or $P = V^2/R$, where i is the current in the resistor and V is the potential difference across it.

III. NOTES

Chapter 28
CIRCUITS

I. BASIC CONCEPTS

The concept of electromotive force (emf) is introduced in this chapter. Seats (or sources) of emf are used to maintain potential differences and to drive currents in electrical circuits. You will use this concept, along with those of electric potential, current, resistance, and capacitance, learned earlier, to solve both simple and complicated circuit problems. You will also learn about energy balance in electrical circuits.

Here are some topics you might review: work from Chapter 7, charge from Chapter 22, work done by an electric field and electric potential from Chapter 25, capacitance from Chapter 26, and current and electrical resistance from Chapter 27.

Emf. An emf device (or seat of emf) performs two functions: it maintains a potential difference between its terminals and, to do that, it moves charge from one terminal to the other inside the device. An ideal emf device contains only a seat of emf; a real emf device, such as a battery, also contains internal resistance.

In electrical circuits an ideal emf device is indicated by the symbol _____. The arrow with the small circle at its tail points from the _____ terminal toward the _____ terminal. An emf tends to drive current in the direction of the arrow but you should remember that, for any circuit, the actual direction of the current through an emf device depends on other elements in the circuit.

An emf, denoted by $\mathcal{E}$, is defined in terms of the work it does as it transports charge. If it does work dW on charge dq then,

$$\mathcal{E} =$$

Emf has an SI unit of _____ .

An emf device does positive work if positive charge moves inside the device from the _____ to the _____ terminal and negative work if positive charge moves in the other direction. It does positive work if negative charge moves inside the device from the _____ to the _____ terminal and negative work if negative charge moves in the other direction. This can be summarized by saying the device does positive work on the charge if the current inside it is from the negative to the positive terminal and negative work if the current is in the other direction.

Positive work done by an emf device results in a decrease in the store of energy of the device, chemical energy in the case of a battery. Negative work results in an increase in the energy of the device. If the device is a battery, then in the first case it is said to be discharging and in the second it is said to be charging.

The potential difference across an ideal emf device with emf $\mathcal{E}$ is _____, with the positive terminal being at the higher potential. This statement is true no matter what the direction of the current through the device.

Single loop circuits. To solve single loop circuits you must be able to express the potential differences across emf devices and resistors in terms of the emf's, resistances, and currents. For each of the circuit elements shown below, assume the potential is V_a at the left side and write an expression for the potential V_b at the right side in terms of V_a, $\mathcal{E}$, R, and i, as appropriate.

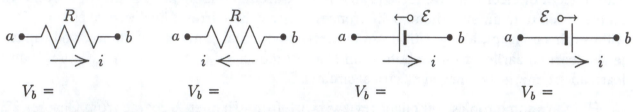

$V_b =$ $V_b =$ $V_b =$ $V_b =$

You will use <u>Kirchhoff's</u> <u>loop</u> <u>rule</u> to solve both single-loop and multiple-loop circuits. Write this rule in words: _____

_____

In general, the rules for solving a single loop circuit are: pick a direction for the current and draw a current arrow on the circuit diagram. For a single-loop circuit, the current is the same in every circuit element. Go around the circuit and, as you do, add the changes in electric potential for the various circuit elements, with appropriate signs. For a resistor assume the current is in the direction of the arrow you drew. The diagrams above should help you. Then, equate the sum to zero and solve for the unknown quantity.

Consider the single-loop circuit shown on the right and use Kirchhoff's loop rule to develop an equation that relates the emf $\mathcal{E}$, resistance R, and current i. In terms of these quantities, the potential difference $V_b - V_a$ is _____ and the potential difference $V_c - V_b$ is _____. Now, sum the potential differences around the loop. Since $V_c = V_a$, you should obtain $\mathcal{E} - iR = 0$. This equation can be solved for any one of the quantities appearing in it. If, for example, $\mathcal{E}$ and R are known, then $i = \mathcal{E}/R$ gives the current.

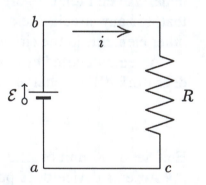

Notice that the value of i is positive, no matter what the values of $\mathcal{E}$ and R. This is because the current arrow was picked to be in the direction of the actual current. In other circuits, the direction of the current might not be so obvious and you might not pick the proper direction for the current arrow. If you pick the wrong direction, the value you obtain for i will be negative. Suppose, for example, the current arrow for the circuit above was picked in the opposite direction, upward through the resistor. Then, the loop equation would be _____ and the expression for the current in terms of $\mathcal{E}$ and R would be $i =$ _____. The value for i is obviously negative now.

A real battery contains an internal resistance in addition to a seat of emf. Although the internal resistance is distributed inside the battery, you may think of it as being in series with the seat. You must distinguish between the emf and terminal potential difference of a real battery. The latter includes the potential difference of the internal resistance. To give an example, suppose a battery is characterized by emf $\mathcal{E}$ and internal resistance r. If the current in the battery is zero then the terminal potential difference is $V =$ _____; if the battery is discharging, with current i in the direction of the emf, then $V =$ _____; and if the battery is charging, with current i opposite the emf, then $V =$ _____.

Suppose a single-loop circuit consists of a battery with emf $\mathcal{E}$ and internal resistance r connected to an external resistance R. Then, the current in the circuit is given by $i =$ _____. Study Sample Problem 28–1 of the text for an example of a circuit with two batteries. Note that the battery with the larger emf is discharging and determines the direction of the current while the battery with the smaller emf is charging.

Energy considerations. An emf with current in the direction of the emf does work on the moving charges and increases their potential energy. If the emf is $\mathcal{E}$ and the current is i, then the rate at which energy is supplied is given by $P =$ _____. An emf with current opposite the direction of the emf removes energy from the moving charges. The rate at which energy is removed is _____. A resistor removes energy from moving charges in collisions between the charges and atoms of the resistor. If the resistance is R and the current is i, then the rate at which energy is removed is given by _____. A discharging battery, being a combination of an emf and a resistor, both supplies and removes energy.

For any circuit in steady state, the rate at which energy is supplied by discharging batteries exactly equals the rate at which energy is removed by resistors and charging batteries.

Multiloop circuits. To analyze multiloop circuits, another rule, called <u>Kirchhoff's junction rule</u>, is needed in addition to the loop rule. State the junction rule in words: _____ _____ _____

The diagrams below show two junctions and the current arrows associated with them. For each, write the equation that follows from the junction rule.

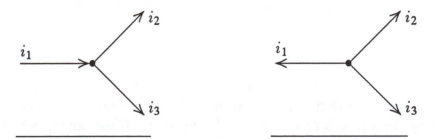

The form of the junction equation does not depend on the directions of the actual currents, only on the directions of the current arrows. Note that all of the currents in the second example above cannot have the same sign.

Successful analysis of a multiloop circuit depends on your ability to identify <u>junctions</u> and <u>branches</u>. A junction is a point where _____ or more wires are joined. A branch is a portion of a circuit with junctions at its ends and none between. You must place a current arrow in each branch and label the arrows in different branches with different current symbols.

Consider the circuit shown to the right and assume the battery has negligible internal resistance. Identify the four junctions by writing the letters associated with them: _____
Identify the six branches by writing the letters corresponding to their end points: _____

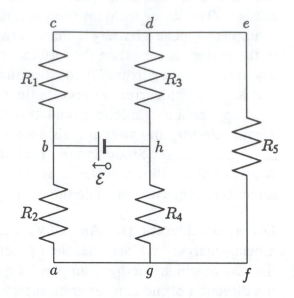

The currents in the various branches may be different. On the diagram, draw a current arrow for each branch and label them i_1 through i_6, with i_j being the current in resistor R_j and i_6 being the current in the battery. The directions of the arrows are immaterial. In general, for any circuit with N junctions, there will be $N-1$ independent junction equations. Pick any three junctions for the circuit shown and write the junction equations here:

The total number of independent equations equals the number of branches, so the number of independent loop equations you will use equals the number of branches minus the number of junction equations. For the circuit shown, you will need $6 - 3 = 3$ loop equations. Pick three loops. You must be a little careful here. Every branch of the circuit must contribute to at least one of the loop equations. You cannot pick $abhg$, $bcdh$, and $acdg$, for example, since this set leaves out the branch deg. Substitute a loop containing R_5 for one of the three loops just mentioned. Now, write the corresponding loop equations:

The six equations can be solved for six unknowns. If the emf and resistances are known, then the equations can be solved for the currents, for example. If the numerical value of a current is positive, the actual current is in the direction of the corresponding arrow on the diagram and electrons move opposite to the direction of the arrow. If the numerical value of a current is negative, the actual current is opposite the direction of the arrow and electrons move in the direction of the arrow.

Resistors in series and parallel. Be sure you can distinguish between parallel and series combinations of two or more resistors. As for any circuit elements, the potential difference is the same for all the resistors in a _____ combination and the current is the same for all resistors in a _____ combination. If neither the potential difference nor the current is the same, then the resistors do not form either a parallel or series combination.

In the space below draw two resistors connected in parallel and two connected in series. Assume a potential difference V is applied to each combination and indicate on the diagrams where it is applied. Draw a current arrow for each resistor and label them to show which resistors have the same current and which have different currents.

PARALLEL   SERIES

In each case, the resistors can be replaced by a single resistor with resistance R_{eq} such that the current is the same as the total current into the original combination when the potential difference is the same as the potential difference across the original combination. If the resistances are R_1 and R_2, then the value of R_{eq} for a parallel combination is given by

$$R_{eq} =$$

and the value of R_{eq} for a series combination is given by

$$R_{eq} =$$

For a parallel combination, the equivalent resistance is less than the least resistance in the combination; for a series combination, the equivalent resistance is greater than the greatest resistance in the combination.

RC circuits. You will now study a circuit with a time-dependent current: a single-loop containing a capacitor, a resistor, and an emf. Kirchhoff's loop rule is still valid and the circuit equation can be derived by summing the potential differences around the loop.

The diagram on the right shows the circuit; it might be used to charge the capacitor. $\mathcal{E}$ is the emf, R the resistance, and C the capacitance. Assume a switch (not shown) is closed at time $t = 0$, when the capacitor is uncharged. Let $i(t)$ be the current at time t, positive in the direction of the arrow; let $q(t)$ be the charge on the upper plate of the capacitor at time t. The potential difference from a to b is $V_b - V_a =$ _____; the potential difference from b to c is $V_c - V_b =$ _____; and the potential difference from c to d is $V_d - V_c =$ _____. Thus, $\mathcal{E} - iR - q/C = 0$.

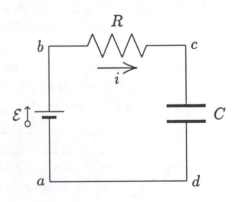

Both i and q are unknown, but since they are related, you can eliminate one in favor of the other to obtain an equation in one unknown. Use $i = dq/dt$ to eliminate i and show that

$$R\frac{dq}{dt} + \frac{q}{C} - \mathcal{E} = 0.$$

This is a differential equation to be solved for $q(t)$.

The relationship between i and q depends in sign on the direction of the current arrow and the selection of q to represent the charge on the *upper* plate. For the choices made above, positive i is equivalent to the statement that positive charge is flowing onto the upper plate and this, in turn, means dq/dt is positive. For a current arrow in the other direction, $i = -dq/dt$.

The solution to the differential equation that obeys the initial condition $q(0) = 0$ is

$$q(t) =$$

This expression is differentiated with respect to time to obtain an expression for the current:

$$i(t) =$$

On the axes below, sketch graphs of $q(t)$ and $i(t)$:

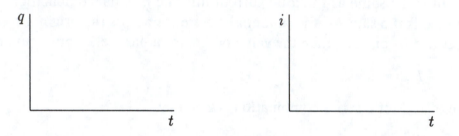

According to this result, the current just after the switch is closed, at $t = 0$, is given by $i(0) = $ _____, just as if there were no capacitor. The results also predict that after a long time, the charge on the capacitor is $q(\infty) = $ _____ and the current is $i(\infty) = $ _____. These conclusions make physical sense. At $t = 0$, the charge on the capacitor is zero so the potential difference across that element is _____. Since the loop equation reduces to $\mathcal{E} = iR$, we expect $i = $ _____. As charge builds up on the capacitor, the potential difference across it increases and, as a consequence, the potential difference across the resistor _____ (they must always sum to $\mathcal{E}$). Since the potential difference across the resistor is iR, this means the current _____ with time. When the capacitor is fully charged, $dq/dt = 0$ and the loop equation becomes $\mathcal{E} = q/C$. Then, the current is _____ and the charge on the capacitor is _____.

The quantity $\tau = RC$ is called the <u>capacitive</u> <u>time</u> <u>constant</u> of the circuit. It has units of _____ and controls the time for the charge on the capacitor to reach any given value. Describe what happens to the curves you drew above if τ is made longer (by increasing C or R): _____

_____

Suppose now that a capacitor C, initially with charge q_0, is discharged by connecting it to a resistor R. In the space to the right, draw a circuit diagram with a current arrow. In terms of the charge q and current i, the loop equation is _____. Use the relationship between q and i to eliminate i and write the resulting differential equation for q:

The solution to the differential equation is

$$q(t) =$$

Note that $q(0) = q_0$ and that after a long time, q becomes zero. The current as a function of time is

$$i(t) =$$

Initially the current is $i(0) =$ _____; after a long time, the current is _____. On the axes below, sketch graphs of $q(t)$ and $i(t)$:

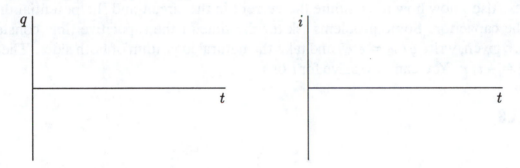

II. PROBLEM SOLVING

You should understand the definition of emf in terms of the work done on a charge as it moves through an emf device. You should also know that real emf devices have both emf's and internal resistances and, because they do, the terminal potential difference is not the same as the emf if charge is passing through the device. Be aware in each instance if the current is in the direction of the emf or in the opposite direction. The terminal potential difference depends on the relative direction of the current.

You should be able to write down Kirchhoff's loop equation for a single loop and solve it for an unknown (the current, a resistance, or an emf). If the current is known, you should be able to compute the potential difference between any two points on the circuit. You should also be able to compute the rate with which an emf supplies energy and the rate with which a resistor dissipates energy. Remember that for any electrical device the rate with which energy is transferred to or from the device is the product of the current through it and the potential difference across it.

You should know how to write and solve Kirchhoff's junction and loop equations for multiloop circuits. Draw a current arrow in each branch of the circuit and label each arrow with a different symbol. Write the junction equations first. The number of equations will be one less than the number of junctions. Then write the loop equations. The number will be such that the total number of equations is equal to the number of branches in the circuit. Solve the junction and loop equations for the unknowns.

Some problems ask you to calculate the equivalent resistance when two or more resistors are connected in parallel or series. You should also know how to use equivalent resistances to solve circuit problems. Remember that for a series connection, the current is the same in all the resistors and is the same as the current in the equivalent resistor. The potential difference across the equivalent resistor is the sum of the potential differences across the resistors of the combination. For a parallel connection, the potential difference across all the resistors is the same and is the same as the potential difference across the equivalent resistor. The current in the equivalent resistor is the sum of the currents in the resistors of the combination.

Some problems deal with a resistor and capacitor in a series circuit, with the capacitor either charging or discharging. You should know how to solve problems using the expression for the charge on the capacitor as a function of time: $q(t) = C\mathcal{E}(1 - e^{-t/\tau})$ for a charging capacitor and $q(t) = q_0 e^{-t/\tau}$ for a discharging capacitor. Here $\tau = RC$ is the capacitive time constant. Also know how to compute the current in the circuit and the potential difference across the capacitor. Some problems ask for the time or the capacitive time constant. If q and q_0 are given, write $q/q_0 = e^{t/\tau}$ and take the natural logarithm of both sides. The result is $\ln(q/q_0) = -t/\tau$. You can now solve for t or τ.

III. NOTES

Chapter 29
MAGNETIC FIELDS

I. BASIC CONCEPTS

Charged particles, when moving, exert magnetic as well as electric forces on each other. Magnetic fields are used to describe magnetic forces and, although the geometry is a little more complicated, the idea is much the same as the idea of using electric fields to describe electric forces. In the next chapter, you will learn how a magnetic field is related to the motion of the charges that produce it. Here you will learn about the force exerted by a magnetic field on a moving charge and on a wire carrying an electric current. In each case, pay particular attention to what determines the magnitude and direction of the force. As you read the chapter, note similarities and differences in electric and magnetic forces.

This chapter makes use of the following topics that were covered in previous chapters: velocity and acceleration in Chapter 4, force in chapter 5, uniform circular motion in Chapters 4 and 6, torque in Chapters 11 and 12, charge in Chapter 22, electric field and electric field lines in Chapter 23, electric potential in Chapter 25, and current and drift velocity in Chapter 27. You might review these topics.

Magnetic force on a moving charge. _____ charges exert magnetic forces on other _____ charges. The view taken is that a moving charge creates a <u>magnetic field</u> in all of space and this magnetic field exerts a (necessarily magnetic) force on any other moving charge. Charges at rest do not produce magnetic fields nor do magnetic fields exert forces on them. A moving charge does not exert a magnetic force on itself.

The force exerted by the magnetic field $\vec{B}$ on a charge q moving with velocity $\vec{v}$ is given by

$$\vec{F}_B =$$

Notice that the force is written in terms of the vector product $\vec{v} \times \vec{B}$. You might review Section 3–7 of the text if you do not remember how to determine the magnitude and direction of a vector product. The magnitude of the magnetic force on the moving charge is given by $F_B =$ _____, where θ is the angle between $\vec{v}$ and $\vec{B}$ when they are drawn with their tails at the same point. The direction of $\vec{v} \times \vec{B}$ is determined by the right-hand rule: _____

_____

Carefully note that the magnetic force is always perpendicular to both the magnetic field and to the velocity of the charge. Because the force is perpendicular to the velocity, a magnetic field cannot change the speed or _____ energy of a charge. It can, however, change the direction of motion.

Also note that the sign of the charge is important for determining the direction of the force. A positive charge and a negative charge moving with the same velocity in the same magnetic field experience magnetic forces in _____ directions.

If a proton is traveling in the positive x direction in a region in which the magnetic field is in the positive z direction, the magnetic force on it is in the _____ direction. If an electron is traveling in the same direction in the same region, the force on it is in the _____ direction. If either particle is traveling parallel to the magnetic field, the force on it is _____ .

The magnetic field can be defined in terms of the force on a moving positive test charge. First, the electric force is found by measuring the force on the test charge when it is _____ . This force is subtracted vectorially from the force on the test charge when it is moving in order to find the magnetic force. Second, the direction of the magnetic field is found by causing the test charge to move in various directions and seeking the direction for which the magnetic force is _____ . Lastly, the test charge is given a velocity perpendicular to the field and the magnetic force on it is found. If its charge is q_0, its speed is v, and the magnitude of the force on it is F_B, then the magnitude of the magnetic field is given by $B =$ _____ .

The SI unit for a magnetic field is _____ and is abbreviated _____ . In terms of the units coulomb, kilogram, and second, $1\,\mathrm{T} = 1$_____ . Another unit in common use is the gauss. $1\,\mathrm{T} =$ _____ gauss.

Magnetic field lines.   Magnetic field lines are drawn so that at any point the field is tangent to the field line through that point and so that the magnitude of the field is proportional to the number of lines through a small area perpendicular to the field. Recall that electric field lines are drawn in the same manner.

The diagram on the right shows the magnetic field lines in a certain region of space. On the diagram, label a region of high magnetic field and a region of low magnetic field. Suppose an electron is moving out of the page at the point marked with the symbol • and draw an arrow to indicate the direction of the magnetic force on it.

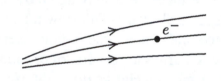

Magnetic field lines form closed loops. They continue through the interior of a magnet, for example. The end of a magnet from which they emerge is called a _____ pole; the end they enter is called a _____ pole. If the magnet is free to rotate in Earth's magnetic field, the north pole of the magnet will tend to point roughly toward the _____ geographic pole.

Crossed electric and magnetic fields.   If both a magnetic field $\vec{B}$ and an electric field $\vec{E}$ act simultaneously on a charge q moving with velocity $\vec{v}$, the total force on the charge is given by

$$\vec{F} =$$

An electric force can be used to balance a magnetic force on a charge, but the magnitude and direction used depend on the velocity of the charge. In particular, if a charge is in a magnetic field $\vec{B}$ and has velocity $\vec{v}$, then the total force is zero if $\vec{E} =$ _____ . Notice that this result does not depend on either the sign or magnitude of the charge. Also notice that the balancing electric field is perpendicular to both the magnetic field and the velocity of the charge. An important special case occurs if the velocity of the charge is perpendicular to the magnetic field. Then, the magnitude of the electric field is given by $E =$ _____ .

Perpendicular electric and magnetic fields form the basis for a <u>velocity</u> <u>selector</u>, used to select charges with a given speed from among a group of charges with a variety of speeds. Uniform fields $\vec{E}$ and $\vec{B}$, with known values and perpendicular to each other, are established in a region of space and the charges are incident in a direction that is perpendicular to both fields. Those with a speed given by $v =$ _____ continue straight through the region without being deflected while those with other speeds are deflected. By changing the ratio of the field strengths, different sets of charges can be selected.

Crossed electric and magnetic fields can be used to measure the mass-to-charge ratio m/q of an electron (or other charged particle). A beam of fast electrons, emitted from a hot filament, is formed by accelerating them in an electric field and passing them through a slit. They then enter a region of crossed electric and magnetic fields. The beam hits a fluorescent screen and produces a spot where it hits. The steps of the measurement are:

1. The fields are turned off and the position of the spot is noted.

2. Only the electric field is turned on and the deflection y of the beam is computed. Review Sample Problem 23–4 to see how this is done. In terms of the length L of the plates, the mass m, charge q, and speed v of the electron, and the electric field E, it is given by

$$y =$$

Notice that y is the deflection of the beam as it exits the region between the plates. It is not the deflection of the spot on the screen but it can be calculated from the spot deflection and knowledge of the geometry. The expression for y cannot be used directly to calculate m/q because the speed v is not known.

3. The magnetic field is now adjusted until the spot returns to its original position. Then, the magnetic and electric forces cancel and, in terms of E and B, the speed is $v =$ _____ . Substitute this expression into the equation for y and solve for m/q. The result is

$$m/q =$$

All of the quantities on the right side can be determined experimentally.

The Hall effect.   A Hall effect experiment is one of the few experiments that can be carried out to determine the sign of the charge carriers in a current-carrying sample. The diagram shows a rectangular sample of width d and thickness ℓ, carrying a current i, in a uniform magnetic field that points out of the page. Assume the current consists of *positive* charges moving in the direction of the current arrow. The magnetic field pushes them to one side of the sample — mark that side with a series of + signs and the opposite side with a series of − signs.

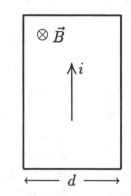

As charge accumulates at the sample sides, it creates an electric field in the sample, transverse to the current. For the situation described above, the electric field points from the _____ side to the _____ side. Charge continues to accumulate until the electric force on charges in the current is exactly the right strength to balance the magnetic force. In terms of the drift speed v_d of the particles and the magnetic field B, the final electric field strength is given by $E =$ _____. Once this condition is met, usually in times of about 10^{-13} or 10^{-14} s, charge in the current is no longer deflected and no additional charge accumulates at the sample sides.

A potential difference is associated with the transverse electric field. Since the charge carriers are positive, a point on the _____ side of the sample is at a higher potential than the point directly opposite on the other side.

If the current consists of *negative* charges going in the direction opposite to the current arrow, they are forced to the _____ side by the magnetic field. The final electric field now points from the _____ side toward the _____ side and the _____ side is at the higher potential. Thus, the sign of the charge carriers can be found by noting the sign of the deflection of a voltmeter attached across the sample width.

In the usual experiment, the transverse potential difference, the current, and the magnetic field are measured and the data is used to calculate the carrier concentration n. No matter what the sign of the carriers, the magnitude of the transverse potential difference is given by $V = Ed$, where E is the magnitude of the transverse electric field. Recall from your study of Chapter 27 that if only one type of carrier is present in the current, the current is given by $i = qnAv_d$, where A ($= \ell d$) is the cross-sectional area of the sample. In the space below, substitute $E = V/d$ and $v_d = i/qn\ell d$ into $E = v_d B$ and solve for n.

You should have obtained $n = iB/qV\ell$. If i, B, V, and ℓ are measured, n can be calculated.

Cyclotron motion. Since a magnetic force is always perpendicular to the velocity of the charge on which it acts, it can be used to hold a charge in a circular orbit. All that is required is to fire the charge perpendicularly into a uniform field. Suppose a region of space contains a uniform magnetic field with magnitude B and a particle with charge of magnitude q is given a velocity $\vec{v}$ perpendicular to the field. Equate the magnitude of the magnetic force (qvB) to the product of the mass and acceleration (mv^2/r) and solve for the radius r of the orbit. In terms of q, m, v, and B, the result is

$$r =$$

The time taken by the charge to go once around its orbit does not depend on its speed because the radius of the orbit and, hence, the distance traveled are directly proportional to _____. You can see how this comes about by developing an expression for the period. In terms of the radius r and speed v, the period is given by $T =$ _____. Now, substitute the expression for r in terms of the magnetic field, mass, charge, and speed. The result, $T = m/qB$, is independent of v. The number of times the charge goes around per unit time, or the frequency of the motion, is given by $f = 1/T =$ _____. The angular frequency is given by $\omega = 2\pi f =$ _____.

If the particle velocity is not perpendicular to the field, but has both parallel and perpendicular components, the trajectory is a _____. The radius is determined by the _____ component and the pitch is determined by the _____ component of the velocity.

These results are important for the operation of a cyclotron, a type of particle accelerator. The two dees of a cyclotron are diagramed to the right. A uniform magnetic field is everywhere out of the page and an electric field is in the region _____.
A positive charge enters the cyclotron near the center. The _____ field causes the charge to travel in a circular orbit. The _____ field causes the speed of the charge to increase and when it does, the charge moves to an orbit of larger radius. Assume the charge goes around four times before leaving the accelerator and draw the path on the diagram. In reality, charges circulate many thousands of times.

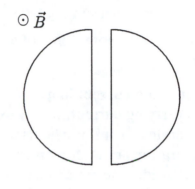

For the electric field to increase the speed of a positive charge each time the charge encounters it, it must be in the direction of motion. For it to increase the speed of a negative charge, it must be opposite the direction of motion. In either case, its direction must be reversed twice each orbit. This is fairly easy to do since the time between reversals does not change as long as the particle speed is significantly less than the speed of light. At speeds near the speed of light, the period does depend on the particle speed and the interval between reversals of the electric field must be _____ as the charge speeds up. Accelerators that do this are called _____.

Magnetic force on a current. A current is just a collection of moving charges and so experiences a force when a magnetic field is applied. If the current is in a wire, the force is transmitted to the wire itself.

To calculate the force on a wire carrying current i, divide the wire into infinitesimal segments, calculate the force on each segment, then vectorially sum the forces. Let dL be the length of an infinitesimal segment of wire with cross-sectional area A. If n is the concentration of charge carriers and each carrier has charge q, then the total charge in the segment is given by $qnA\,dL$. The magnetic force on the segment is the product of this and the force on a single charge. That is, $d\vec{F}_B = qnA\vec{v}_d \times \vec{B}\,dL$, where $\vec{v}_d$ is the drift velocity. Notice that the combination $qnAv_d$ appears in the expression for the magnitude of the force. This is the current i. If we take the vector $d\vec{L}$ to be in the direction of $q\vec{v}_d$ (i.e. in the direction of the current arrow), then the force can be written in terms of i, $d\vec{L}$, and $\vec{B}$: $d\vec{F}_B =$ _____.

For a finite wire, the resultant force is given by the integral

$$\vec{F}_B =$$

Check to be sure you have the correct order for the factors in the vector product. For a straight wire of length L in a uniform field $\vec{B}$, perpendicular to the wire, the magnitude of the force is $F_B =$ _____. Explain how the direction of the force is found: _____

_____

For a wire of any shape in a *uniform* magnetic field, the magnetic force is given by $\vec{F}_B = i\vec{L} \times \vec{B}$, where $\vec{L}$ is the vector from the end where the current enters to the end where the current exits the wire. Since the field is uniform, it can be factored from the integral for the force, with the result $\vec{F}_B = i\left(\int d\vec{L}\right) \times \vec{B}$. The integral $\int d\vec{L}$ is just $\vec{L}$. For a closed loop, $\vec{L} = 0$ and the total force of a uniform field is zero. The force of a non-uniform field is not zero.

Torque on a current loop.   Although a uniform field does not exert a net force on a closed loop carrying current, it may exert a torque. Thus, the center of mass of the loop does not accelerate in a uniform field but if the magnetic torque is not balanced, the loop has an angular acceleration about the center of mass.

Consider the rectangular loop of wire shown, in the plane of the page and carrying current i. A uniform magnetic field points from left to right. The force on the upper segment is zero since the current there is parallel to the field. The torque on it is also zero. Both the force and torque on the lower segment are zero for the same reason. The force on the left segment has magnitude iaB and is directed into the page. The torque on this segment about the center of the loop has magnitude $iabB/2$ and is directed toward the bottom of the page. The force on the right segment has magni-

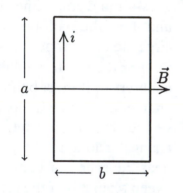

tude _____ and is directed toward _____. The torque on this segment about the center of the loop has magnitude _____ and is directed toward _____. The total force is _____ and the total torque has magnitude _____ and is directed toward _____. The loop tends to turn so its _____ side comes out of the page and its _____ side goes into the page. Use a dotted line to show the axis of rotation on the diagram. Label it. If the wire forms a coil with N turns the magnitude of the torque on it is _____.

For the orientation shown above, the magnitude of the torque is a maximum. It decreases as the loop turns. The diagram on the right shows the view looking down on the loop from the top of the page after it has turned through the angle ϕ. The magnitude of the torque on it is now given by _____. The torque always tends to orient the loop so it is _____ to the magnetic field and when it has this orientation, the torque is _____.

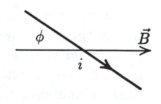

The torque exerted by a uniform field on any planar current loop is easily expressed in terms of the <u>magnetic dipole moment</u> $\vec{\mu}$ of the loop. For a loop with area A, having N turns, and carrying current i, the magnitude of the dipole moment is given by $\mu =$ _____. The direction of the dipole moment is determined by the right-hand rule: _____

A magnetic dipole moment is not the same as an electric dipole moment. Do not confuse them.

The torque exerted by a field $\vec{B}$ on a loop with magnetic dipole moment $\vec{\mu}$ is given by

$$\vec{\tau} =$$

Its magnitude is given by _____, where θ is the angle between $\vec{\mu}$ and $\vec{B}$ when they are drawn with their tails at the same point.

The magnitude of the dipole moment of the loop diagramed above is $\mu =$ _____. On the diagram above, draw a vector representing the dipole moment and label the angle θ. The vector product $\vec{\mu} \times \vec{B}$ has magnitude _____ and direction _____. Your answers should be in agreement with the results of the direct calculation of the torque.

Although magnetic fields are not conservative and a potential energy cannot be associated with a charge in a magnetic field, a potential energy can be associated with a magnetic dipole $\vec{\mu}$ in a magnetic field $\vec{B}$. It is given by

$$U =$$

The potential energy is a minimum when the dipole moment is _____ to the magnetic field and is in the _____ direction. It is a maximum when the dipole moment is _____ to the magnetic field and is in the _____ direction. In both these cases, the plane of the loop is _____ to the field. If the loop is in the plane of the page and the field is pointing toward you, then the direction of the current is _____ at a minimum of potential energy and is _____ at a maximum.

Suppose a dipole initially makes the angle θ_i with the magnetic field, but then rotates so it makes the angle θ_f. During this rotation, the work done on the dipole by the field is given by $W =$ _____ and the work done by the agent turning the dipole is given by _____ if the dipole is initially and finally at rest.

II. PROBLEM SOLVING

You should know how to compute the magnetic force on a moving charge. Use $F = qvB \sin \theta$ to compute the magnitude and use the right-hand rule to find the direction. Draw a diagram to help find the angle θ between the vectors $\vec{v}$ and $\vec{B}$. In other cases, the velocity and field might be given in unit vector notation. Use $\hat{\imath} \times \hat{\jmath} = \hat{k}$, $\hat{\jmath} \times \hat{k} = \hat{\imath}$, and $\hat{k} \times \hat{\imath} = \hat{\jmath}$. You also need to know that the sign of the vector product reverses if the factors are interchanged: $\hat{\jmath} \times \hat{\imath} = -\hat{k}$, for example. Remember that the vector product of two parallel vectors is 0: $\hat{\imath} \times \hat{\imath} = 0$, for example.

Some problems deal with the simultaneous influence of electric and magnetic fields. Use $\vec{F} = q(\vec{E} + \vec{v} \times \vec{B})$. In many cases, the charge is not accelerated but continues in a straight line with constant speed. Then, $\vec{E} + \vec{v} \times \vec{B} = 0$. This equation is often used to find the value of one of three quantities that appear in it.

Several problems deal with the Hall effect. Here the electric and magnetic forces cancel but now you might need to relate the electric field to the transverse potential difference ($E = V/d$, where d is the width of the sample) and the carrier speed to the current density ($v_d = J/ne$, where n is the carrier concentration) or current ($v_d = i/neA$, where A is the cross-sectional area).

Some problems deal with a charged particle that moves around a circular orbit, under the influence of a uniform magnetic field. You should know the relationship between the particle speed, the orbit radius, and the magnetic field. The period of the motion is given by $T = 2\pi m/qB$, where m is the mass of the particle, Q is its charge, and B is the magnitude of the magnetic field. The radius of the orbit is given by $r = mv/qB$, where v is the particle speed.

You should know how to calculate the force of a magnetic field on a current-carrying wire and the torque of a uniform magnetic field on a current-carrying loop. Often the easiest way to compute the torque is by using $\vec{\tau} = \vec{\mu} \times \vec{B}$. You will need to know how to compute the dipole moment $\vec{\mu}$, both magnitude and direction, of a current loop.

III. NOTES

Chapter 30
MAGNETIC FIELDS DUE TO CURRENTS

I. BASIC CONCEPTS

You will learn to calculate the magnetic field produced by a current using two techniques, one based on the Biot-Savart law and the other based on Ampere's law. When using the first, you will sum the fields produced by infinitesimal current elements. Carefully note how the directions of the current and the position vector from a current element to the field point influence the direction of the field. Ampere's law relates the integral of the tangential component of the field around a closed path to the net current through the path. Pay attention to the role played by symmetry when you use this law to find the field.

Before you begin studying this chapter, you might review the following concepts from previous chapters: conductors and insulators (Chapter 22), electric field (Chapter 23), current (Chapter 27), and magnetic field, magnetic force, and magnetic dipole (Chapter 29).

The Biot-Savart law. The diagram on the right shows a portion of a current-carrying wire. Suppose you wish to calculate the magnetic field it produces at point P. First, mark an infinitesimal element $d\vec{s}$ of the wire, in the same direction as _____. Now draw the displacement vector $\vec{r}$ from the selected current element to P. Draw an arrow head ($\odot$) or tail ($\otimes$) near P to represent the (infinitesimal) field produced there by the element. Be careful about the direction. Finally, write the Biot-Savart law in vector form for the (infinitesimal) magnetic field produced at P by the current element:

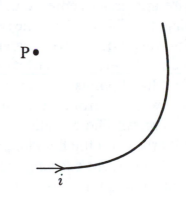

$$d\vec{B} =$$

Notice that the field of the element is perpendicular to both $d\vec{s}$ and $\vec{r}$. To find the total field at P, sum the contributions from all elements of the wire. The result is the vector integral

$$\vec{B} =$$

You should recognize that the integrand is a vector and you must actually carry out three integrations, one for each component. In many cases, however, you can choose the coordinates so only one component does not vanish.

The Biot-Savart law contains the constant μ_0. It is called the _____ constant and its value is _____ T·m/A. Do not confuse the symbol with that for the magnitude of a magnetic dipole moment (μ without the subscript 0).

In Section 30–1 of the text, the Biot-Savart law is used to find an expression for the magnitude of the magnetic field produced by a long straight wire carrying current i. Each infinitesimal element of the wire produces a field in the same direction so the magnitude of the total field is the sum of the magnitudes of the fields produced by all the elements. Go over the calculation carefully. The diagram shows an infinitesimal element of the wire at x, a distance r from point P. In terms of i and r, the magnitude of the field it produces at P is

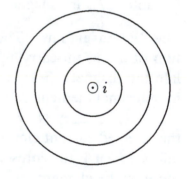

$$dB =$$

The direction of the field is _____ the page. In terms of x and R, $r =$ _____ and $\sin \theta =$ _____, so

$$dB =$$

Integrate in x from $-\ell$ to $+\ell$, then take the limit as $\ell \to \infty$. The result is

$$B =$$

You should know that the field lines around a long straight wire are circles centered on the wire. Three are shown in the diagram. The field has the same magnitude, given by the equation you wrote above, at all points on any given circle. The direction is determined by a right-hand rule: if the thumb of the right hand points in the direction of the current, then the fingers will curl around the wire in the direction of the magnetic field lines. The current in the wire shown is out of the page. Put arrows on the field lines to show the direction of the field.

For current-carrying wires that are not straight, the magnitude of the field is not uniform along a field line. You should be able to make a symmetry argument to show that it is for a straight wire.

You should also know how to compute the magnetic field produced at its center by a circular arc of current. Consider the wire shown at the right. It has radius R, subtends the angle ϕ, and carries current i. Other wires, not shown, carry the current into and out of the arc. The magnetic field at the center C due to a small segment of length ds has the magnitude

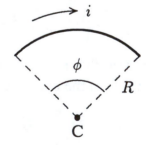

$$dB =$$

and is directed into the page.

The values of i and R are the same for all segments and the fields due to all segments are in the same direction, so the magnitude of the total field at C is given by

$$B = \frac{\mu_0}{4\pi} \int \frac{i \, ds}{R^2} =$$

where s is the length of the arc. If ϕ is measured in radians, then $s = R\phi$ and

$$B =$$

For a semicircle, $\phi = \pi$ and $B = $ _____. For a complete circle, $\phi = 2\pi$ and $B - $ _____ .

The expression for the field of a long straight wire can be combined with the expression for the magnetic force on a wire, developed in the last chapter, to find an expression for the force exerted by one wire on another, parallel to the first. The magnetic field produced by one wire at a point on the other wire is _____ to the second wire. If the wires are separated by a distance d and carry currents i_1 and i_2, then the magnitude of the force per unit length of one on the other is given by

$$F/L =$$

If the currents are in the same direction, the wires _____ each other; if the currents are in opposite directions, they _____ each other. Notice that the forces of the wires on each other obey Newton's third law: they are equal in magnitude and opposite in direction.

Ampere's law.   Ampere's law is given by Eq. 30–16. Copy it here:

The integral on the left side is a path integral around any *closed* path. The current i on the right side is the net current through a surface that is bounded by the path. For example, if the path is formed by the edges of this page, then i is the net current through the page. You should recognize that the Amperian path need not be the boundary of any physical surface and, in fact, may be purely imaginary. You should also recognize that the surface need not be a plane. The sides and bottom of a wastepaper basket form a valid surface with the rim as the boundary.

You should know how to evaluate the right side of the Ampere's law equation when the current is distributed among several wires. First, choose a direction, clockwise or counter-clockwise, to be used in evaluating the integral on the left side of the Ampere's law equation. The choice is immaterial but it must be made since it determines the direction of $d\vec{s}$. If the tangential component of the field is in the direction of $d\vec{s}$, then the integral is positive; if it is in the opposite direction, the integral is negative.

Now curl the fingers of your right hand around the loop in the direction chosen. Your thumb will point in the direction of positive current. Examine each current through the surface and algebraically sum them. If a current arrow is in the direction your thumb pointed, it enters the sum with a positive sign; if it is in the opposite direction, it enters with a negative sign. If the net current through the path is positive, then the average tangential component of $\vec{B}$ is in the direction chosen for $d\vec{s}$. If the net current is negative, then the average tangential component of $\vec{B}$ is in the opposite direction.

A current outside the path does not contribute to the right side of the Ampere's law equation. You should be aware that a current outside the path produces a magnetic field at every point on the path but the integral $\oint \vec{B} \cdot d\vec{s}$ of the field it produces vanishes.

Only the tangential component of the magnetic field enters. When Ampere's law is used to calculate the magnetic field of a current distribution, the path is taken, if possible, to be either parallel or perpendicular to field lines at every point. Then, the integral reduces to $\int B\,ds$, along those parts of the path that are parallel to field lines. If, in addition, the magnitude of the field is constant along the path, then the integral is Bs, where s is the total length of those parts of the path that are parallel to field lines. This is the case for most examples considered in this chapter.

Ampere's law can be used to find an expression for the magnetic field produced by a long straight wire carrying current i. The Amperian path used is a circle in a plane perpendicular to the wire and centered at the wire. Why? _____

_____

_____

Since the magnetic field is tangent to the circle and has the same magnitude at all points around the circle, the integral $\oint \vec{B} \cdot d\vec{s}$ is _____, where r is the radius of the circle. In the space below, equate this expression to $\mu_0 i$ and solve for B:

Your result should be $B = \mu_0 i/2\pi r$, as before.

Ampere's law can also be used to find the magnetic field *inside* a long straight wire. The diagram shows the cross section of a cylindrical wire of radius R carrying current i, uniformly distributed throughout its cross section. Take the Amperian path to be the dotted circle of radius r. The magnetic field is tangent to the circle and has constant magnitude around the circle, so the integral $\oint \vec{B} \cdot d\vec{s}$ is _____. Not all the current in the wire goes through the dotted circle. In fact, the fraction through the circle is the ratio of the circle area to the wire area: r^2/R^2. Thus the right side of the Ampere's law equation is _____. Equate the two sides and solve for B. The result is

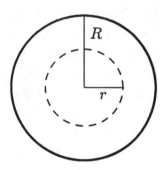

$$B =$$

The magnetic field at the center of the wire is _____. The field has its largest value at _____ and this value is given by

$$B =$$

Ampere's law can be applied to a solenoid, a cylinder tightly wrapped with a thin wire. For an ideal solenoid (long, with the current approximated by a cylindrical sheet of current rather than a wrapped wire), the magnetic field outside is negligible and the field inside is uniform and is parallel to _____ .

The diagram shows the cross section of a solenoid along its length, with the current in each wire coming out of the page at the top and going into the page at the bottom. Use dotted lines to draw the rectangular path you will use to evaluate the integral on the left side of the Ampere's law equation. The integral is $\oint \vec{B} \cdot d\vec{s} =$ _____ , where h is the length of the rectangle. Label it on the diagram.

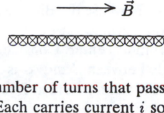

If the solenoid has n turns of wire per unit length, then the number of turns that pass through the surface bounded by the Amperian path is _____ . Each carries current i so the right side of the Ampere's law equation is _____ . The equation can be solved for the magnitude of the field, with the result $B =$ _____ .

Ampere's law can also be applied to a toroid, with a core shaped like a doughnut and wrapped with a wire, like a solenoid bent so its ends join.

The diagram to the right shows a cross section with the wire omitted for clarity. The magnetic field is confined to the interior of the core ($a < r < b$) and the field lines are concentric circles centered at the center of the hole. Draw the Amperian path you will use to find an expression for the field a distance r from the center. The integral $\oint \vec{B} \cdot d\vec{s}$ is _____ and if there are N turns of wire, each carrying current i, the total current through the surface bounded by the path is _____ . Thus the Ampere's law equation is _____ = _____ and

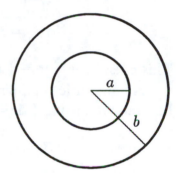

$$B =$$

Consider an Amperian path in the form of a circle, centered at the center of the hole and inside the hole. The net current through the path is _____ and the field in the "doughnut hole" is _____ . Consider a similar path outside the toroid. The net current through this path is _____ and the field outside the toroid is _____ .

The magnetic field of a circular current loop. This calculation is an excellent illustration of how the Biot-Savart law is used. Pay careful attention to it and notice particularly that the vector integral for the field is evaluated one component at a time. First, write the expression for the magnitude dB of the field produced by an infinitesimal element of the current. This involves the distance r from the current element to the field point and the sine of the angle θ between the element $d\vec{s}$ and the vector $d\vec{r}$. Next, determine the direction of the field associated with $d\vec{s}$ and let α be the angle between the field and one of the coordinate axes. $dB \cos \alpha$ is the component of the infinitesimal field along the chosen axis. Integrate $dB \cos \alpha$ over the current to find the component of the total field along the axis. The quantities r, θ, and α may be different for different segments of the current. If they are, you must use geometry write them in terms of a single variable of integration. Finally, repeat the process for the other coordinate axes.

Two angles enter the calculation: the angle θ between $d\vec{s}$ and $\vec{r}$ and the angle α between the field and a coordinate axis. The first is used to find an expression for the magnitude of the field produced by an infinitesimal element of current; the second is used to find a component, once the magnitude is known. Be careful to keep them straight.

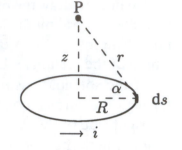

Now use the Biot-Savart law to develop an expression for the field on the axis of a circular loop, a distance z from its center. Consider the element $d\vec{s}$ shown. The angle between $\vec{r}$ and $d\vec{s}$ is _____ so the magnitude of the infinitesimal field at P reduces to

$$dB =$$

Draw a vector at P to indicate the direction of $d\vec{B}$ and verify that it makes the angle α with the z axis. The expression for the component of $d\vec{B}$ along the axis is

$$dB_{\parallel} =$$

This must be integrated around the loop.

All quantities in the integral are the same for all segments of the loop, so the integral can be evaluated using $\oint ds = 2\pi R$. The result is

$$B_{\parallel} =$$

In terms of z and R, $r =$ _____ and $\cos\alpha =$ _____. Make these substitutions to obtain

$$B_{\parallel} =$$

A symmetry argument can be made to show all other components vanish.

The field of a magnetic dipole. An important result can be derived from this equation. For points far away from the loop ($z \gg R$), the expression for the field can be written in terms of the magnetic dipole moment μ of the loop. $\vec{B}$ is in the direction of μ for points above the loop, in the direction of the dipole moment, and is in the opposite direction for points below the loop. Write the vector expression for the magnetic field:

$$\vec{B} =$$

This expression is valid for the magnetic field of *any* plane loop, regardless of its shape, for points far away along the axis defined by the direction of the dipole moment.

You should be able to estimate the direction of the field produced in its plane by a current-carrying loop of wire. Near any segment of the wire, you may treat the segment as a long straight wire and use the right-hand rule given above. Since the field decreases with distance, take the direction at any point to be roughly in the same direction as the field produced by the nearest segment of the loop, perhaps adjusted for other nearby segments.

For example, the circle on the right represents a current loop, with current i as shown. Points A, B, C, and D are in the plane of the loop. The magnetic field is directed _____ the page at A, _____ the page at B, _____ the page at C, and _____ the page at D. At points far away from the loop, the field closely resembles that of a bar magnet (both are dipole fields). The _____ pole is at the upper surface of the loop, where magnetic field lines leave the "magnet" and the _____ pole is at the lower surface of the loop, where field lines enter the "magnet".

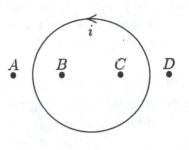

II. PROBLEM SOLVING

A few of the early problems deal with the field of a long straight wire. You should be able to find the magnitude and direction of the field at any point in space, given the current in the wire. Use $B = \mu_0 i/2\pi r$ to find the magnitude and the right-hand rule to find the direction. You may be asked for the total filed of two or more long straight wires. You will then need to carry out vector addition.

Once you have found the magnetic field you may be asked for the force it exerts on a moving charge. Use $q\vec{v} \times \vec{B}$. Don't forget to include the sine of the angle between $\vec{v}$ and $\vec{B}$ when you calculate the magnitude. Know how to use the right-hand rule to find the direction of the force. Be sure to take into account the sign of the charge.

Some problems ask you to use the Biot-Savart law to compute the magnetic field of a current. Divide the current into infinitesimal elements, write the expression for the field of an element, then integrate each component over the current. You will need to write the integrand in terms of a single variable. If the wire is straight place it along the x axis, say, and use the coordinate x of points along the wire as the variable of integration. If the wire is a circular loop use the angle made by a radial line with a coordinate axis as the variable of integration.

In many cases you may think of an electrical circuit as composed of finite straight-line and circular-arc segments, each of which produces a magnetic field. You can then calculate the field produced by each segment and vectorially sum the individual fields to find the total field. Use the result given in Problem 11 if you need to find the field on the perpendicular bisector of a finite straight wire. To find the field at some other point you sill need to integrate the Biot-Savart law. Use the procedure given in Section 1 for an infinite wire but replace the limits of integration with finite values. Use $B = \mu_0 i\phi/4\pi R$ for the field of a circular arc at its center of curvature.

Some problems ask you to use what you learned in the last chapter to calculate the force that one wire exerts on another.

Ampere's law problems take three forms. The most straightforward give the currents and ask you to find the value of $\oint \vec{B} \cdot d\vec{s}$ around a given path. You must pay attention to the directions of the currents as they pierce the plane of the path and carefully observe which currents are encircled by the path and which are not. Other problems ask you to use Ampere's law to calculate the magnetic field. Carefully choose the Amperian path you will use and pay attention to the evaluation of the Ampere's law integral in terms of the unknown field. Still other problems give you the magnetic field as a function of position and ask for the current through a given region. Carry out the line integral of the tangential component of the field around the boundary of the region and equate the result to $\mu_0 i$, then solve for i.

You should know how to compute the magnetic fields of some special current configurations (in addition to a long straight wire): a solenoid, a toroid, and a magnetic dipole. The field of an ideal solenoid is given by $B = \mu_0 n i$ inside the solenoid and by $B = 0$ outside. The field inside is along the cylinder axis. The field of a toroid is given by $B = \mu_0 i N / 2\pi r$ at a point inside, a distance r from the center. The field inside the hole and outside the toroid is zero. The field lines inside are circles that at concentric with the toroid. The field of a magnetic dipole is given by $\vec{B} = \mu_0 \vec{\mu} / 2\pi z^3$ at a point on the axis defined by the direction of the dipole moment, a distance z from the dipole. You should recall from the last chapter how to find the dipole moment of a current loop (both magnitude and direction) and how to compute the torque of a uniform magnetic field on a dipole.

III. NOTES

Chapter 31
INDUCTION AND INDUCTANCE

I. BASIC CONCEPTS

As the magnetic flux through any area changes, an emf is generated around the boundary and, if the boundary is conducting, a current is induced. Faraday's law tells us the relationship between the emf and the rate of change of the flux. You will learn that an emf can be generated by changing the magnetic field or by moving a physical object through a magnetic field. You should concentrate on how to calculate the magnetic flux and the emf in each case. In addition, you will learn that a nonconservative electric field is always associated with a changing magnetic field and it is this field that is responsible for the emf when the magnetic field is changing.

You will also learn about inductive circuit elements, which produce emfs via Faraday's law when their currents are changing and which store energy in magnetic fields. Pay attention to the definition of inductance and learn to calculate its value for solenoids and toroids. Learn about the influence of inductance on the current in a circuit. Also learn what factors determine the energy stored in the magnetic field of an inductor and learn how to calculate the stored energy.

Here are some important topics that have been covered in previous chapters and are used in this chapter: velocity in Chapter 4, charge in Chapter 22, electric field and electric field lines in Chapter 23, flux in Chapter 24, electric potential in Chapter 25, current and electrical resistance in Chapter 27, emf in Chapter 28, and magnetic field and magnetic torque in Chapter 29, solenoid, toroid, and permeability constant in Chapter 30.

Faraday's law. To understand Faraday's law you must first understand <u>magnetic flux</u>. The magnetic flux through any area is proportional to the number of magnetic field lines through that area. Recall that the number of field lines through a small area perpendicular to the field is proportional to the magnitude of the field and that the number of lines through *any* small area is proportional to the component B_n along a normal to the area.

The diagram shows an area A bounded by the curve C. If a magnetic field pierces the plane of the page, then there is magnetic flux through the area. It is the integral over the area of the normal component of the magnetic field. Divide the area into a large number of small elements, each with area ΔA, and for each element, calculate the quantity $B_n \Delta A$, then sum the results. The magnetic flux is the limit of this sum as the area elements become infinitesimal. That is, the flux through the area is given by the integral $\Phi_B = \int B_n \, dA$.

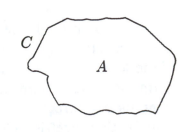

Usually the infinitesimal area is assigned a direction normal to the surface and written as the vector $\mathrm{d}\vec{A}$. Then, $B_n\mathrm{d}A = \vec{B} \cdot \mathrm{d}\vec{A}$ and the integral for the flux becomes

$$\Phi_B =$$

If the field is uniform over the entire area, the flux is given by

$$\Phi_B =$$

where θ is the angle between _____ and _____. The SI unit of magnetic flux is _____ and is abbreviated _____.

Faraday's law tells us that whenever the flux through any area is changing with time, then an emf is generated around the boundary of that area. Symbolically, the law is

$$\mathcal{E} =$$

where $\mathcal{E}$ is the induced emf. The induced emf is distributed around the boundary and is not localized as is the emf of a battery. Its direction is specified as clockwise or counterclockwise around a loop in the plane of the page. The minus sign that appears in Faraday's law is closely related to the direction of the emf.

Notice that the direction of $\mathrm{d}\vec{A}$ is ambiguous since two directions are normal to any surface. For the area shown above, the normal directions are into and out of the page. Φ_B has the same magnitude for both choices but is positive for one and negative for the other. You may choose either but you may find thinking about Faraday's law a bit easier if you pick the one for which Φ_B is positive. If you point the thumb of your right hand in the direction chosen for $\mathrm{d}\vec{A}$, then your fingers will curl around the boundary in the direction of positive emf. If $\mathcal{E}$, calculated using Faraday's law including the minus sign, turns out to be positive, then the emf is in the direction of your fingers. If it turns out to be negative, then it is in the opposite direction. If the boundary of the region is conducting and no other emf's are present, the induced current will be in the direction of the induced emf.

Some problems deal with a coil of wire consisting of several identical turns, like a solenoid, with the same magnetic field through each turn. If Φ_B is the flux through each turn and there are N turns, then the total emf induced around the coil is given by

$$\mathcal{E} =$$

The flux through the interior of a loop can be changed by changing the magnitude or direction of the magnetic field, by changing the area of the loop, or by changing the orientation of the loop. No matter how the emf is generated, it represents work done on a unit test charge as the test charge is carried around the loop. If the magnetic field is changing in either magnitude or direction, an electric field is generated and the emf is the work per unit charge done by that electric field. If the magnetic field is constant and the loop changes area or orientation, work must be done by an external agent, the agent that is changing the loop.

Lenz's law.   Lenz's law provides another way to determine the direction of an induced emf. Imagine that the bounding curve is a conducting wire so current flows in it when an emf is induced, the current being in the direction of the emf. The induced current produces a magnetic field everywhere but you should concentrate on the field it produces inside the loop and the flux associated with that field. According to Lenz's law, when the flux of the externally applied field changes, current flows in the loop in such a way that the flux it produces in the interior of the loop counteracts the change.

Suppose a bar magnet is held above the loop shown above, with its north pole toward the loop. Then, the magnetic field points into the page through the loop. If the magnet is moved toward the loop, the magnitude of its field in the plane of the loop increases (the field of a magnet is stronger near the magnet). According to Lenz's law, current will flow in the loop in such a way that its field in the interior of the loop is directed _____ the page. If the magnet is moved away from the loop, the magnitude of the applied field decreases and, as it does, current flows in the loop in such a way that its magnetic field in the interior of the loop is directed _____ the page. Thus the law is used to determine the direction of the field produced by the induced current.

Once you have determined the direction of the magnetic field produced by the induced current, you can determine the direction of the current itself and, hence, that of the emf. Very near any segment of the loop, the field is quite similar to the field of a long straight wire; the lines are nearly circles around the segment. You can use the right-hand rule explained in the last chapter: curl your fingers around the segment so they point in the direction of the field in the interior of the loop. That is, they should curl upward through the loop if the field is upward through the loop and they should curl downward if the field is downward. Your thumb will then point along the segment in the direction of the current.

For the situation being considered, the current and the emf are clockwise around the loop if the induced field inside the loop is _____ the page and are counterclockwise if the induced field inside the loop is _____ the page.

Motional emf.   An emf is also generated if all or part of the loop moves in a manner that changes the flux through it. You should realize that if the loop above is moved through a uniform field, the flux through it does not change and no emf is generated. If, however, the field is not uniform, then the flux changes as the loop moves and an emf is generated. You must find an expression for the flux as a function of the position of the loop, then differentiate it with respect to time to obtain $d\Phi_B/dt$. The emf clearly depends on the velocity of the loop. Either Lenz's law or the sign convention associated with Faraday's law can be used to find the direction of the emf.

If part of the loop moves in such a way that the area changes, an emf is generated even if the field is uniform. The flux through the loop is a function of time because the area is a function of time. In terms of the rate dA/dt with which the area changes in a uniform field B, the emf is given by $\mathcal{E} = $ _____ . It is generated only along the moving part, not around the whole loop.

Suppose a wire forms three sides of a rectangle and a rod, free to move on the wire, forms the fourth side, as shown. If W is the fixed dimension of the rectangle, x is the variable dimension, and the magnetic field is uniform and perpendicular to the rectangle, then the flux through it is Φ_B = _____ and the emf generated around it is _____. Note that the emf is proportional to the speed.

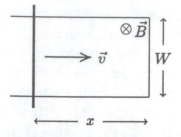

The loop need not consist entirely of physical objects, such as wires. Part of it may be imaginary. An emf is generated along an isolated rod moving in a magnetic field, for example. To calculate it, you may complete the loop with imaginary lines. If the rod is conducting, electrons move in the direction opposite to the emf when the emf first appears. They collect at one end of the rod, leaving the other end positively charged, and as a result, an electric field exists along the rod. Very quickly the field prevents further build-up of charge and the current becomes zero. A steady current cannot flow in an isolated rod.

Do not confuse the electric field created by charge at the ends of a moving rod with an electric field associated with a time-varying magnetic field. In the first case, the magnetic field is not changing and no electric field is associated with it.

An emf is also generated around a loop that is changing orientation so the angle between its normal and the magnetic field is changing with time. Suppose a circular loop with radius R is rotated about a diameter in a uniform magnetic field, perpendicular to the axis of rotation. The flux through the loop is given by $\Phi_B = BA\cos\theta = \pi R^2 B \cos\theta$, where θ is the angle between the normal to the plane of the loop and the magnetic field. If θ is changing according to $\theta = \omega t$, then $d\Phi_B/dt$ = _____.

You should be able to find the direction of the induced emf (and current if the loop is conducting). When ωt is between 0 and $\pi/2$, the flux is positive and so is the emf. Positive flux tells us that the direction we picked for the normal makes an angle of less than 90° with the magnetic field. If we look from a point where the magnetic field is pointing toward us from the loop, the emf is counterclockwise. When ωt is between 90° and 180°, the flux is _____ and the emf is _____. When viewed from the same point, the emf is _____. When ωt is between 180° and 270°, the flux is _____ and the emf is _____. When viewed from the same point, the emf is _____. When ωt is between 270° and 360°, the flux is _____ and the emf is _____. When viewed from the same point, the emf is _____.

Energy considerations. When a current is induced, energy is transferred, via the emf, to the moving charges. Recall that the rate at which a seat of emf $\mathcal{E}$ does work on a current i is given by P = _____. When the emf is associated with a changing magnetic field, the energy comes from the agent changing the field; when the emf is motional, the energy comes from the agent moving the loop or from the kinetic energy of the loop. In either case, the energy is dissipated in the resistance of the loop. You will learn more about the mechanism of energy transfer by a changing magnetic field in the next chapter. Here you learn about energy transfer by a motional emf.

Consider the example above in which one side of a rectangular loop moves in a uniform magnetic field and recall that the emf generated is given by $\mathcal{E} = BWv$. If the resistance of the loop is R, then the current is given by $i = $ _____ , so in terms of B, W, v, and R, the emf transfers energy at the rate $P = $ _____ . This is precisely the rate at which the external agent does work on the rod to keep it moving at a constant speed.

In terms of i, B, and W, the external field $\vec{B}$ exerts a force of magnitude $F = $ _____ on the rod. The direction of the current is _____ in the diagram and the magnetic field is into the page, so the magnetic force on the moving rod is directed to the _____ . If the rod is to move with constant velocity, an external agent must apply a force of equal magnitude but in the opposite direction. Thus the force of the agent is directed to the _____ . Since the rod is moving to the right with speed v, the rate at which the agent is doing work is $P = Fv$. In terms of B, v, W, and R, this is $P = $ _____ .

The rate at which energy is dissipated in the resistance of the loop is given by $P = i^2 R$ and, in terms of B, v, W, and R, this is $P = $ _____ . All of the energy supplied by the agent is dissipated. The magnetic field provides the mechanism of energy transfer but it supplies no energy.

If the agent stops pushing, the magnetic field exerts a force that slows the moving rod and eventually stops it. The kinetic energy of the rod is then dissipated in the resistance of the loop. This is the basis of magnetic braking.

Induced electric fields. An electric field is always associated with a changing magnetic field and this electric field is responsible for the induced emf. The relationship between the electric field $\vec{E}$ and the emf around a closed loop is given by the integral

$$\mathcal{E} = $$

Note that the integral is zero for a conservative field, such as the electrostatic field produced by charges at rest. The electric field induced by a changing magnetic field, however, is non-conservative and the integral is not zero.

Suppose a cylindrical region of space contains a uniform magnetic field, directed along the axis of the cylinder, as shown. The field is zero outside the region. If the magnetic field changes with time, the lines of the electric field it produces form circles, concentric with the cylinder. You should be able to derive an expression for the magnitude of the electric field at points inside and outside the cylinder.

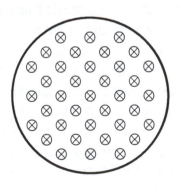

First, consider a point inside the cylinder, a distance r from the center. Draw a circle through the point, concentric with the cylinder cross section. The magnetic flux through the circle is given by _____ and in terms of dB/dt, its rate of change is given by _____ .

The circle coincides with an electric field line and the electric field has uniform magnitude around the circle. In terms of the magnitude of the electric field and the radius of the circle, the emf is given by $\mathcal{E} = $ _____ . Solve the Faraday's law equation for the magnitude of the electric field as a function of r. It is $E = $ _____ .

Now repeat the calculation for a point outside the cylinder. Draw a circle of radius r. The magnetic flux through this circle is given by _____, where R is the radius of the cylinder. The magnitude of the emf around the circle is given by _____ and the magnitude of the electric field is given as a function of r by $E =$ _____.

If the magnitude of the magnetic field is increasing, the emf (either inside or outside the cylinder) is in the _____ direction and so is the electric field. If the magnitude of the magnetic field is decreasing, the emf is in the _____ direction and so is the electric field.

Inductance. Current in a circuit produces a magnetic field and magnetic flux through the circuit. Both are proportional to the current. If the circuit consists of N turns and the flux is the same through all of them, then its inductance L is defined by

$$L =$$

where i is the current that produces flux Φ through each turn. L does not depend on the current or its rate of change. In general terms, it does depend on

_____

_____

The SI unit of inductance is called the _____ (abbreviated _____). The quantity $N\Phi$ is called the _____.

When the current changes, the flux changes and an emf in induced in the circuit. If the circuit has inductance L, the induced emf is given in terms of the rate of change of the current by

$$\mathcal{E}_L =$$

Any circuit has an inductance, usually small, but there are electrical devices, called inductors, that are used expressly for the purpose of adding a given amount of inductance to a circuit. They usually consist of a coil of wire, like a solenoid. The symbol for an inductor is

_____ .

A solenoid is an excellent example of an inductor because it concentrates the field in its interior and the magnetic flux through each turn is large. For all practical purposes, the emf induced by a changing current in a circuit containing a solenoid appears in its entirety along the wire of the solenoid.

The diagram shows an inductor carrying current i (the rest of the circuit is not shown). The direction of positive current is from a toward b. If the current is increasing, then the emf induced in the inductor is from _____ toward _____ ; if it is decreasing, then the emf is from _____ toward _____ .

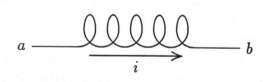

In either case, the potential V_b at point b is given by $V_b = V_a - L\,di/dt$, where V_a is the potential at point a and L is the inductance.

For an ideal solenoid of length ℓ and cross-sectional area A, with n turns per unit length and carrying current i, the magnetic field in the interior is given by $B =$ _____, the flux through each turn is given by $\Phi_B =$ _____, and the inductance is given by

$$L =$$

Notice that the inductance depends on the geometry (the length, area, and number of turns per unit length) but not on the current.

Toroids are also often used as inductors. Consider a toroid with N turns, a square cross section, inner radius a, and outer radius b. If the current in the toroid is i, the magnetic field a distance r from the center, within the toroid, is $B =$ _____ and the flux through a cross section is $\Phi =$ _____ . The inductance is therefore

$$L =$$

An LR circuit. The diagram on the right shows a series LR circuit. Take the current to be positive in the direction of the arrow and develop the loop equation in terms of L, R, $\mathcal{E}$, i, and di/dt. Take the electric potential to be 0 at point a. Then, the potential at point b is _____ and the potential at point c is given by _____ . The potential at point a is this value minus iR and the result must, of course, be zero. Thus the loop equation is

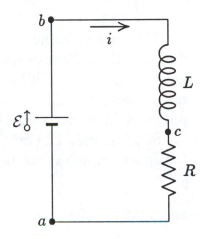

Assume the source of emf $\mathcal{E}$ is connected to the circuit at time t = 0, at which time the current is 0. The solution to the loop equation is then

$$i(t) =$$

Note that this equation predicts $i = 0$ at $t = 0$. It also predicts that long after the emf $\mathcal{E}$ is connected, the current is $i =$ _____ . The rate at which the current increases to its final value is controlled by the inductive time constant τ_L, which in terms of L and R, is $\tau_L =$ _____ . The larger the time constant, the slower the rate. A large time constant can be obtained by making _____ large and _____ small. When $t = \tau_L$, the current is about _____ percent of its final value.

The potential difference across the inductor is given by

$$V_L(t) = L\frac{di}{dt} =$$

and the potential difference across the resistor is given by

$$V_R(t) = iR =$$

On the axes to the left below, draw a graph of the current as a function of time. On the time axis, mark the approximate position of $t = \tau_L$. On the axes to the right below, draw graphs of the potential differences across the resistor and the inductor, all as functions of time. Here's some information that might help you: $V_R = V_L$ at $t \approx 0.7\tau_L$ and each of these potential differences have the value $\mathcal{E}/2$ at that time. Label the graphs V_R and V_L, as appropriate.

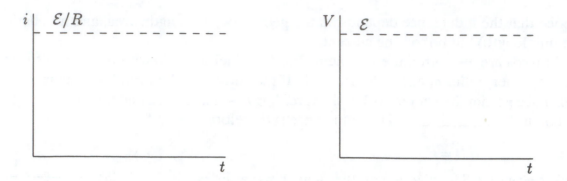

The current is the least but its rate of change is the greatest at $t = $ _____. Then, the potential difference across the resistor is _____ and the potential difference across the inductor is _____. A long time after the emf $\mathcal{E}$ is connected, the current is _____ and its rate of change is _____. Then, the potential difference across the resistor is _____ and the potential difference across the inductor is _____.

Suppose that the emf $\mathcal{E}$ is replaced by a wire when the circuit has a steady current i_0. As time goes on, we expect the current to decay to 0. The loop equation is

and its solution is

$$i(t) = $$

where the emf was replaced by the wire at time $t = 0$. According to this expression, at $t = 0$ the current is _____ and its rate of change is _____. Then, the potential difference across the resistor is _____ and the potential difference across the inductor is _____. A long time after the wire is inserted, the current is _____ and its rate of change is _____. Then, the potential difference across the resistor is _____ and the potential difference across the inductor is _____. Again the time dependence is controlled by the inductive time constant. When $t = \tau_L$, the current is about _____ percent of its initial value.

Magnetic energy. Energy must be supplied to build up the magnetic field in an inductor, perhaps by a source of emf. The energy may be considered to be stored in the magnetic field and can be retrieved when the current and field decrease. If current i is in an inductor with inductance L, the energy stored is given by

$$U_B = $$

The loop equation for a series LR circuit, when multiplied by the current, can be written $i\mathcal{E} = Li\, di/dt + i^2R$. This equation tells us that the rate with which energy is supplied by the source of emf $\mathcal{E}$ equals the sum of the rates with which it is dissipated in the resistor and stored in the magnetic field of the inductor. Identify each of the terms:

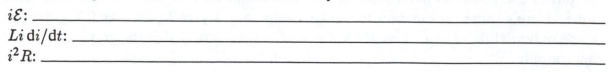

$i\mathcal{E}$: _____

$Li\, di/dt$: _____

i^2R: _____

The energy density (energy per unit volume) stored in a magnetic field $\vec{B}$ is given by

$$u_B =$$

The total energy stored in a field is given by the volume integral

$$u_B = \frac{1}{2\mu_0} \int B^2 \, dV \, .$$

These two equations are universally valid. They hold for every magnetic field, not just the field of a solenoid.

Mutual induction. The current in one circuit influences the current in another, although the two are not connected electrically. The current in the first circuit produces a magnetic field at all points in space and thus is responsible for a magnetic flux through the second. When the current changes, the flux changes and an emf is induced in the second circuit.

If the second circuit has N_2 turns and Φ_{21} is the flux produced through it by the current i_1 in the first circuit, then

$$M_{21} =$$

is the mutual inductance of circuit 2 with respect to circuit 1. Similarly, the mutual inductance of circuit 1 with respect to circuit 2 is

$$M_{12} =$$

where N_1 is the number of turns in circuit 1, i_2 is the current in circuit 2, and Φ_{12} is the flux through circuit _____ produced by the current in circuit _____ . The two mutual inductances M_{12} and M_{21} are always equal and the subscripts are not needed.

When the current i_1 in circuit 1 changes at the rate di_1/dt, the emf induced in circuit 2 is

$$\mathcal{E}_2 =$$

and when the current i_2 in circuit 2 changes the emf, induced in circuit 1 is

$$\mathcal{E}_1 =$$

.

II. PROBLEM SOLVING

You should know how to compute the flux through a given area. In some cases, the magnetic field is uniform and you simply multiply the perpendicular component by the area. In other cases, the field is not uniform and you must carry out an integration of the perpendicular component over the area. In some problems, the magnetic field is given while in others the current that produces the field is given and you must use what you learned in the last chapter to write an expression for it as a function of position. For example, the field may be produced by current in a long straight wire.

Once an expression for the magnetic flux is found, you differentiate it with respect to time to find the emf around the boundary of the region. Look carefully at the expression for the flux and decide what is changing with time. It might be the magnitude of the magnetic field, the magnitude of the area of the region, or the orientation of the area with respect to the field.

In some problems, the boundary of the region is a conductor and you are asked to find the current in it. First, note all the emfs around the boundary. One of them is the emf associated with the changing magnetic flux but there may be others, produced by batteries or generators, for example. Add them with their correct signs and divide the total by the resistance in the loop. Obviously, you must be very careful about the signs of the various emfs here.

Some problems ask for the electric field associated with a changing magnetic field. Use Faraday's law in the form $\oint \vec{E} \cdot d\vec{s} = -d\Phi/dt$. All of the problems have cylindrical symmetry, with the electric field lines forming circles around the cylinder axis. For them, you integrate the tangential component of the electric field around one of the field lines, with result $\oint \vec{E} \cdot d\vec{s} = 2\pi r E$, where r is the radius of the circle. Find the rate of change of the magnetic flux through the circle and solve the Faraday's law equation for E.

Some problems ask you to use the definition of self-inductance ($L = N\Phi/i$) to compute one of the quantities that appear in it. To carry out this task, you may need to compute the magnetic flux Φ of the magnetic field, perhaps by carrying out an integration. Other problems deal with the emf generated by an inductor when the current changes. Use $\mathcal{E} = -L\, di/dt$.

Some problems deal with RL series circuits. If a source of emf, such as a battery, is in the circuit and the current is 0 at time $t = 0$ (because, for example, a switch is closed then), the current is given by $i(t) = (\mathcal{E}/R)(1 - e^{-t/\tau_L})$, where the inductive time constant is $\tau_L = L/R$. If the current is i_0 at $t = 0$ and no source of emf is in the circuit, Calculate $i_0 e^{-t/\tau_L}$.

You may be asked to compute the potential difference across the resistor (iR) or across the inductor ($L\, di/dt$). Be sure you can tell which end is at the higher potential. You may also be asked for the rate with which the source of emf is supplying energy ($i\mathcal{E}$), the rate at which the resistor is dissipating energy ($i^2 R$), or the rate at which the inductor is storing energy ($Li\, di/dt$). If you are asked for the energy densities use $u_E = \frac{1}{2}\epsilon_0 E^2$ and $u_B = B^2/2\mu_0$.

Some problems deal with the calculation of mutual inductance. Here you assume a current in one circuit and compute the magnetic flux it produces in another circuit.

III. NOTES

Chapter 32
MAGNETISM OF MATTER; MAXWELL'S EQUATIONS

I. BASIC CONCEPTS

No particles have been observed that produce magnetic fields with lines that begin or end on them as electric field lines begin and end on electric charge. This statement is codified in Gauss' law for magnetism, one of the fundamental laws of electromagnetic theory. You should understand both the mathematical statement and the important ramifications of the law. Magnetic dipoles, not monopoles, are the fundamental magnetic entities responsible for the magnetic properties of matter. You should understand that the motions of electrons in matter produce atomic dipole moments and that the dipole moment of an atom is intimately related to its angular momentum. You should be able to describe the magnetic properties of paramagnetic, diamagnetic, and ferromagnetic materials and you should understand how the behavior of atomic dipoles leads to these properties.

The four equations that tell us about the electric and magnetic fields produced by any source are collectively known as Maxwell's equations and are presented together in this chapter so you can see how they complement each other. Most of this material should be a review for you. The new topic deals with displacement current. You should pay close attention to its definition and learn how to calculate it and the magnetic field it produces.

The following topics have been discussed in previous chapters and are used in this chapter: angular momentum in Chapter 12, charge in Chapter 22, electric field in Chapter 23, electric flux and Gauss' law for electricity in Chapter 24, current in Chapter 27, magnetic field in Chapter 29, magnetic dipole in Chapters 29 and 30, permeability constant in Chapter 30, and magnetic flux and Faraday's law in Chapter 31.

Gauss' law for magnetism.   Suppose the normal component of the magnetic field is integrated over a *closed* d over a *closed* surface (one that completely surrounds a volume). The result is _____ no matter what closed surface is considered and no matter what the distribution of current. Write the mathematical statement of the law, Eq. 32–1 of the text, here:

The direction of the infinitesimal element of area $d\vec{A}$ is _____ and $\vec{B} \cdot d\vec{A} = B_n \, dA$, where B_n is the normal component of the magnetic field.

The law does not necessarily mean that the magnetic field is zero at any point on the surface, only that the total magnetic flux through any closed surface is zero. This usually happens because the field is essentially outward over some portions and essentially inward over others. No field lines start or stop in the interior of any surface: every line that enters any volume also leaves the volume.

The law can be interpreted to mean that magnetic monopoles do not exist, or at any rate are so rare that their influence has not be detected. Explain what a magnetic monopole is:

_____

_____

If monopoles existed and if a closed surface surrounded one of them, the right side of the equation would not be 0. Magnetic field lines would start and stop at monopoles so a net magnetic flux would pass through the surface. On the other hand, magnetic dipoles do exist: a charge circulating around a small loop is an example. They do not violate Gauss' law for magnetism and are, in fact, the fundamental sources of magnetism in matter.

The magnetic field in the exterior of a permanent bar magnet can be closely approximated by the field that would be produced by a positive monopole (a north pole) at one end and a negative monopole (a south pole) at the other, but the field does not actually arise from single monopoles but rather from magnetic dipoles associated with electron motion. As proof that the field is not due to monopoles, the magnet can be cut in half with the result that _____

_____ .

This process can be continued to the atomic level, with the same result.

The magnetic field of Earth.   The magnetic field of Earth can be approximated by the field of a magnetic dipole with moment μ = _____ J/T, located at Earth's center. The dipole is not along the axis of rotation but is tilted slightly from that direction. It points from _____ to _____ , so magnetic field lines leave Earth in the _____ hemisphere and enter it in the _____ hemisphere. Thus, the north geomagnetic pole is actually a _____ pole of Earth's magnetic dipole.

The direction of Earth's magnetic field is often specified in terms of its <u>declination</u> and <u>inclination</u>. The declination is _____

_____ .

The inclination is _____

_____ .

If the field has horizontal component B_h and vertical component B_v, then the inclination ϕ_i is given by $\tan \phi_i$ = _____ .

Dipole moments and angular momentum.   Review some properties of magnetic dipoles:

1. If current i is in a loop that bounds area A, the magnitude of its dipole moment is given by μ = _____ . The direction of the moment is given by a right hand-rule. Curl the fingers of your right hand in the direction of the _____ ; then your thumb points in the direction of the _____ .

2. On the line defined by the moment, the magnitude of the magnetic field produced by a magnetic dipole is given by B = _____ , where r is the distance from _____ to _____ . If a dipole is in the page with its moment normal to the plane of the page, pointing upward out of the page, then the direction of the magnetic field at points above the dipole is roughly _____ and the direction of the magnetic field at points below the dipole is roughly _____ .

3. An external magnetic field $\vec{B}$ exerts a torque on a magnetic dipole. The torque is given by $\vec{\tau} =$ _____ . Its magnitude is given by $\tau =$ _____ , where θ is the angle between _____ and _____ .

4. The potential energy of a magnetic dipole with moment $s\vec{\mu}$ in an external magnetic field $\vec{B}_{ext}$ is given by $U =$ _____ .

The magnetic dipole moment of a circulating charge is closely related to its angular momentum. Suppose a charge q is traveling with speed v around a circle of radius r. The period of its motion is given by $T =$ _____ , the current is given by $i = q/T =$ _____ , and the magnitude of its dipole moment is given by $\mu_{orb} = iA =$ _____ . If its mass is m, then the magnitude of its angular momentum is $L_{orb} =$ _____ . In terms of L_{orb}, $\mu_{orb} =$ _____ . The vector relationship between $\vec{\mu}_{orb}$ and $\vec{L}_{orb}$ is:

$$\vec{\mu}_{orb} =$$

If q is positive, then $\vec{\mu}_{orb}$ and $\vec{L}_{orb}$ are in the same direction; if q is negative, they are in opposite directions. Suppose an electron is traveling in a counterclockwise direction around a circular orbit in the plane of this page. The direction of its orbital angular momentum vector is _____ the page and the direction of its magnetic dipole moment is _____ the page.

Only one component of the orbital angular momentum, the z component, say, can be measured. Quantum mechanically, the z component is given by

$$L_{orb,z} =$$

where h is the Planck constant and $m_\ell = 0, \pm 1, \pm 2, \ldots, \pm(\text{limit})$. The z component of the orbital magnetic dipole moment is given by

$$\mu_{orb,z} =$$

Every electron has an intrinsic angular momentum, often called its spin angular momentum or simply its spin. It is denoted by $\vec{S}$. An intrinsic magnetic dipole moment is associated with spin and it is related to the spin angular momentum by

$$\vec{\mu}_S =$$

Note that the relationship differs by a factor of two from the analogous relationship for orbital angular momentum. The dipole moment and spin angular momentum are in opposite directions for electrons.

Only one component of the spin dipole moment, the z component, say, can be measured. Since the z component of the spin angular momentum is given by $S_z =$ _____ , the z component of the dipole moment is given by

$$\mu_{S,z} =$$

where m_S is _____ .

For a collection of electrons, as in an atom, the individual electron dipole moments sum vectorially to produce the total dipole moment of the atom. For most atoms, the individual moments cancel and the total moment is zero; some atoms, however, have a net moment. Whether the atoms of a material have net dipole moments is important for the determination of the magnetic properties of the material.

Particle and atomic magnetic moments are often measured in units of what is called a Bohr magneton and denoted μ_B. In terms of the Planck constant h and the mass m of an electron, a Bohr magneton is given by μ_B = _____ and its SI value is _____ J/T. In units of a Bohr magneton the magnitude of the intrinsic dipole moment of an electron is _____ .

Magnetic dipole moments are also associated with the spins of protons and neutrons. The moments, however, are extremely small compared with that of an electron because _____ _____ .

In spite of the smallness of the magnetic moments of protons and neutrons, nuclear magnetism has found important applications in medicine and elsewhere.

Magnetization. In words, the magnetization of a uniformly magnetized object is _____ _____ per unit _____ . If the object is not uniformly magnetized, you may need the definition of the magnetization at a point. It is the limiting value of the expression you wrote as the volume shrinks to zero around the point.

Some materials, called paramagnetic and diamagnetic, become magnetized only when an external magnetic field is applied; others, called ferromagnetic, can be permanently magnetized. In any event, the dipoles of a magnetized object produce a magnetic field of their own and, to find the total magnetic field at any point, this field must be added vectorially to any applied field that is present. If a magnetic field $\vec{B}_0$ is applied to magnetic material, the total field is given by $\vec{B} = \vec{B}_0 + \vec{B}_M$, where $\vec{B}_M$ is the field produced by the magnetic dipoles of the material. Since the field $\vec{B}_M$ depends on the geometry of the sample and the uniformity of the magnetization, it is often difficult to calculate.

Paramagnetism. Atoms of paramagnetic materials have permanent dipole moments. When no external field is applied, however, the magnetization is zero because _____ _____ .

When an external field is applied, the material becomes magnetized because _____ _____ .

When the applied field is removed, the magnetization quickly reduces to zero because _____ _____ .

When an external field is applied, the direction of the field produced by dipoles of the material is _____ the direction of the applied field, so the total field in the material is _____ in magnitude than the applied field.

According to Curie's law (Eq. 32–19), the magnetization at any point is directly proportional to the _____ at that point and inversely proportional to the _____ . Symbolically,

$$M =$$

where C is called the _____ constant for the material. The relationship is valid only for small values of B/T. Explain qualitatively why the magnetization increases if the local magnetic field is increased without changing the temperature: _____

_____

Explain qualitatively why the magnetization decreases if the temperature is increased without changing the applied field: _____

_____

For large magnetic fields, the magnetization is no longer proportional to the field and, in fact, is less in magnitude than a proportional relationship predicts. For sufficiently high fields, the magnetization is independent of the field. The magnetization is then said to be <u>saturated</u>. This occurs when all the dipoles are _____.

On the axes to the right, draw a graph of the magnetization as a function of the magnetic field for a typical paramagnetic material at any given temperature. Label the saturation magnetization M_{max} and indicate the linear region where Curie's law is valid.

If the sample contains N dipoles per unit volume, each with dipole moment μ, then the saturation value of the magnetization is given by

$$M_{max} =$$

When the orientation of a magnetic dipole changes with respect to a magnetic field, its energy changes. The potential energy of a dipole $\vec{\mu}$ in a magnetic field $\vec{B}$ is given by $U =$ _____. Suppose a dipole originally aligned with a field is turned end for end. Its initial potential energy is _____, its final potential energy is _____, and the energy required to turn the dipole is _____. To see if the thermal energy of the material is adequate to inhibit paramagnetism, this energy is compared with the mean kinetic energy of the atoms of the material. For an ideal gas of atoms, this is _____ at absolute temperature T.

Diamagnetism. In the absence of an applied field, the atoms of a diamagnetic substance have no magnetic dipole moments, but moments are induced when a field is applied. As an external field is turned on, the orbits of the electrons change so the electrons produce an opposing field. The direction of the magnetization (and the induced dipole moments, on average) is _____ that of the local magnetic field. The total magnetic field in a diamagnetic material is _____ in magnitude than the applied field.

The effect occurs for all materials, but if the atoms have permanent dipole moments, the effect of their alignment with the field dominates and the material is _____ rather than _____.

Ferromagnetism. The atoms of ferromagnetic materials have permanent dipole moments but unlike the atomic moments of paramagnetic materials they *spontaneously* align with each other. List some ferromagnetic materials: _____

_____

You should realize that the spontaneous alignment of dipoles in a ferromagnet is *not* due to the magnetic torque exerted by one magnetic dipole on another. These torques are not sufficiently strong to overcome thermal agitation that tends to randomize the dipole directions. The torques that align the dipoles have their source in the quantum mechanics of electrons in solids. Above a certain temperature, called its _____ temperature, a ferromagnetic object becomes paramagnetic. For iron, this temperature is about _____ K.

Ferromagnetic materials exhibit <u>hysteresis</u>. On the axes to the right draw a curve that shows the field B_M due to atomic dipoles for a typical ferromagnet as an applied field B_0 increases from 0 to B_{0f}. Assume the material is unmagnetized to start. Next, draw the curve as the applied field decreases from B_{0f} to 0. The magnetization does not retrace the original curve and it does not become 0 when the applied field is 0. The ferromagnet is magnetized although there is no external field.

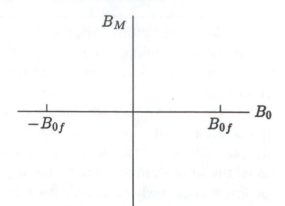

Now suppose the applied field is reversed (has negative values on the graph) and its magnitude is increased to B_{0f}. Draw the curve. Note that a field must be applied opposite to the magnetization in order to reduce the magnetization to 0. Now draw the curve that shows what happens if the applied field is reduced to 0. Finally, draw the curve that shows what happens if the applied field is reversed again and increased to B_{0f}.

Hysteresis is a direct result of the existence of ferromagnetic <u>domains</u>. In a domain, the dipole moments are _____ to each other, while dipoles in a neighboring domain are in _____ direction. The field B_M you plotted above is the vector sum of the fields due to all domains. If a ferromagnet is unmagnetized, the vector sum of the dipole moments of the domains is _____. When an external field is applied, domains with dipoles parallel to the field tend to _____ in size while domains in other directions tend to _____. This amounts to changes in the orientations of dipoles near domain boundaries. Some reorientation of dipoles within a domain may also take place.

Hysteresis comes about because the growth and shrinkage of domains is not reversible. When an external field is applied to an unmagnetized sample and then turned off, the domains do not spontaneously revert to their original sizes.

Magnetic Induction.   Faraday's law tells us that an electric field is associated with a changing magnetic field. The symmetric statement is also true: a magnetic field is associated with a changing electric field. If Φ_E is the electric flux through some area and $\oint \vec{B} \cdot d\vec{s}$ is the integral of the tangential component of the magnetic field around the boundary, then

$$\oint \vec{B} \cdot d\vec{s} =$$

This relationship is called Maxwell's law of induction.

You should be able to find the direction of the magnetic field lines. First, chose a direction for $d\vec{A}$. It is in one of the two directions that are perpendicular to the surface. This choice determines the signs of Φ_E and $d\Phi_E/dt$. Point the thumb of your right hand in the direction chosen. Your fingers will curl in the direction you *must* use for $d\vec{s}$ when you evaluate $\oint \vec{B} \cdot d\vec{s}$. Now evaluate $d\Phi_E/dt$. If it is positive, the magnetic field lines follow your fingers; if it is negative, they go the opposite way.

Magnetic fields arise from currents and from changing electric flux. The first is described by Ampere's law and the second by Maxwell's law of induction. The two can be combined into the Ampere-Maxwell law:

$$\oint \vec{B} \cdot d\vec{s} =$$

Consider a parallel-plate capacitor, with circular plates. Take the region of interest to be a circle of radius r, between the plates and centered on the line through the centers of the plates. Take $r < R$, where R is the plate radius. Let E be the magnitude of the (uniform) electric field in the plane of the circle. The electric flux is then $\Phi_E =$ _____ .

While the charge on the capacitor plates is changing, the electric field and electric flux are changing and there is a magnetic field between the plates. Its lines are circles that are parallel to the plates with their centers on the line through the plate centers. The magnitude of the field is uniform around any field line. If B is the magnitude of the field, then $\oint \vec{B} \cdot d\vec{s} =$ _____, in terms of the radius r. Substitute into Maxwell's law of induction to obtain

$$B =$$

in terms of the rate of change of the electric field. Note that the magnetic field is zero on the line joining the plate centers and increases in proportion to r.

Now consider a circle outside the plates ($r > R$). The electric flux through the circle is $\Phi_E =$ _____. The integral $\oint \vec{B} \cdot d\vec{s}$ is still $2\pi r$. The Maxwell law of induction now yields

$$B =$$

for the magnitude of the magnetic field. It decreases in proportion to $1/r$.

The displacement current. Look at the Ampere-Maxwell law and note that $\epsilon_0 \, d\Phi_E/dt$ plays the same role as the current i in producing a magnetic field. In fact, the quantity $i_d =$ _____ is called the <u>displacement</u> <u>current</u> through the area. You should realize that a displacement current is emphatically NOT a true current, which consists of moving charge. A displacement current exists in any region of space that contains a changing electric field, even if charge is not moving in the region. True currents exist only in regions that contain moving charge.

Suppose a cylindrical region with radius R contains a uniform electric field $\vec{E}$ parallel to its axis. A cross section is shown to the right, with the field pointing into the page. First consider the whole cross section. If $d\vec{A}$ is also into the page, the electric flux through the circle of radius R is given by $\Phi_E =$ _____, where E is the magnitude of the field. If the magnitude is changing at the rate dE/dt, then the displacement current through the entire cross section is given by $i_d =$ _____ .

Now consider the circle of radius r ($< R$). The electric flux through this circle is given by $\Phi_E = $ _____ and the displacement current through it is given by $i_d = $ _____.

Finally, consider a circle of radius r ($> R$). The electric flux through this circle is given by $\Phi_E = $ _____ and the displacement current through it is given by $i_d = $ _____.

A magnetic field is associated with a displacement current just as a magnetic field is associated with moving charge. For the cylindrical region considered above, the displacement current is uniformly distributed over every cross section, so the cylinder acts like a long straight wire of radius R with uniform current density. The magnetic field lines are _____ centered at the _____ of the cylinder.

Take the left side of the Ampere-Maxwell law to be an integral around a circle in the plane of a cross section and centered at the cylinder axis. Then, in terms of the magnitude B of the magnetic field and the radius r of the circle, $\oint \vec{B} \cdot d\vec{s} = $ _____. Let i_d be the total displacement current through a cross section of the cylinder. Then, if $r < R$, the displacement current through the amperian circle is given by _____. In terms of i_d, the magnitude of the magnetic field is given by $B = $ _____. You should be able to show that this result agrees with the expression you derived above for B in terms of $d\Phi_E/dt$.

If $r > R$, the displacement current through the amperian circle is the same as that through the entire cross section of the cylinder; that is, it is i_d. For this case, the magnitude of the magnetic field is given in terms of i_d by $B = $ _____. You should be able to show that this result agrees with the expression you derived above for B in terms of $d\Phi_E/dt$.

Maxwell's equations. Complete the following statements of Maxwell's equations (see Table 32–1 of the text):

I. $\oint \vec{E} \cdot d\vec{A} = $

II. $\oint \vec{B} \cdot d\vec{A} = $

III. $\oint \vec{E} \cdot d\vec{s} = $

IV. $\oint \vec{B} \cdot d\vec{s} = $

Additional terms must be added to some of these equations if magnetic or dielectric materials are present but otherwise they are completely general.

You should understand what each of the Maxwell equations says. Equations I and II contain surface integrals on their left sides and the surface must be specified to evaluate the integral. In each case, $d\vec{A}$ is a vector whose magnitude is an infinitesimal element of _____ and whose direction is _____ to the surface. In equation I, the _____ component of the electric field is integrated over a _____ surface. The symbol q on the right side represents the net charge _____ by the surface. In equation II, the right side is zero because there are no _____.

Earlier in this course you used equation I (Gauss' law for electricity) to find the electric field of a point charge, a large plate of charge, a line of charge, and other charge configurations. You also used it to show that static charge on a conductor resides on its surface. The equation tells us that electric field lines can be drawn so they start on positive charge and end on negative charge. Equation II (Gauss' law for magnetism) tells us that magnetic field lines form closed loops.

Equations III and IV contain path integrals on their left sides. In each case, $d\vec{s}$ is a vector whose magnitude is an infinitesimal element of _____ and whose direction is _____ to the path. On the left side of equation III, the _____ component of the electric field is integrated around a _____ path. This integral gives the _____ around the path. On the left side of equation IV, the _____ component of the magnetic field is integrated around a _____ path. On the right side of equation III, Φ_B is the magnetic flux through the _____ bounded by the path. On the right side of equation IV, i is the net _____ through the area and Φ_E is the _____ through the area.

Equation III is Faraday's law. You used it to compute the emf in a loop that is rotating in a magnetic field, the emf generated in a loop by the changing field of a nearby long straight wire, and other emf's generated by changing magnetic fluxes.

Equation IV is the Ampere-Maxwell law. It has been used to find the magnetic field of a long straight wire and the magnetic fields inside solenoids and toroids. For these situations, the fields are produced by currents alone and $d\Phi_E/dt = 0$. It has also been used to compute the magnetic field inside a charging or discharging capacitor. There $i = 0$ but $d\Phi_E/dt \neq 0$.

Although these equations have been used in specific examples, chosen in most cases for ease in computation, they are generally valid. The first two are true for *any* closed surface, the second two are true for *any* closed path.

Take special care to distinguish between the various equations. Remember that a changing magnetic field produces an electric field and that *both* a changing electric field and a current produce a magnetic field. Remember that the fields produced appear on the left sides of Maxwell's equations, the sources of the field appear on the right sides. Look at the equations and explain to yourself what each describes.

How can you generate an electric field in a region without having a net charge anywhere?

_____

_____

_____

For this situation, which of the Maxwell equations have zero for the value of their right sides? _____ Which do not? _____

How can you generate an electric field in a region without having a current anywhere?

_____

_____

_____

For this situation, which of the Maxwell equations have zero for the value of their right sides? _____ Which do not? _____

How can you generate a magnetic field in a region without having a net charge anywhere?

_____

_____

_____

For this situation, which of the Maxwell equations have zero for the value of their right sides? _____ Which do not? _____

II. PROBLEM SOLVING

Many problems that deal with Gauss' law for magnetism give you a closed surface such that you can easily calculate the magnetic flux through part of it. You are then asked for the flux through the rest of the surface. You must know that the flux through the complete surface is zero.

Earth's magnetic field can be approximated by the field of magnetic dipole located at the center of Earth. Some problems ask you to use this approximation to calculate the magnitude of Earth's field or else its inclination angle. You should know that the magnetic field of a magnetic dipole is given by $\vec{B} = \mu_0 \vec{\mu}/2\pi z^3$, where $\vec{\mu}$ is the magnetic dipole moment and z is the distance from the dipole.

Some problems deal with the interaction of electrons with magnetic fields. Remember that the energy of a magnetic dipole in a magnetic field is given by $U = -\vec{\mu} \cdot \vec{B}$.

Some problems deal with paramagnetic, diamagnetic, and ferromagnetic materials. To solve them, you should understand the concept of magnetization the net dipole moment per unit volume. You should also know the Curie law for paramagnetic materials.

You should be able to calculate the magnetic field induced by a uniform time-varying electric field in a cylindrical region. Remember that although the electric field exists only inside the region, a magnetic field is induced at all points, both inside and outside.

You should know how the displacement current and the rate of change of the electric flux are related: $i_d = \epsilon_0 \, d\Phi_E/dt$. You should also know how to calculate the electric flux and the displacement current when the electric field is uniform. If the field is uniform and normal to the surface, then $\Phi_E = EA$, where A is the area of the surface. In this case, $i_d = \epsilon_0 A \, dE/dt$. In some cases, you must find the electric field from a given electric potential function. Recall, for example, that the electric field between the plates of a capacitor is related to the potential difference across the plates by $E = V/d$, where d is the plate separation.

Also remember that the right side of the Ampere-Maxwell law contains both true current and displacement current. In most cases, one or the other of the currents is likely to be zero but you must think about both when you solve a problem.

III. NOTES

Chapter 33
ELECTROMAGNETIC OSCILLATIONS
AND ALTERNATING CURRENT

I. BASIC CONCEPTS

The combination of a charged capacitor and an inductor connected in series produces a sinusoidally varying current. Learn what factors determine the frequency of oscillation. When a resistor is added, the amplitude decays exponentially with time. Energy is an important theme of this chapter. Pay careful attention to the form it takes, sometimes as energy stored in the electric field of the capacitor and sometimes as energy stored in the magnetic field of the inductor. In the first circuit, electrical energy is conserved but when a resistor is added, it is converted to thermal energy.

You also study a series circuit consisting of a resistor, an inductor, and a capacitor, driven by a sinusoidal emf. The current is an alternating current (ac): it is sinusoidal and periodically reverses direction. Pay close attention to the relationship between the potential difference across each circuit element and the current in the element. Learn how to compute the current amplitude and phase in terms of the generator emf. You will be able to apply what you learn to many other circuits.

The following are some topics that were discussed in earlier chapters and are used in this one: oscillation of a mass on an ideal spring and angular frequency (Chapter 16), charge (Chapter 22), electric potential (Chapter 25), capacitance and electrical energy (Chapter 26), current, electrical resistance, and thermal energy generated by a resistor (Chapter 27), emf (Chapters 28 and 31), and inductance and magnetic energy (Chapter 31).

An LC circuit. The circuit shown in the diagram to the right consists of an inductor with inductance L and a capacitor with capacitance C, connected in series. As you will see, the charge on either plate of the capacitor oscillates sinusoidally, being positive for half a cycle and negative for the other half. The current also oscillates, being in one direction for half a cycle and in the other direction for the other half cycle. Finally, the energy also oscillates between the capacitor and the inductor. You should understand the behavior of the charge, current, and energy.

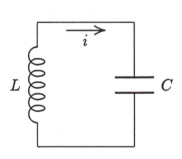

The first step is to develop and solve the loop equation for the circuit. Let q represent the charge on the upper plate of the capacitor and take the current to be positive in the direction of the arrow. The potential difference across the capacitor is given by _____ , with the _____ plate at the higher potential if q is positive. The potential difference across the inductor is given by _____ , with the _____ end at the higher potential if i is increasing. In terms of q and di/dt, the loop equation is _____. Substitute $i = dq/dt$ to obtain the loop equation in terms of q:

Suppose that at time $t = 0$, the current is zero and charge Q is on the capacitor, with the upper plate positive. Then, the solution to the loop equation is

$$q(t) =$$

and the current is given by

$$i(t) = \frac{dq}{dt} =$$

Both q and i oscillate with an angular frequency that is determined by the values of L and C: $\omega =$ _____. Solutions with other phase constants, of the form $q(t) = Q\cos(\omega t + \phi)$, are valid for other initial conditions. Initial conditions, however, do not influence ω.

No matter what the value of the phase constant, the maximum magnitude I of the current is _____ and $i = \pm I$ when $q =$ _____. When q is either $+Q$ or $-Q$, the current is _____.

Energy is stored in the electric field of the capacitor and in the magnetic field of the inductor. As functions of time the energy in the capacitor is given by

$$U_E(t) = \frac{1}{2}\frac{q^2}{C} =$$

and the energy in the inductor is

$$U_B(t) = \frac{1}{2}Li^2 =$$

Note that the energy oscillates back and forth between the inductor and the capacitor. The energy is entirely magnetic when the _____ is a maximum and the _____ is zero; it is entirely electric when the _____ is a maximum and the _____ is zero. The sum, $U = U_E + U_B$, is constant and can be written in terms of the current amplitude I as $U =$ _____ or in terms of the charge amplitude Q as $U =$ _____.

Study Fig. 33–1 carefully. In (a), the sign of the charge on the upper plate of the capacitor is _____ and the current through the inductor is _____. At this stage, all the energy is stored in the _____ field of the _____. As soon as this situation is set up, charge begins to flow. Why? _____

_____

In (b), the charge on the upper plate of the capacitor is _____ (the current is away from that plate). A magnetic field is building up in the inductor and the energy is shared between the inductor and capacitor. In (c), the discharge of the capacitor has been completed. The current and the magnetic field in the inductor are now at their maximum values and all the energy is stored in the _____ field of the _____.

Since the current continues as before, positive charge collects on the _____ plate of the capacitor. Because the charge on the plate repels other charge of the same sign, the current in (d) is _____. This means the magnetic field and the energy stored in the inductor are _____. In (e), the capacitor is maximally charged but now the _____ plate is positive. The current is _____ and all energy is stored in the _____. Now the capacitor begins discharging. In (f), the charge on the capacitor is _____, the current is _____, and its direction is _____ through the inductor. The energy is again shared by the capacitor and inductor. In (g), the discharge is completed, the current is a maximum, and the energy is all stored in the _____. The current continues and in (h), the capacitor is again charging, this time with the _____ plate positive. The current is _____ and the energy is shared. In (a), the capacitor is fully charged, the current is zero, and the cycle begins again.

Electromagnetic oscillations of an LC circuit are quite similar to the mechanical oscillations of a block-spring system. If a block of mass m moves in one dimension on a frictionless horizontal surface, acted on by a spring with spring constant k, its coordinate obeys the differential equation

$$m \frac{d^2}{dt^2} + kx = 0.$$

This can be changed into the differential equation for the charge on a capacitor plate in an LC series circuit by making the following replacements:

$$x \rightarrow$$

$$m \rightarrow$$

and

$$k \rightarrow$$

Although the velocity of the block does not appear in the differential equation, in other equations it is replaced by _____.

The potential energy stored in the spring is given by $U = \frac{1}{2}kx^2$. When the replacements are made, this becomes _____, which is the energy stored in the _____. The kinetic energy of the block is given by $K = \frac{1}{2}mv^2$. When the replacements are made, this becomes _____, which is the energy stored in the _____.

The coordinate and velocity of the block oscillate with a phase difference of 90° just like the charge and current in the electrical circuit. The energy of the block-spring system remains constant but changes from potential to kinetic and back again. The energy of the LC circuit also remains constant. It changes from electrical energy stored in the capacitor to magnetic energy stored in the inductor and back again.

Such analogies between mechanical and electrical systems help us think about the systems. Since you have studied the block-spring system in detail, you can use your knowledge to help answer questions about an LC series circuit. Engineers who need to study complicated mechanical systems often build the analogous electrical circuit (at a savings in cost) and study it experimentally.

RLC circuits. If a resistor with resistance R is added in series to the LC circuit shown above, the charge and current amplitudes decay exponentially with time. The loop equation becomes and its general solution is

$$q(t) =$$

where the angular frequency of oscillation is now

$$\omega' =$$

Each time the charge on the capacitor reaches a maximum, it is less than the previous time. Does this violate the principle of charge conservation? _____ Explain: _____

_____

If the resistance is small, so it may be neglected in the expression for the angular frequency, then $\omega' =$ _____ .

The sum of the electric and magnetic energies is not constant but decreases exponentially because _____

_____ .

In terms of the current i and resistance R, the rate of change of the energy is given by

$$\frac{dU}{dt} =$$

An ac circuit. Consider the circuit diagramed on the right. The ac generator is symbolized by _____ and the emf it produces is given by $\mathcal{E} = \mathcal{E}_m \sin(\omega_d t)$, where ω_d is its angular frequency. Its frequency is given by $f =$ _____ . $\mathcal{E}$ is taken to be positive if the upper terminal of the generator is positive and negative if the upper terminal is negative. The arrow on the diagram shows the direction of positive current.

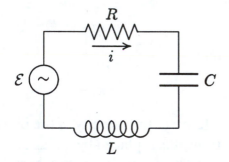

After transients die out, the current is given by $i(t) = I \sin(\omega_d t - \phi)$, where I is the current amplitude and ϕ is a phase constant. Note that the current has the same frequency as the generator emf but because the circuit contains a capacitor and an inductor, it may not be in phase with the generator emf. If ϕ is between 0 and 90°, the current is said to _____ the emf. If it is between 0 and −90° the current _____ the emf.

Quantities with sinusoidal time dependence, such as the generator emf and the current, can be represented by <u>phasors</u>. Tell what a phasor is: _____

_____

The diagram on the right shows the phasor associated with the generator emf at some instant of time. Its length is $\mathcal{E}_m$ and it rotates in the counterclockwise direction. Its projection on the vertical axis is $\mathcal{E}(t) = \mathcal{E}_m \sin(\omega_d t)$. Label the appropriate angle $\omega_d t$. Assume ϕ is about 40° and draw the phasor associated with the current. Label its length I. Label the angle $\omega_d t - \phi$ and the angle ϕ between the two phasors.

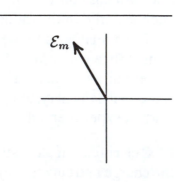

The potential difference v_R across the resistor is given by

$$v_R = iR = IR\sin(\omega_d t - \phi).$$

The potential difference v_R and the current i are in phase. The phasor associated with v_R is along the same line as the phasor associated with _____ and its length is

$$V_R =$$

The potential difference v_L across the inductor is given by

$$v_L = L\frac{di}{dt} = \omega_d LI\cos(\omega_d t - \phi).$$

Use the trigonometric identity $\sin(A + \pi/2) = \cos(A)$ to write this

$$v_L = \omega_d LI\sin(\omega_d t - \phi + \pi/2).$$

v_L is said to *lead* the current by $\pi/2$ radians. Its amplitude is given by

$$V_L =$$

When i is positive and increasing, the _____ end of the inductor in the circuit diagram is at a higher potential than the _____ end.

The phasor associated with v_L is $\pi/2$ radians ahead of (more counterclockwise than) the phasor associated with the current. On the axes to the right, draw the phasors associated with the current and with the potential difference across the inductor at the instant for which the previous phasor diagram was drawn. Label their lengths I and V_L, respectively. Also label the right angle between them.

The potential difference v_C across the capacitor is given by $v_C = q/C$. Now

$$q = \int i\,dt = -\frac{I}{\omega_d}\cos(\omega_d t - \phi),$$

so

$$v_C = -\frac{I}{\omega_d C}\cos(\omega_d t - \phi).$$

Use the trigonometric identity $\sin(A - \pi/2) = -\cos(A)$ to write this

$$v_C = \frac{I}{\omega_d C}\sin(\omega_d t - \phi - \pi/2).$$

v_C is said to *lag* the current by $\pi/2$ radians. Its amplitude is given by

$$V_C =$$

When q is positive, the _____ plate of the capacitor in the circuit diagram is at a higher potential than the _____ plate.

The phasor associated with v_C is $\pi/2$ radians behind (less counterclockwise than) the phasor associated with the current. On the axes to the right, draw the phasors associated with the current and with the potential difference across the capacitor at the instant for which the previous phasor diagrams were drawn. Label their lengths I and V_C, respectively. Also label the right angle between them.

The relationship between the current and potential amplitude is often written

$$V_C = IX_C$$

for a capacitor and

$$V_L = IX_L$$

for an inductor. Here

$$X_C =$$

and

$$X_L =$$

X_C is called the <u>capacitive reactance</u> and X_L is called the <u>inductive reactance</u>. These relations and $V_R = IR$ for a resistor hold no matter what circuit the elements are in. In addition, the potential difference across an inductor always leads the current in the inductor by $\pi/2$ radians and the potential difference across a capacitor always lags the current into the capacitor by $\pi/2$ radians.

Reactances are not resistances but they play a similar role: they relate the amplitude of the current through a circuit element to the amplitude of the potential difference across the element. X_L, X_C, and R all have SI units of _____.

Carefully note that the reactances depend on the generator frequency. As the frequency increases, X_C _____ and X_L _____. If the frequency is somehow increased without changing I, the maximum potential difference across the capacitor and the maximum charge on the capacitor both decrease. Explain why this makes sense physically: _____

_____

If the frequency is increased without changing I, the maximum potential difference across the inductor increases. Explain why this makes sense physically: _____

_____

Now you are ready to put the pieces together and solve for the current amplitude I and phase constant ϕ, given the generator emf. On the axes to the left below draw the phasors for the potential differences across the resistor, capacitor, and inductor at the time for which the previous phasor diagrams were drawn. Label them.

According to the loop equation for the circuit, $\mathcal{E}(t) = v_R(t) + v_L(t) + v_C(t)$. This means the phasors associated with v_R, v_L, and v_C must add like vectors to produce the phasor associated with $\mathcal{E}$. Since the phasors for v_L and v_C are parallel to each other they can be summed easily. Assume $V_L > V_C$ and on the axes to the right below, draw phasors associated with v_R and $v_L + v_C$. The first has length V_R and the second has length $(V_L - V_C)$. Now draw the "vector" sum of the two phasors with its tail at the origin. This must be the emf phasor. Label the angle ϕ between it and the phasor associated with v_R. This is also the angle between the emf and current phasors.

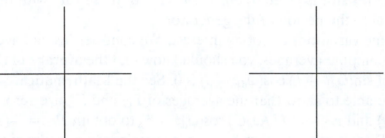

The phasors associated with v_R and $v_L + v_C$ form two sides of a right triangle with a hypotenuse of length $\mathcal{E}_m$, so $\mathcal{E}_m^2 = V_R^2 + (V_L - V_C)^2$. Substitute $V_R = IR$, $V_L = IX_L$, and $V_C = IX_C$, then solve for I. The result can be written

$$I = \frac{\mathcal{E}_m}{Z},$$

where

$$Z =$$

Z is called the _____ of the circuit.

If the impedance of a circuit is increased without changing the generator emf, the current amplitude _____. The impedance can be increased by increasing the resistance or by increasing the difference between the inductive and capacitive reactances. Suppose $X_L > X_C$. Then, as the frequency increases, the impedance _____. If $X_C > X_L$ and the frequency increases, then the impedance initially _____.

According to the phasor diagram, the phase constant for the current is given in terms of the potential differences by

$$\tan\phi =$$

and in terms of the resistance and the reactances by

$$\tan\phi =$$

Notice from your phasor diagram that ϕ must be between $-90°$ and $+90°$. Since R is positive, the value given for ϕ by your calculator is correct. You will never need to add 180°.

Give the relation between X_L and X_C for which each of the following occurs:

the current lags the generator emf $(0 < \phi < 90°)$ _____

the current leads the generator emf $(-90° < \phi < 0)$ _____

the current is in phase with the generator emf $(\phi = 0)$ _____

Note that the angular frequency for which the current is in phase with the generator emf is the resonance angular frequency $\omega = \sqrt{1/LC}$. For a given emf, the current amplitude has its greatest value for this angular frequency.

Power.   The rate with which energy is supplied by the generator is given by $P_\mathcal{E} = i\mathcal{E}$, the rate with which energy is stored in the capacitor is given by $P_C = iv_C$, the rate with which energy is stored in the inductor is given by $P_L = iv_L$, and the rate with which thermal energy is generated in the resistor is given by $P_R = iv_R$. All of these vary with time, being periodic with a period equal to the period of the generator.

Usually the time variations are of no interest. We consider instead averages over a cycle of oscillation. To compute averages, you should know that the average of $\sin^2(\omega_d t - \phi)$ is $1/2$ and the average of $\sin(\omega_d t - \phi)\cos(\omega_d t - \phi)$ is 0. See the Mathematical Skills section.

You should be able to show that the averages of P_C and P_L are zero. For example, use $i = I\sin(\omega_d t - \phi)$ and $v_C = -(I/\omega_d C)\cos(\omega_d t - \phi)$ to obtain $P_C = -(I^2/\omega_d C)\sin(\omega_d t - \phi)\cos(\omega_d t - \phi)$. This averages to zero because it contains the product of the sine and cosine of $\omega_d t - \phi$. In the space below, show that the average of P_L is also zero:

The power supplied by the generator is

$$P_\mathcal{E} = i\mathcal{E} = I\mathcal{E}_m \sin(\omega_d t - \phi)\sin(\omega_d t).$$

The first sine function can be expanded as $\sin(\omega_d t)\cos\phi - \cos(\omega_d t)\sin\phi$. When the first term is multiplied by $\sin(\omega_d t)$ and averaged, the result is $\frac{1}{2}\cos\phi$. When the second term is multiplied by $\sin(\omega_d t)$ and averaged, the result is 0. Thus the average power supplied is

$$P_{\text{av}} = \frac{1}{2}I\mathcal{E}_m \cos\phi.$$

This can also be written

$$P_{\text{av}} = \frac{1}{2}(\mathcal{E}_m^2/Z)\cos\phi$$

and as

$$P_{\text{av}} = \frac{1}{2}I^2 Z \cos\phi,$$

where $\mathcal{E}_m = IZ$ was used. The quantity $\cos\phi$ is called the _____.

In the space below, use $\tan\phi = (X_L - X_C)/R$ and the trigonometric identity $\cos^2 A = 1/(1 + \tan^2 A)$ to show that $\cos\phi = R/Z$:

Thus the average power supplied by the generator is also given by

$$P_{av} = \frac{1}{2}I^2R.$$

The rate with which thermal energy is generated in the resistor is

$$P_R = i^2R = I^2R\sin^2(\omega_d t - \phi)$$

and its average value is

$$P_{av} = \frac{1}{2}I^2R,$$

the same as the power supplied by the generator. Once transients have died out, all energy supplied by the generator is converted to thermal energy in the resistor.

For a given emf, the greatest rate of thermal energy generation occurs when the power factor has the value $\cos\phi =$ _____. For this to occur, the impedance must be $Z =$ _____, which means the inductive and capacitive reactances must be related by $X_L =$ _____. This is the condition for resonance. If L and C are fixed, it occurs if the angular frequency has the value given by $\omega_d =$ _____. Often a circuit is designed to deliver as much power as possible to a resistive load. L and C are then adjusted so $X_L = X_C$ for the frequency of the generator.

Average power is often expressed in terms of root-mean-square quantities instead of amplitudes. The meaning of the term "root-mean-square" (rms) is _____
_____.

Because the average over a cycle of $\sin^2(\omega_d t + \phi)$ is 1/2, the rms value of a sinusoidal function of time is the amplitude divided by $\sqrt{2}$. For example, the rms value of $\mathcal{E}(t) = \mathcal{E}_m\sin(\omega_d t)$ is $\mathcal{E}_{rms} =$ _____. In terms of $\mathcal{E}_{rms}$,

$$P_{av} =$$

and in terms of i_{rms},

$$P_{av} =$$

Resonance The angular frequency of the current and of the potential difference across any of the circuit elements is the same as the angular frequency of the generator but the current and potential difference amplitudes depend on the difference between the angular frequency of the generator and the natural angular frequency of the circuit ($1/\sqrt{LC}$). See Fig. 33–13. As the generator frequency approaches the natural frequency from either side, the current amplitude _____ and reaches its maximum value when $\omega_d =$ _____. This is the condition for _____. At resonance, the _____ reactance and the _____ reactance are equal. The resonance peak can be made sharper by reducing the _____ in the circuit.

You should realize that at resonance, thermal energy is being generated in the resistor but it is being supplied by _____ at the same rate.

The transformer. Power lines transport electrical energy at low current, with a high potential difference across the ends. A low current is advantageous because it reduces _____. On the other hand, a low potential difference is advantageous at both the generator and the consumer ends. A transformer is used to change the potential difference.

A transformer consists of two coils wrapped around the same iron core. See Fig. 33–15 of the text. Changing magnetic flux produced by changing currents in the coils induces emfs in the coils. Consider an ideal transformer (negligible resistance and hysteresis loss) with N_p turns in the primary coil and N_s turns in the secondary coil. The potential difference V_p across the primary coil is related to the potential difference V_s across the secondary coil by

$$\frac{V_p}{N_p} =$$

because, according to Faraday's law, the emf generated in each turn of primary coil is the same as the emf generated in each turn of secondary coil.

If $N_s > N_p$, then the potential differences are related by _____ > _____ and the transformer is called a _____ transformer. If $N_s < N_p$, then the potential differences are related by _____ > _____ and the transformer is called a _____ transformer.

Since there are no losses in an ideal transformer, the rate with which energy enters the primary coil equals the rate with which energy leaves the secondary coil and

$$I_p V_p =$$

These are rms values.

If the secondary circuit has resistance R and no inductance, the current in the secondary is given by $I_s = V_s/R$. In terms of the potential difference across the primary coil, the current in the secondary circuit is

$$I_s =$$

and the current in the primary circuit is

$$I_p =$$

The current in the primary circuit does not change if the transformer and the secondary circuit are replaced by a resistor with resistance $R_{eq} =$ _____. Transformers are often used to match the impedances of generators (or amplifiers) and their loads. That is, they are used to make R_{eq} the same as the generator resistance. Maximum energy transfer occurs if this condition is met.

II. PROBLEM SOLVING

You should know the sequence of events in an LC circuit oscillation:

1. capacitor maximally charged (plate A positive, say), current zero
2. capacitor uncharged, maximum current in one direction
3. capacitor maximally charged (plate B positive), current zero
4. capacitor uncharged, maximum current in the other direction

The cycle then repeats. These events are one-quarter cycle apart ($T/4$, where T is the period of the oscillation). You may need to use $T = 2\pi/\omega_d$, where ω_d ($= 1/\sqrt{LC}$) is the angular frequency of the oscillation. Also remember that the electrical energy stored in the capacitor is a maximum when the charge on the capacitor is a maximum and the magnetic energy stored in the inductor is a maximum when the current is a maximum.

Some problems deal with the analogy between an LC electrical circuit and a block-spring mechanical system. Remember that $q \leftrightarrow x$, $L \leftrightarrow m$, $C \leftrightarrow 1/k$, and $i \leftrightarrow v$.

You should know the relationship between the maximum charge on the capacitor and the maximum current: $I = \omega_d Q$. You should also know that the total energy of the circuit can be written in terms of either of these: $U = Q^2/2C = LI^2/2$. Some problems ask you to compute the maximum charge on the capacitor, given the maximum potential difference across the plates: $Q = CV_{max}$.

Some problems deal with damped circuits. You should know that the frequency of oscillation changes with the addition of resistance and that in succeeding cycles, the maximum charge on the capacitor decreases exponentially, the relevant factor being $e^{-Rt/2L}$.

You should know how to calculate the reactances and the impedance for an RLC series circuit: $X_L = \omega_d L$, $X_C = 1/\omega_d C$, and $Z = \sqrt{R^2 + (X_L - X_C)^2}$, where ω_d is the angular frequency. Some problems give the frequency f_d and you must use $\omega_d = 2\pi f_d$.

You should know how to compute the current given the generator emf. If $\mathcal{E} = \mathcal{E}_m \sin \omega_d t$, then $i = I \sin(\omega_d t - \phi)$, where $I = \mathcal{E}_m/Z$ and $\tan \phi = (X_L - X_C)/R$. Think of ϕ as a phase *difference*. That is, if $\mathcal{E} = \mathcal{E}_m \sin(\omega_d t + \alpha)$, then $i = I \sin(\omega_d t + \alpha - \phi)$.

Given the current, you should know how to compute the potential differences across the individual circuit elements: $v_R = iR$, $v_C = IX_C \sin(\omega_d t - \pi/2)$, and $v_L = IX_L \sin(\omega_d t + \pi/2)$.

Finally, you should know that the generator supplies energy at the rate $P_\mathcal{E} = i\mathcal{E}$, the inductor stores energy at the rate $P_L = iv_L$, the capacitor stores energy at the rate $P_C = iv_C$, and the resistor generates thermal energy at the rate $P_R = i^2 R$. These are all time dependent quantities.

Sometimes the rms values of current and potential differences are given or requested. You should remember that if $i = I \sin(\omega_d t - \phi)$, then $i_{rms} = I/\sqrt{2}$. Similar expressions hold for the potential differences. You should also recognize that equations such as $v_L = IX_L \sin(\omega_d t + \pi/2)$ lead to $V_{L\,rms} = i_{rms} X_L$.

III. MATHEMATICAL SKILLS

You should be able to prove that the average over a cycle of $\sin^2(\omega_d t)$ is $1/2$ and the average over a cycle of $\sin(\omega_d t) \cos(\omega_d t)$ is 0. First, consider the average of $\sin^2(\omega_d t)$, given by the integral

$$I = \frac{1}{T} \int_0^T \sin^2(\omega_d t)\, dt\,,$$

where T is the period of oscillation. Use the trigonometric identity $\cos(2\omega_d t) = \cos^2(\omega_d t) - \sin^2(\omega_d t) = 1 - 2\sin^2(\omega_d t)$, where $\cos^2(\omega_d t) + \sin^2(\omega_d t) = 1$ was used. Thus $\sin^2(\omega_d t) =$

$\frac{1}{2}[1 - \cos(2\omega_d t)]$ and

$$I = \frac{1}{2T} \int_0^T [1 - \cos(2\omega_d t)] \, dt = \frac{1}{2} - \frac{1}{4\omega_d T} \sin(2\omega_d T).$$

Now $2\omega_d T = 4\pi f T = 4\pi$ and $\sin(2\omega_d T) = \sin(4\pi) = 0$, so $I = 1/2$.

The average of $\sin(\omega_d t) \cos(\omega_d t)$ is somewhat easier to find. You must evaluate the integral

$$I = \frac{1}{T} \int_0^T \sin(\omega_d t) \cos(\omega_d t) \, dt$$

and this is

$$I = \frac{1}{2\omega_d T} \sin^2(\omega_d t) \Big|_0^T = 0.$$

The last result follows from $\sin(\omega_d T) = \sin(4\pi) = 0$.

IV. NOTES

Chapter 34
ELECTROMAGNETIC WAVES

I. BASIC CONCEPTS

Maxwell's equations predict the possibility of electric and magnetic fields that propagate in free space, with the changing magnetic field producing changes in the electric field, via Faraday's law, and the changing electric field producing changes in the magnetic field, via the displacement current term in the Ampere-Maxwell law. Pay close attention to the relationship between the amplitudes of the fields, to the relationship between their phases, and to the relationship between their directions. One important consequence of the equations is that they produce an expression for the wave speed, the speed of light, in terms of ϵ_0 and μ_0.

You will also learn what happens when light is incident on the boundary between two materials and learn to determine the propagation directions of the reflected and refracted light. These are given by the laws of reflection and refraction. First, you should understand what geometrical optics is and when it is valid.

The following are the most important topics that have been discussed in previous chapters and are used in this chapter: pressure in Chapter 15, angular frequency in Chapter 16, angular wave number and wave speed in Chapter 17, charge in Chapter 22, electric field in Chapter 23, electric dipole in Chapters 23 and 25, electrical energy in Chapter 26, current in Chapter 27, magnetic field in Chapter 29, magnetic flux, Faraday's law, and magnetic energy in Chapter 31, and displacement current in Chapter 32.

The electromagnetic spectrum. Electromagnetic waves exist for all wavelengths and frequencies, from very large to very small. Various ranges of wavelengths have been named and you should be familiar with the ranges and their names.

1. <u>Gamma</u> radiation extends from the very shortest wavelengths (highest frequencies) to a wavelength of about _____ m (a frequency of about _____ Hz).

2. <u>X-ray</u> radiation extends from a wavelength of about _____ m (a frequency of about _____ Hz) to a wavelength of about _____ m (a frequency of about _____ Hz).

3. <u>Ultraviolet</u> radiation extends from a wavelength of about _____ m (a frequency of about _____ Hz) to a wavelength of about _____ m (a frequency of about _____ Hz).

4. <u>Visible</u> radiation extends from a wavelength of about _____ m (a frequency of about _____ Hz) to a wavelength of about _____ m (a frequency of about _____ Hz).

5. <u>Infrared</u> radiation extends from a wavelength of about _____ m (a frequency of about _____ Hz) to a wavelength of about _____ m (a frequency of about _____ Hz).

6. <u>Microwave</u> radiation extends from a wavelength of about _____ m (a frequency of about _____ Hz) to a wavelength of about _____ m (a frequency of about _____ Hz).

7. Above this in wavelength lies the TV and radio portion of the spectrum, which extends from a wavelength of about _____ m (a frequency of about _____ Hz) to the longest wavelengths (lowest frequencies).

All these electromagnetic radiations are exactly alike except for their frequencies and wavelengths. They are all electromagnetic. That is, they all consist of traveling electric and magnetic fields. They all travel in free space with the same speed, the speed of light $c =$ _____ m/s.

Electromagnetic waves   Classically, electromagnetic waves are produced by charges that are accelerating. Charges at rest or moving with constant velocity do not radiate. In the quantum mechanical picture, electromagnetic radiation is produced when a charge (perhaps an electron in an atom or an atomic nucleus) changes its state or in reactions of fundamental particles.

Fig. 34–5 shows the electric and magnetic fields of a plane traveling electromagnetic wave. Look at this figure and identify the important characteristics of electromagnetic radiation (the x axis is positive to the right and the z axis is positive out of the page):

1. The electric and magnetic fields are perpendicular to each other.

2. The wave travels in a direction that is perpendicular to both the electric and magnetic fields.

3. The electric and magnetic fields are in phase. Explain what this means: _____
   _____

4. The wave is linearly polarized. Explain what this means: _____
   _____

Very far from the source, the wave becomes a plane wave: the electric and magnetic field lines are very close to straight lines at the detector. Suppose the detector is on the x axis so that waves reaching it are traveling in the positive x direction. The diagram shows the fields in the wave near the detector. The electric field is parallel to the y axis and the magnetic field is parallel to the z axis. Mathematically, the fields are given by

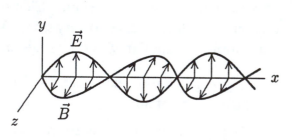

$$E(x,t) =$$

$$B(x,t) =$$

where $k = 2\pi/\lambda$, λ is the wavelength, and ω is the angular frequency. These quantities are related by $\omega/k =$ _____, where c is the speed of the wave. The phase constants are the same for the two fields.

The electric field is different at a point with coordinate x and a point with coordinate $x + dx$, an infinitesimal distance away, because a changing magnetic flux penetrates the region between these points. The magnetic field is different at x and $x + dx$ because a changing electric flux penetrates the region between these points. Differences in the fields at different points can be computed using the Faraday and Ampere-Maxwell laws.

When these laws are applied, you find that the amplitudes of the fields are related by

$$\frac{E_m}{B_m} =$$

and that the wave speed is determined by the constants ϵ_0 and μ_0:

$$c =$$

You should understand how the laws of electromagnetism lead to these relationships. The diagram on the right shows a small region of space in which a wave is traveling toward the right. The electric fields at two infinitesimally separated points, x and $x + \Delta x$, are shown, as is the magnetic field in the region between. Apply Faraday's law to this situation. To calculate the magnetic flux, take $d\vec{A}$ to be out of the page and to calculate the emf, traverse the path shown in the counterclockwise direction. Clearly, $\Phi_B = Bh\Delta x$ and $\oint \vec{E} \cdot d\vec{s} = h[E(x + \Delta x) - E(x)]$. Why does $E(x + \Delta x)$ enter with a plus sign and $E(x)$ enter with a minus sign? _____

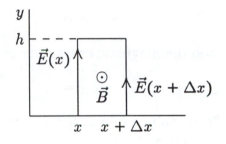

As Δx becomes small, $[E(x + \Delta x) - E(x)]/\Delta x$ becomes $\partial E/\partial x$ and Faraday's law yields

$$\frac{\partial E}{\partial x} =$$

The diagram on the right shows the view looking up from underneath the previous diagram. The magnetic field at the two points and the electric field between are shown. Apply the Ampere-Maxwell law to this situation. Take $d\vec{A}$ to be into the page and traverse the path in the clockwise direction. Clearly, $\Phi_E = Eh\Delta x$ and $\oint \vec{B} \cdot d\vec{s} = h[B(x) - B(x + \Delta x)]$. Why does $B(x)$ enter with a plus sign and $B(x + \Delta x)$ enter with a minus sign?

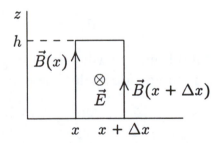

_____

_____

As Δx becomes small, $[B(x) - B(x + \Delta x)]/\Delta x$ becomes $-\partial B/\partial x$ and the Ampere-Maxwell law yields

$$\frac{\partial B}{\partial x} =$$

Substitute $E = E_m \sin(kx - \omega t)$ and $B = B_m \sin(kx - \omega t)$ into these two equations and show that $kE_m = \omega B_m$ and that $kB_m = \mu_0\epsilon_0\omega E_m$:

Now use $\omega/k = c$ to show that $E_m = cB_m$ and $c^2 = 1/\mu_0\epsilon_0$:

Energy transport. Electromagnetic waves carry energy. The electrical energy per unit volume associated with an electric field $\vec{E}$ is given by

$$u_E =$$

and the magnetic energy per unit volume associated with a magnetic field $\vec{B}$ is given by

$$u_B =$$

You can use $B = E/c$ and $c = 1/\sqrt{\epsilon_0\mu_0}$ to show that for a plane wave,

$$u_E = u_B = \frac{EB}{2\mu_0 c}$$

and that the total energy density is

$$u = \frac{EB}{\mu_0 c}.$$

This energy moves with the wave, with speed c. Consider a region of space with infinitesimal width dx and cross-sectional area A, perpendicular to the direction of travel of a plane electromagnetic wave. The volume of the region is $A\,dx$ so, in terms of the energy density u, the electromagnetic energy in the region is

$$dU =$$

All this energy will pass through the area A in time dt, given by $dt = dx/c$. Substitute $dx = c\,dt$ and divide by dt. The energy passing through the area per unit time is

$$\frac{dU}{dt} =$$

and the energy per unit area passing through per unit time is

$$\frac{dU}{A\,dt} =$$

In terms of the field magnitudes E and B, this is

$$\frac{dU}{A\,dt} =$$

The transport of energy is described in terms of the <u>Poynting vector</u> $\vec{S}$, defined by the vector product

$$\vec{S} = \underline{}$$

Since $\vec{E}$ and $\vec{B}$ are perpendicular to each other, the magnitude of the Poynting vector is given by $S = \underline{}$. In terms of the magnitude S of the Poynting vector, the energy passing through a surface of area A, perpendicular to the direction of travel, per unit time, is given by $dU/dt = \underline{}$ and the energy passing through per unit area per unit time is given by $dU/A\,dt = \underline{}$.

For most sinusoidal electromagnetic waves of interest, the fields oscillate so rapidly that their instantaneous values cannot be detected or else are not of interest. Energies and energy densities are then characterized by their average over a period of oscillation. The average value of E^2, for example, is $\frac{1}{2}E_m^2$, where E_m is the amplitude.

The <u>intensity</u> of a wave is the magnitude of the Poynting vector averaged over a period of oscillation. In terms of the electric field amplitude E_m, it is

$$I = \underline{}$$

and in terms of the magnetic field amplitude B_m, it is

$$I = \underline{}$$

For sinusoidal waves, the intensity is often written in term of the rms value E_{rms} of the electric field:

$$I = \underline{}$$

The direction of $\vec{S}$ is the same as the direction in which the wave is traveling. Recall that this direction is intimately connected with the relative signs of the terms kx and ωt in the argument of the trigonometric function that describes the fields. If the wave travels in the negative x direction, you write $E_y(x,t) = E_m \sin(kx + \omega t)$ and $B_z(x,t) = \underline{}$. Be sure you have selected the appropriate signs so $B_z(x,t)$ represents a wave traveling in the negative x direction and so that $\vec{E} \times \vec{B}$ is also in the negative x direction. Carefully note that $B_z = (E_m/c)\sin(kx + \omega t)$ is *wrong* since this means $\vec{E} \times \vec{B}$ is in the positive x direction.

Electromagnetic waves also carry momentum. If ΔU is the energy in any infinitesimal portion of a wave, then

$$\Delta p = \underline{}$$

is the magnitude of the momentum in that portion. The momentum is in the direction of $\vec{S}$, the direction of propagation. If u is the energy per unit volume, then $\underline{}$ is the momentum per unit volume and if I is the energy transported through an area per unit area per unit time, then $\underline{}$ is the momentum transported through the area per unit area per unit time.

When an electromagnetic wave interacts with a charge, the electric field component of the wave does work on the charge and energy is transferred from the wave to the charge. Since the fields exert forces on the charge, momentum is also transferred. When electromagnetic radiation is absorbed by a material object, its energy and momentum are transferred to the object. Consider a plane wave with time averaged energy density u, incident normally on the plane surface of an object. If the area of the surface is A and the wave is completely absorbed, then averaged over a period the energy transferred in time Δt is given by _____ and the momentum transferred is given by _____. The momentum transferred per unit time is the force of the radiation on the object and the force per unit area is the radiation pressure. In terms of u, the force is given by _____ and the radiation pressure is given by _____.

If the wave is reflected without loss back along the original path, the energy transferred is _____ and the momentum transferred is _____. The radiation pressure is now _____, twice what it would be if the radiation were absorbed. Why? _____

Polarization. The electric field of a <u>linearly</u> <u>polarized</u> electromagnetic wave is always parallel to the same _____. The <u>direction of polarization</u> is parallel to the _____ field. For maximum signal, the wires in an electric dipole antenna used to detect polarized waves must be aligned with the _____ direction.

You should be able to contrast polarized and unpolarized electromagnetic radiation. Describe the orientation of the electric field of unpolarized radiation: _____

Light from an incandescent bulb or from the Sun is not polarized. The direction of its electric field changes rapidly in a random fashion. Explain why this comes about: _____

You should recognize that any linearly polarized wave can be treated as the sum of two waves, polarized in any two mutually orthogonal directions that are perpendicular to the direction of propagation. Suppose a wave with an electric field amplitude E_m is traveling out of the page and is polarized with its electric field along a line that makes an angle θ with the x axis, as shown. You can consider the electric field to be the vector sum of two fields, one with amplitude _____, polarized along the x axis and one with amplitude _____, polarized along the y axis.

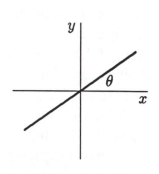

Polarized radiation can be produced by shining unpolarized radiation through a sheet of Polaroid. These sheets contain certain long-chain molecules that are aligned in the manufacturing process. Radiation with its electric field vector parallel to the chains is preferentially _____ while radiation with its electric field perpendicular to the chains is preferentially _____. A line perpendicular to the molecules is said to be along the <u>polarizing direction</u> of the sheet.

Suppose that linearly polarized radiation with amplitude E_m and intensity I_0 is incident on an ideal polarizing sheet and that its electric field is along a line that makes an angle θ with the polarizing direction. Then, the amplitude of the transmitted radiation is given by $E =$ ___ and its intensity is given by

$$I =$$

This is called the law of Malus. Note that the transmitted amplitude is the component of the incident amplitude along the polarization direction of the sheet. You simply resolved the incident electric field into components parallel and perpendicular to the polarization direction. The parallel component is transmitted while the perpendicular component is absorbed.

If unpolarized radiation with intensity I_0 is incident on an ideal polarizing sheet, the intensity of the transmitted radiation is given by

$$I =$$

Be sure you understand the conditions for which each of these expressions for the transmitted intensity applies. In both cases the transmitted radiation is linearly polarized with its electric field parallel to _____ .

Geometrical and wave optics.   In physical optics (or wave optics), the wave nature of light is used to describe and understand optical phenomena. Wave optics is close to the fundamental principles, Maxwell's equations, and is valid for all optical phenomena outside the quantum realm. However, when any obstacles to light or any openings through which light passes are large compared to the wavelength of the light, then details of its wave nature are not important. What is important is the direction of travel of the light. This is the realm of geometrical optics (or ray optics).

Geometrical optics uses rays to describe the path of light. A ray is a line in the direction of _____ of a light wave. The propagation direction changes when light is reflected or enters a region where the wave speed is different.

The laws of reflection and refraction.   When light in one medium encounters a boundary with another region in which the wave speed is different, some light is reflected back into the region of incidence and some is transmitted into the second region. Two laws, the law of reflection and the law of refraction, describe the directions of propagation of the reflected and transmitted light.

The diagram shows a ray in medium 1 incident on the boundary with medium 2. It makes the angle θ_1 with the normal to the boundary. Reflected and refracted rays all lie in the plane of incidence, defined by the incident ray and the normal to the surface. The reflected ray makes the angle θ_1' with the normal and the transmitted ray makes the angle θ_2 with the normal. The law of reflection is

$$\theta_1' =$$

and the law of refraction is

$$n_2 \sin \theta_2 =$$

where n_1 is the index of refraction for medium 1 and n_2 is the index of refraction for medium 2. The angle θ_1 is called the angle of _____, θ_1' is called the angle of _____, and θ_2 is called the angle of _____. The law of refraction is also known as Snell's law.

The index of refraction of a medium is the ratio of the speed of light in a vacuum to the speed of light in the medium. Specifically, if v is the speed of light in the medium, then

$$n =$$

Look at Table 34–1 of the text for some values for yellow sodium light. Of the materials listed, the one with the largest index of refraction is _____, with $n =$ _____. The speed of light is _____ in this material than in any other listed. The index of refraction depends on the wavelength; light of different wavelengths, incident at the same angle, is refracted through different angles. See Fig. 34–19 of the text.

You should understand Snell's law: $n_1 \sin \theta_1 = n_2 \sin \theta_2$. Recall that $\sin \theta$ increases as θ increases from 0 to 90°. If n_1 is greater than n_2, then θ_1 is less than θ_2. Light incident from a medium with a small index of refraction bends toward the normal as it enters a medium with a higher index of refraction. Light incident from a medium with a large index of refraction bends away from the normal as it enters a medium with a smaller index of refraction. When you shine light into water from air, it bends _____ the normal. Light from an underwater source bends _____ the normal as it enters the air.

<u>Total</u> <u>internal</u> <u>reflection</u> can occur when light travels in an optically dense medium (large index of refraction) toward a less optically dense medium (smaller index of refraction). If the angle of incidence is greater than a certain critical value denoted by θ_c, no light is transmitted into the less optically dense medium. All the light incident on the boundary is reflected. Suppose $n_1 > n_2$ and light is incident from medium 1 on the boundary with medium 2. Then, $\theta_1 = \theta_c$ if $\theta_2 =$ _____ and $\sin \theta_2 =$ _____. In terms of n_1 and n_2,

$$\theta_c =$$

Polarization by reflection. If unpolarized light is incident on a boundary between two different materials, both the reflected and refracted waves are partially polarized for most directions of incidence. There is, however, a special angle of incidence, called Brewster's angle and denoted by θ_B, for which the reflected wave contains only one polarization component. If θ_r is the angle of refraction for incidence at Brewster's angle, then $\theta_B + \theta_r =$ _____. If n_1 is the index of refraction for the medium of incidence and n_2 is the index of refraction for the medium of refraction, then Snell's law leads to

$$\tan \theta_B =$$

This is Brewster's law for the polarizing angle.

The direction of polarization for light reflected at the polarizing angle is normal to the plane determined by the _____ and the _____. Carefully note that even at Brewster's angle, the *refracted* light is partially, not completely, polarized. It contains light with all polarization directions but is deficient in light with the polarization of the reflected light.

II. PROBLEM SOLVING

Many problems deal with the relationships between various characteristics of plane electromagnetic waves. You should know that the angular wave number k and the wavelength λ are related by $k = 2\pi/\lambda$, that the angular frequency ω and the frequency f are related by $\omega = 2\pi f$, and that the wavelength and frequency are related by $\lambda f = c$, where c is the speed of light. This last relationship leads to $\omega/k = c$.

You should also know that the electric and magnetic fields are in phase, that their amplitudes are related by $B_m = cE_m$, that $\vec{E}$ and $\vec{B}$ are perpendicular to each other, and that $\vec{E} \times \vec{B}$ is in the direction of propagation. The last statement implies that both $\vec{E}$ and $\vec{B}$ are perpendicular to the direction of propagation. You should also remember that the direction of propagation determines the argument of the trigonometric function in the expressions for $\vec{E}$ and $\vec{B}$. That is, for example, $E_m \sin(kx - \omega t)$ is propagating in the positive x direction and $E_m \sin(kx + \omega t)$ is propagating in the negative x direction.

You should know that the magnitude of the Poynting vector, given for a plane wave by $S = E^2/c\mu_0$, is the energy per unit area per unit time that crosses an area that is perpendicular to the direction of propagation. Its average over a cycle is called the intensity and is given by $I = E_m^2/2c\mu_0$ or by $I = E_{rms}^2/c\mu_0$. For a point source that radiates uniformly in all directions, the intensity at a point is proportional to the inverse square of the distance from the source to the point.

You should also know that when an object absorbs electromagnetic radiation, it receives momentum: $\Delta p = \Delta U/c$, where ΔU is the energy absorbed. When the radiation is reflected back along the path of incidence, the momentum received is $\Delta p = 2\Delta U/c$, where ΔU is the reflected energy. For total absorption, the radiation pressure is $p_r = I/c$ and the force on the object is $F = IA/c$, where A is the area struck by radiation. If the radiation is reflected back on its path of incidence, the radiation pressure and force are both twice as great.

Some problems deal with polarization. You should know how to identify the direction and plane of polarization. You should also know how to compute the intensity of radiation transmitted by a polarizing sheet. Remember that the exiting radiation is polarized in the polarizing direction of the sheet.

Many problems deal with refraction at a single plane surface and with total internal reflection. For refraction problems, use Snell's law: $n_1 \sin\theta_1 = n_2 \sin\theta_2$. Remember to measure the angle from the surface normal. For total internal reflection, remember that $n_1 \sin\theta > n_2$, where n_1 is the index of refraction for the medium of incidence and n_2 is the index of refraction for the medium beyond the surface. Other problems ask you to calculate the Brewster angle. Use $\tan\theta_B = n_2/n_1$, where n_1 is the index of refraction for the medium of incidence and n_2 is the index of refraction for the medium of the refracted light. In some cases, you will also need to use geometry to trace rays.

III. NOTES

Chapter 35
IMAGES

I. BASIC CONCEPTS

The law of reflection is applied to spherical mirrors and the law of refraction is applied to spherical refracting surfaces. Each type surface forms images of objects placed in front of it and you will learn the relationship between the position of the object, the position of the image, and the radius of curvature of the surface. You will also apply what you learn to lenses with spherical faces and to systems of lenses, such as those used in telescopes, microscopes, and cameras. Pay careful attention to ray tracing techniques. They will help you to visualize image formation and to understand the position and size of the image. When using the equations relating object and image distances, be sure to concentrate on the signs of the variables. Knowing how to choose the signs of given quantities and how to interpret the signs of results is vital for obtaining correct answers.

Some of the topics you should understand in connection with this chapter are: angular frequency from Chapter 16, wave speed and angular wave number from Chapters 17 and 34, electromagnetic waves, polarization, and intensity of electromagnetic waves from Chapter 34.

Images in mirrors. When you view light from an object after it has been reflected by a plane mirror, the light appears to come from points behind the mirror and, in fact, you see an image that appears to be behind the mirror.

The diagram shows a point source P in front of a mirror and two rays emanating from it. The reflected rays are drawn according to the law of reflection: the angles of incidence and reflection are the same. Draw dotted lines to extend the reflected rays to behind the mirror and use P' to label the point where they intersect. This is the image of the source. When your eyes view the light reflected from the mirror, it appears to come from a point source at P'.

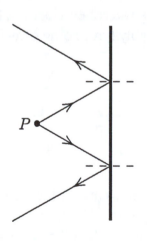

You should realize that many other rays from the source to the mirror could have been drawn. All reflected rays, extended backward to the other side of the mirror, intersect at P'. When you look at the mirror, only a very narrow bundle of light enters your eyes, but all rays in the bundle diverge from the image.

The image point P' is on the line through the source P that is _____ to the plane of the mirror. It is the same distance behind the mirror as _____ is in front. If you want to know if you can see the image when your eye is in any given position, draw a line from the image to your eye. If the line intersects the mirror, then light reflected from the mirror gets to your eye and you see the image. If the line does not intersect the mirror, you do not see the image.

If the source is extended, you may think of each point on it as a point source, emitting light in all directions. Only the light that reaches the mirror is reflected, of course. For each point that sends light to the mirror, an image point is formed. In many examples, an extended source is made up of one or more straight lines. To find the image of a line, simply find the images of each end and connect them with a straight line. Clearly the image of an extended source is the same size as the source.

You should be familiar with some of the terms used. A source of light is often called an object. An image may be real or virtual. In both cases, light diverges from the image but if the image is real, the light actually _____ , while if it is virtual, it does not. Images formed by plane mirrors are _____ .

The distance from a point source to a mirror is denoted by p. It is positive. The image position is denoted by i. It is negative (for virtual images) and its magnitude is the distance from the image to the mirror. In terms of these quantities, the statement that the image is as far behind the mirror as the object is in front is written

$$i =$$

Spherical mirrors. The diagrams below show two spherical mirrors, with their centers labelled c and their centers of curvature labelled C. A point source of light S is in front of each mirror. The line through the center of curvature and the center of the mirror is called the _____ . Usually angles are measured with respect to this line and distances are measured along it. Light from the source is reflected by the mirror.

On each diagram, draw a ray from the source to the mirror, about halfway to its upper edge. Use a dotted line to draw the normal to the mirror at the point where the ray strikes, then draw the reflected ray. The rays should obey the law of reflection at the mirror, but you will probably have some drawing error.

In some cases, reflected rays converge to a point on the source side of the mirror. Reflected light then forms a _____ image. In other cases, the reflected rays diverge as they leave the mirror but they follow lines that would pass through a single point behind the mirror if they were extended to that region. That is, they form a _____ image. Strictly speaking, sharp images are formed only by light whose rays make small angles with the central axis.

The object distance p is the distance from the source to the mirror, measured along the central axis. It is positive for a source in front of a single mirror. Indicate p on each of the diagrams above. The image distance i is the distance from the mirror to the image position, measured along the central axis. A sign is associated with it. It is positive for a _____ image, located _____ the mirror and negative for a _____ image, located _____ the mirror.

The mirror equation relates the object distance p, the image distance i, and the radius of curvature r. It is

$$\frac{1}{p} + \frac{1}{i} =$$

This is often written in terms of the <u>focal length</u>, defined by $f = $ _____. Then,

$$\frac{1}{p} + \frac{1}{i} =$$

Signs are associated with the radius of curvature and focal length as well as with object and image distances. A mirror that is <u>concave</u> with respect to the source has a _____ radius of curvature and focal length; a mirror that is <u>convex</u> with respect to the source has a _____ radius of curvature and focal length. On each of the preceding diagrams, indicate whether the mirror is concave or convex with respect to S and give the sign of r and f.

The <u>focal point</u> of a mirror is important for tracing rays and graphically finding the position of an image when the object is not on the central axis. It is on the central axis, a distance $|f|$ from the mirror. It is in front of a _____ mirror and behind a _____ mirror. For each of the mirrors, place a dot at the focal point and label it F.

Incident light with rays that are parallel to the central axis is reflected so its rays are along lines that pass through _____. If the focal point is in front of the mirror, the rays actually pass through it. If the focal point is behind the mirror, they do not but their extensions do. Incident light with rays along lines that pass through the focal point is reflected so its rays are along lines that are _____.

The image is located at the intersection of all small-angle reflected rays. The two special rays mentioned above are easy to locate and draw, so they are often used to locate an image. Another that is used for the same reason is a ray along a line through the center of curvature. Its reflection is along a line that passes through _____.

The diagram shows a spherical mirror with its center of curvature at C. Locate the focal point and mark it F. Locate the image of the arrow at O by drawing two rays both before and after reflection. The first is from the head of the arrow to the mirror and is parallel to the axis. The reflected ray goes through _____. The second is from the head of the arrow through C to the mirror. The reflected ray goes through _____. Draw the image of the arrow with its head at the point where these rays intersect and its tail on the axis.

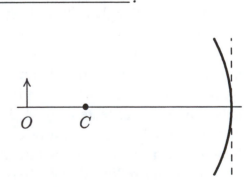

You might also draw a ray from the arrow head through the focal point to the mirror. The reflected ray is parallel to the axis and passes through the intersection of the rays you have already drawn. All of the rays you draw may not intersect at precisely the same point and none of the intersections may agree exactly with the image position as calculated by the mirror equation. Discrepancies arise because the rays you have drawn do not make sufficiently small angles with the central axis. You can obtain a single intersection point and agreement with the mirror equation if you draw the rays from the object to the dotted line shown, rather than to the actual mirror surface.

Note that the image is real. Light actually passes through points of the image. It is also inverted. If you move the object arrow to the left, away from the mirror, the image moves _____ the mirror. To verify this, imagine the object after it is moved to the left and note what happened to the ray through the center of curvature and its intersection with the ray that is parallel to the axis. The latter does not change as the object moves.

Now suppose the object is placed between the focal point and the mirror, as shown. Draw a line from the center of curvature through the head of the arrow to the mirror. The portion between the arrow head and the mirror is a ray. It is reflected through _____. Draw a ray from the arrow head to the mirror, parallel to the axis. It is reflected through _____. Use dotted lines to extend the reflected rays backward into the region behind the mirror, then locate the image of the arrow head at their intersection and draw the image of the arrow.

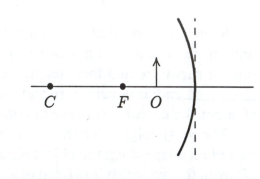

You might also use a line from the focal point through the arrow head to the mirror. The portion between the head and the mirror is a ray and its reflection is parallel to the axis. Its extension passes through the intersection of the dotted lines you have drawn.

Notice that the image is virtual and erect. It is behind the mirror and no light reaches it but reflected light in front of the mirror appears to come from it. If the object arrow is moved closer to the mirror, its image moves _____ the mirror.

The results you found graphically are predicted by the mirror equation. When it is solved for i the result is $i = fp/(p - f)$. If $p > f$, then i is positive: the image is real and is in front of the mirror. If $p < f$, then i is negative: the image is virtual and is behind the mirror. If $p = f$ all small-angle reflected rays are parallel to each other and the image is said to be at infinity.

Now consider a convex mirror. Locate the focal point on the diagram and label it F. Draw a ray from the head of the arrow to the mirror, parallel to the axis. Draw the reflected ray. It is along a line that passes through _____. Draw a ray from the arrow head to the mirror, along the line that passes through C. Draw the reflected ray. It is along a line that passes through _____. Use dotted lines to extend the reflected rays to behind the mirror and draw the image of the arrow at their intersection. It is virtual and erect.

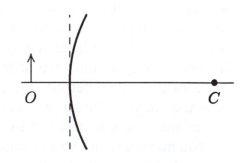

You might also use a line from the head of the arrow through the focal point. The reflected ray is parallel to the axis and its extension behind the mirror should pass through the intersection of the other lines.

The mirror equation also predicts a virtual image. Since f is negative, $i = fp/(p - f)$ is negative for any positive value of p.

The lateral magnification m of a mirror is the ratio of the lateral size of the _____ to the lateral size of the _____. The term *lateral size* means the dimension perpendicular to the central axis. In terms of the object distance p and image distance i, the lateral magnification of a spherical mirror is given by

$$m =$$

The negative sign here has special meaning. Values for p and i are substituted with their signs. If m is positive, then the orientations of the object and image are _____; if m is negative, then the orientations of the object and image are _____. Notice that virtual images of an erect object are erect and real images of an erect object are inverted.

Spherical refracting surfaces. The diagrams below show two spherical refracting surfaces, with their centers of curvature labelled C. Each separates a medium with an index of refraction n_1 from a medium with index of refraction n_2. Light from point source S in front of a surface is refracted at the surface and continues into the medium on the other side. Suppose n_2 is greater than n_1, so the rays are bent toward the normal, and for each diagram draw a ray from the source that strikes the surface about halfway up. Use a dotted line to draw the normal to the surface at the point where the ray strikes and show the ray after refraction.

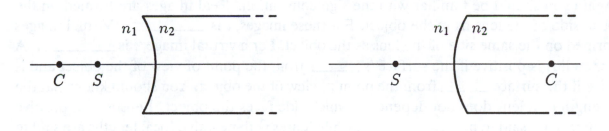

If the rays make small angles with the central axis, the light forms a sharp image of the source. If the image is virtual, it is formed in the region containing the source; if the image is real, it is formed on the side into which light is transmitted.

For a single refracting surface, the object distance p is positive. Indicate it on each diagram above. The image distance is positive if the image is real and negative if it is virtual. For the surfaces and sources of the diagrams above, images with positive values of i are to the _____ of the surface and images with negative values of i are to the _____ .

The law of refraction yields a relationship between the object distance, the image distance, the radius of curvature of the surface, and the indices of refraction for the two sides. It is

$$\frac{n_1}{p} + \frac{n_2}{i} =$$

For this equation to be valid, the appropriate sign must be associated with the radius of curvature. If the surface is concave with respect to the source, r is _____. If the surface is convex with respect to the source, r is _____.

If $n_2 > n_1$, then a convex surface forms a real image of an object that is far from it and a virtual image of an object that is near it. Details depend on the values of the indices of refraction and the radius of curvature. A concave surface, on the other hand, always forms a _____ image. If $n_1 > n_2$, then a convex surface always forms a _____ image and a concave surface forms a _____ image for an object that is far from it and a _____ image for an object that is near it.

Thin lenses. A lens consists of two refracting surfaces. The image formed by the first may be considered the object for the second. If the surrounding medium is a vacuum and the thickness of the lens can be neglected, then the object distance p, image distance i, and focal length f are related by

$$\frac{1}{p} + \frac{1}{i} =$$

where the focal length is given in terms of the radii of curvature by

$$\frac{1}{f} =$$

Here r_1 is the radius of curvature of the surface nearer the object, r_2 is the radius of curvature of the surface farther away, and n is the index of refraction of the lens material. The equation is a good approximation if the surrounding medium is air.

Again you should be familiar with the sign convention. Real images are formed on the opposite side of the lens from the object. For these images, i is _____. Virtual images are formed on the same side of the lens as the object. For a virtual image, i is _____. A surface radius is positive if the surface is _____ from the point of view of the object and is negative if the surface _____ from the point of view of the object. You should know that the focal length of a lens does not depend on which side faces the object. Lenses with positive focal lengths are said to be _____, while lenses with negative focal lengths are said to be _____.

For each of the lenses shown below, give the sign of each radius, the sign of the focal length, and say if the lens is converging or diverging. Assume the object is to the left of the lens.

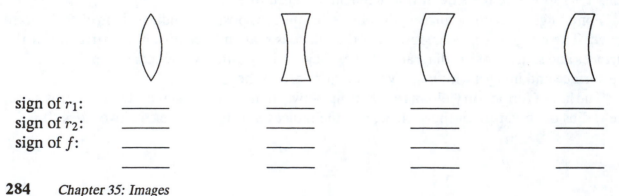

sign of r_1: _____   _____   _____   _____
sign of r_2: _____   _____   _____   _____
sign of f: _____   _____   _____   _____

_____   _____   _____   _____

A lens has two focal points, located equal distances $|f|$ on opposite sides of the lens. The first focal point, denoted by F_1, is on the side of the incident light for a converging lens (f positive) and on the side of the refracted light for a diverging lens (f negative). Light rays that are along lines that pass through F_1 are bent by the lens to become parallel to the axis. For a converging, lens the rays before refraction actually pass through F_1. For a diverging lens, the rays strike the surface, so only their extensions pass through F_1. For each of the lenses shown below, draw an incident ray along a line that passes through F_1. Also draw the refracted ray, parallel to the axis. For the diverging lens, use a dotted line to continue the line of the incident ray to the focal point. Since we have neglected the thickness of the lenses, you may assume refraction takes place at the plane through the lens center, perpendicular to the axis. These planes are shown as dotted lines on the diagrams.

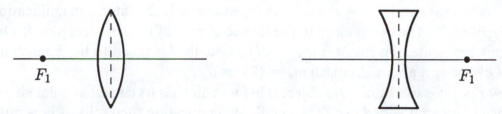

The second focal point, denoted by F_2, is on the side of the refracted light for a converging lens (f positive) and on the side of the incident light for a diverging lens (f negative). Light rays that are parallel to the central axis are bent by the lens to lie along lines that pass through F_2. For a converging lens, the rays after refraction actually do pass through F_2. For a diverging lens, the backward extension of the rays pass through F_2. For each of the lenses shown below, draw an incident ray parallel to the central axis. Also draw the refracted ray, which is along a line through F_2. For the diverging lens, use a dotted line to continue the line of the refracted ray backward to the focal point.

The position of an off-axis image can be found graphically by tracing two or more rays originating at the same point on the object. Two of these might be the rays you drew above: one along a line through F_1 and then parallel to the axis and the other parallel to the axis and then along a line through F_2. The third ray you might use goes through _____ of the lens. It is not refracted.

In terms of the object and image distances, the lateral magnification associated with a lens is given by

$$m =$$

an expression that is identical to the expression for a spherical mirror. If m is negative, the image of an upright object is _____; if m is positive, the image of an upright object is _____. A virtual image formed by single thin lens is always _____; a real image is always _____.

Angular magnification. Because it takes into account the apparent diminishing of the size of an object with distance, the <u>angular</u> <u>magnification</u> is often a better measure of the usefulness of a lens used for viewing than is the lateral magnification.

If an object has a lateral dimension (perpendicular to the central axis) of h and is a distance d from the eye, as in Fig. 35–16(a) of the text, its angular size is given in radians by $\theta =$ _____ . If the object is viewed through a lens and the image is a distance d' from the eye and has a lateral dimension of h', then the angular size of the image is $\theta' =$ _____ . The angular magnification of the lens is the ratio of these, o or $m_\theta = \theta'/\theta =$ _____ , where the last expression gives the angular magnification in terms of h, h', d, and d'.

A single converging lens used as a magnifying glass is usually positioned so the object is just inside the focal point F_1. See Fig. 35–16(c) of the text. The image is then virtual and far away. Its lateral size is $h' = mh = -ih/p$, where m is the lateral magnification. Thus, $m_\theta = -(i/p)(d/d')$. The eye is close to the lens so $d' \approx |i|$. Furthermore, $p \approx f$. Once these substitutions are made, the result is $m_\theta = d/f$. Usually d is taken to be the distance to the near point of the eye, about 25 cm, so $m_\theta = (25\,\text{cm})/f$.

The near point is used since the object is in focus and has its largest angular size when it is this distance from an unaided eye. The angular magnification then tells us how much better the lens is than the best the unaided eye can do.

Lens systems. Telescopes, microscopes, and other optical instruments usually consist of a series of lenses. They can be analyzed graphically by tracing a few rays as they pass through each lens in succession. The lens equation can also be applied to each lens in succession to find the position of the image. Consider a system of two lenses, a distance ℓ apart on the same central axis, as shown below. Let p_1 be the distance from the object O to the first lens struck. This lens forms an image I_1 at i_1, given by $1/p_1 + 1/i_1 = 1/f_1$, where f_1 is the focal length of the lens. You may think of this image as the object for the second lens. The object distance is $p_2 = \ell - i_1$ and the image distance i_2 is given by $1/p_2 + 1/i_2 = 1/f_2$, where f_2 is the focal length of the second lens. On the diagram, identify and label p_1, i_1, p_2, and i_2. Verify that $i_1 + p_2 = \ell$.

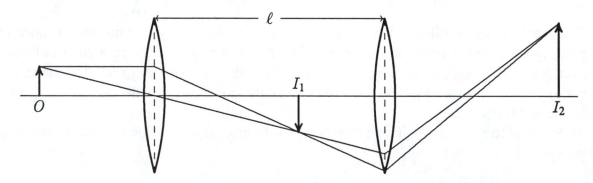

For some systems, the object distance for the second lens may be negative. This occurs if the image formed by the first lens is behind the second lens, so the light exiting the first lens is converging as it strikes the second lens. Such objects are called *virtual* objects. The equation relating object and image distances is still valid — just substitute a negative value for p_2.

A simple microscope consists of an objective lens, near the object, and an eyepiece (or ocular). The focal length of the objective lens is small and the lens is positioned so the object lies just outside the _____ of the lens. The image produced by the objective is real, large, inverted, and far from the lens. If the object has a lateral dimension h, then the image has a lateral dimension $h' = mh - ih/p$. The magnification is great since i is large and p is small.

The distance s between the second focal point of the objective lens and the first focal point of the eyepiece is called the _____ of the microscope. In terms of this quantity, $h' = $ _____, where θ is the angle the ray through the focal point makes with the central axis. Since $h = f_{ob} \tan \theta$, the magnification of the objective lens is given by $m = $ _____.

The eyepiece is positioned so the image produced by the objective lens falls at its _____ focal point. The eye is placed close to the eyepiece, which then acts as a simple magnifying glass with an angular magnification of $25\,\text{cm}/f_{ey}$, where f_{ey} is the focal length of the eyepiece. The overall angular magnification is given by $M = $ _____. The expression again compares the angular size of the image produced by the microscope with the angular size of the object when it is at the near point of an unaided eye.

Telescopes are used to view objects that are far away. Rays entering the objective lens are essentially parallel to the central axis and that lens produces an image that is close to the _____ focal point. To compute the angular magnification of a telescope, we compare the angular size of the object at its far-away position (not the near point) to the angular size of the image produced by the telescope.

Since the length of the telescope is much less than the object distance, the angle subtended by the object at the eye is the same as the angle it subtends at the objective lens. This is the same as the angle subtended at the objective lens by the image produced by that lens, so $\theta_{ob} = h'/f_{ob}$, where h' is the lateral dimension of the image. The angle subtended by the final image at the eye is the same as the angle subtended by the intermediate image at the eyepiece, or $\theta_{ey} = h'/f_{ey}$. The angular magnification of the telescope is given by $m_\theta = \theta_{ob}/\theta_{ey} = $ _____. To obtain a large angular magnification, a telescope should have an objective lens with a _____ focal length and an eyepiece with a _____ focal length.

II. PROBLEM SOLVING

You should be able to calculate the image position for an object in front of a plane or spherical mirror. Use the mirror equations: $i = -p$ for a plane mirror and $(1/p) + (1/i) = (1/f)$, where $f = 2/r$ for a spherical mirror. Be careful about the signs of the various quantities. Be sure you know the location of the image formed by a plane mirror and how to tell if the image is visible to an observer at a given location. Recall that the lateral magnification is given by $m = -i/p$ and that the lateral size h' of the image is related to the lateral size h of the object by $h' = mh$. Know how to interpret signs.

You should know how to find the image formed by a spherical refracting surface and by a lens. If the incoming rays make small angles with the central axis, use $(n_1/p) + (n_2/i) = (n_2 - n_1)/r$ for a single surface and $(1/p) + (1/i) = (1/f)$ for a thin lens in air. You should also know how to compute the focal length, given the radii of the two surfaces of a lens. See Eq. 35–10.

You should understand compound lens systems, such as microscopes and telescopes. Be sure you can calculate the angular magnification for each of these instruments.

III. NOTES

Chapter 36
INTERFERENCE

I. BASIC CONCEPTS

You studied the fundamentals of interference in Chapter 17. Now the results are specialized to light waves and applied to double-slit interference, thin-film interference, and the Michelson interferometer, an important instrument for measuring distances. Pay special attention to the role played by the distances traveled by interfering waves in determining their relative phase.

Be sure you understand the following concepts, covered in earlier chapters: interference in Chapter 17, wavelength in Chapters 17 and 34, phasors in Chapter 33, electromagnetic waves, the intensity of an electromagnetic wave, index of refraction, law of reflection, and law of refraction in Chapter 34.

Huygens' principle. Maxwell's equations lead to a geometrical construction, called Huygens' principle, that shows us how to construct the wavefront for an electromagnetic wave at some time $t + \Delta t$, given the wavefront at an earlier time t. According to the principle, you may think of each point on a wavefront as a point source of spherical waves, called Huygens wavelets. After time Δt, the radius of the wavelets will be $v\Delta t$, where v is the wave speed, and the wavefront will be tangent to the wavelets.

The principle is illustrated by the diagram to the right, which shows a plane wavefront at some instant of time. Use the time t to label it. Three spherical Huygens wavelets are shown. Each is centered on the plane wavefront and each has a radius of $c\Delta t$. The position of the plane wavefront at time $t + \Delta t$ is the common tangent to the spherical wavefronts. Draw it on the diagram and label it with the time $t + \Delta t$.

You will use Huygens' principle in this chapter and the next to understand what happens to light when it passes through one or more slits or passes by the edge of an obstacle. In each case, the light spreads out to enter the region you might think is in shadow. You may think of the spreading Huygens wavelets as being responsible for the phenomenon. To obtain some practice with the principle, first use it to prove the law of refraction.

In the diagram to the right, two wavefronts impinging on a plane boundary between two media are drawn with solid lines; two rays are drawn with dashed lines. Wavefront 1 has reached the boundary at B and wavefront 2 has reached the boundary at A. Point b on wavefront 2 has not yet reached the boundary but

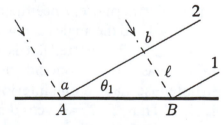

it will arrive at B in time $t = \ell/v_1$, where ℓ is the distance between the wavefronts and v_1 is the wave speed in medium of incidence.

The next diagram shows the two wavefronts in the region on the other side of the boundary, when b has reached B. If the wave speed in this region is v_2, the Huygens wavelet emanating from A has a radius of $v_2 t$, so $\ell' = $ _____. Draw an arc of the wavelet through a. Notice that the line AB is the hypotenuse of a right triangle for which ℓ' is one side. Thus, the length of AB is given by $v_2 t / \sin \theta_2$.

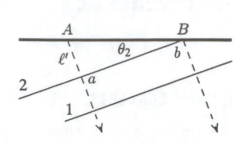

According to the diagram showing the *incident* wavefronts, the length of AB is also given by $v_1 t / \sin \theta_1$. In the space below, set the two expressions for AB equal to each other, replace v_1 with c/n_1, replace v_2 with c/n_2, and obtain the law of refraction:

Phasors. To find the total disturbance when two or more waves are present, you add the waves. Here you consider two sinusoidal waves with the same frequency and traveling in the same direction. The resultant is again a sinusoidal wave. Its amplitude may be as much as the sum of the amplitudes of the individual waves or as small as their difference, depending on their relative phase.

Phasors can be used to sum waves. The diagram on the right shows a phasor used to represent the electric field of a light wave at some point in space. Suppose the field is given by $E = E_0 \sin(\omega t)$, where ω is the angular frequency. The length of the phasor is proportional to the amplitude E_0 and its projection on the vertical axis is proportional to the wave itself. Label the angle ωt and indicate the direction of rotation of the phasor. Its angular speed is _____. Indicate the projection on the vertical axis.

Now draw a second phasor, associated with the field $E = E_0 \sin(\omega t + \phi)$, where ϕ is 75°. Draw it with its tail at the head of the first phasor. Finally draw the phasor that represents the sum of the two waves. Label it E.

The phasors are redrawn to the right. Since two sides of the triangle are the same, the two angles marked β are equal. Furthermore, since the third angle of the triangle is $180° - \phi$ and the angles of any triangle must sum to 180°, $\beta = $ _____. The dotted line is the perpendicular bisector of the phasor E. It is one side of a small right triangle with hypotenuse E_0 and one interior angle β. The other side is $E/2$. Thus, $E/2 = E_0 \cos \beta$ and, in terms of E_0 and ϕ,

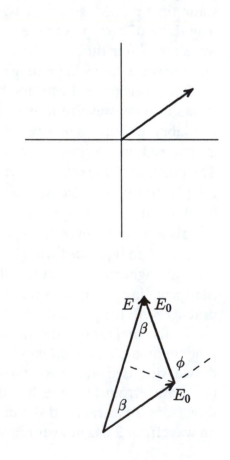

$$E =$$

If $\cos(\phi/2) = $ _____, constructive interference is complete and $E = 2E_0$. This means ϕ is a multiple of _____ rad. Both waves have their maximum values at the same time. If $\cos(\phi/2) = $ _____, destructive interference is complete and $E = 0$ This means ϕ is an odd multiple of _____ rad. At all times, one wave is the negative of the other.

You should recognize that ϕ in the equations above is the phase of one wave *relative* to the other. Any constant can be added to the phase of both waves without changing the amplitude of the resultant. Said another way, if the waves are given by $E_0 \sin(\omega t + \phi_1)$ and $E_0 \sin(\omega t + \phi_2)$, then $\phi = \phi_1 - \phi_2$. Notice that ϕ can be replaced by $-\phi$ without changing the expression for the amplitude. This is because $\cos(\phi/2) = \cos(-\phi/2)$. Thus, you may also take ϕ to be $\phi_2 - \phi_1$. In addition, any multiple of _____ can be added to or subtracted from ϕ without changing the resultant.

Phase differences can come about because two waves, starting in phase, travel different distances to get to the same point. If two waves of the same frequency travel in the same medium, so their speeds and wavelengths are the same, and one goes a distance ΔL farther than the other, the phase difference is $\phi = $ _____, where λ is the wavelength.

If a wave with wavelength λ in vacuum enters a medium with index of refraction n, the frequency does not change but the wave speed does. The wavelength in the medium is $\lambda' = $ _____. Suppose two waves have the same frequency and start in phase. One travels a distance L_1 in a medium with index of refraction n_1 and the other travels a distance L_2 in a medium with index of refraction n_2. At the end their phase difference is _____.

The electric fields of the two waves are vectors and, strictly speaking, the waves should be added vectorially. We assume, however, that the waves are plane polarized and the fields are along the same line. Then, scalar addition can be used. This is a good approximation for most of the situations considered in this chapter.

Double-slit interference. A monochromatic plane wave is incident normally on a barrier with two slits, as diagramed in Fig. 36–8 of the text. You can find the intensity at any point on the screen by summing the Huygens wavelets emanating from points within the slits. Note that wavelets arrive at every point on the screen, not just those directly behind the slits. For now, assume the slits are so narrow that only one wavelet from each slit is required. Furthermore, assume the amplitude of each wavelet is the same at the screen. At the point P in the diagram, the wavelet from the upper slit can be written $E_0 \sin(\omega t - kr_1)$ and the wavelet from the lower slit can be written $E_0 \sin(\omega t - kr_2)$, where $k = 2\pi/\lambda$ and λ is the wavelength. The phase difference is $\phi = k(r_2 - r_1) = 2\pi(r_2 - r_1)/\lambda$.

Carefully note that the phase difference arises because the wavelets travel different distances from the slits to the same point on the screen. Also note that because a plane wave is incident normally on the slits, the phases of the wavelets are the same at the slits. This is changed if the wave is incident at some other angle or if the slits are covered by transparent materials with different indices of refraction. The expression for ϕ must then be modified.

If the screen is far from the slits, then rays from the slits to any point on the screen are nearly parallel to each other and the difference in the distances can easily be written in terms of the angle θ between a ray and a line normal to the barrier. The geometry is shown in the diagram to the right. Since the screen is far away, the distance from the dotted line between the rays to P is the same along each ray. One of the interior angles of the right triangle formed by the dotted line, the barrier, and the lower ray is also θ. Label it. The lower ray is longer than the upper by $d \sin \theta$, where d is the slit separation. Indicate this distance on the diagram and label it with the expression for its length.

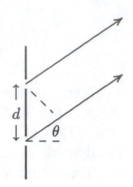

In terms of d, θ, and λ, the phase difference at the screen is

$$\phi =$$

The amplitude of the resultant wave at P is denoted by E. In terms of E_0 and ϕ, it is given by

$$E =$$

The intensity I at P is proportional to the square of the amplitude and, in terms of ϕ, is given by

$$I =$$

where I_0 is the intensity associated with a single wave. The same result is obtained for a nearby screen if a lens is used to focus light on the screen.

You should recognize that light emanates from each slit in all forward directions but only the portions of wavefronts that follow the rays shown get to point P. Other portions of the wavefronts get to other places on the screen and for them, θ has a different value. For different values of θ, the phase difference is different and, as a result, so are the resultant amplitude and intensity. Alternating bright and dark regions (fringes) are seen on the screen. Centers of bright fringes (maxima of intensity) occur at points where the phases of the two wavelets differ by _____, where m is an integer. To reach these places, one wave travels farther than the other by a multiple of _____. That is,

$$d \sin \theta =$$

Centers of dark fringes (minima of intensity) occur at points where the phases of the two wavelets differ by _____, where m is an integer. To reach these places, one wave travels farther than the other by an odd multiple of _____. That is,

$$d \sin \theta =$$

Notice that complete constructive interference occurs at the point on the screen directly in back of the slits, for which $\theta = 0$. The angular separation of the first minima on either side of the central maximum is a measure of the extent to which the intensity pattern is spread on the screen. This is given by $2\theta_0$, where $\sin \theta_0 = \lambda/2d$. As d _____ or λ _____, the angular separation increases and the pattern spreads. If $d =$ _____, then $\theta_0 = 90°$ and no bright fringes appear beyond the central maximum. For any given values of d and λ, the number of maxima that are seen can be found by setting θ equal to _____ and solving $d \sin \theta = m\lambda$ for _____.

Coherence. Two sinusoidal waves are said to be coherent if the difference in their phases is constant. Light is emitted from atoms in bursts lasting on the order of nanoseconds and each burst may have a different phase constant associated with it. Thus, light from atoms emitting independently of each other is not coherent.

If an extended incoherent source, such as an incandescent lamp, is used to produce light that is incident on a double slit barrier, the light from each atom goes through each slit and combines on the other side to form an interference pattern. Light from different atoms, however, form patterns that are shifted with respect to each other, the amount of the shift depending on the separation of the atoms in the source. Describe a technique that can be used to insure that only light from a small region of the source reaches the slits: _____

_____

All the light from another type light source, a _____, is coherent, even though many different atoms are emitting simultaneously. When this light is incident on a double slit, it produces an interference pattern without any additional apparatus.

Thin-film interference. If light is incident normally on a thin film, some is reflected from the front surface and some from the back. These two waves interfere and the resultant intensity may be quite large or quite small, depending on the thickness of the film.

To calculate the intensity for a given wavelength light, you must be able to find the relative phases of the two waves. For normal incidence the wave reflected from the far surface travels _____ farther than the wave reflected from the near surface. Here L is the thickness of the film. In addition, for one or both of the waves, the medium beyond the surface of reflection may have a higher index of refraction than the medium of incidence. On reflection, the wave then suffers a phase change of _____ radians.

The diagram shows a film of thickness L with a plane wave incident normally. You should be able to show that the phase difference for waves reflected from the two surfaces is given by $\phi = 4\pi L n_2/\lambda$ if $n_1 > n_2 > n_3$ or $n_1 < n_2 < n_3$. In the first case, neither wave suffers a phase change on reflection. In the second, they both suffer a phase change of π rad.

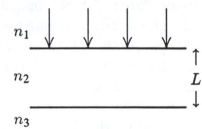

If $n_1 < n_2 > n_3$ or $n_1 > n_2 < n_3$, the phase difference is given by $(4\pi L n_2/\lambda) \pm \pi$. In the first case, the wave reflected from the front surface suffers a phase change of π rad but the wave reflected from the back surface does not. In the second case, the wave reflected from the back surface suffers a phase change of π but the wave reflected from the front surface does not. Notice that the wavelength of the wave in the film must be used to calculate the difference in phases. It is given by λ/n_2, where n_2 is the index of refraction of the film and λ is the wavelength in vacuum. Since you may always add 2π to any phase, it is immaterial whether the sign in front of π is plus or minus.

When white light (a combination of all wavelengths of the visible spectrum) is incident on a thin film, the reflected light is colored. It consists chiefly of those wavelengths for which interference of the two reflected waves produces a maximum or nearly a maximum of intensity. Those wavelengths for which interference produces a minimum are missing. This phenomena accounts for the colors of oil films and soap bubbles, for example.

Suppose a film with thickness L has an index of refraction n and is in air (with a smaller index of refraction). Monochromatic light is incident normally on a surface. There is a change in phase on reflection at the _____ surface but not at the _____ surface. The difference in phase of the two reflected waves is ϕ = _____. Interference maxima occur for wavelengths given by λ = _____ and interference minima occur for wavelengths given by λ = _____, where m is an integer. How are these results changed if the film is deposited on glass with a higher index of refraction? _____

_____

_____

When monochromatic light is incident on a thin film with a varying thickness, like a wedge, interference produces bright and dark bands. Bright bands appear in regions for which the film thickness is such that the phase difference between the two reflected waves is close to _____ rad; dark bands appear in regions for which the film thickness is such that the phase difference is close to _____ rad. Here m is an integer.

The Michelson interferometer. The diagram on the right is a schematic drawing of the instrument. Light from a source S is incident on a half-silvered mirror M, where the beam is split. Half is reflected to mirror M_2 where it is reflected back through M to the eye at E. The other half is transmitted to mirror M_1 where it is reflected back to M. There, half is reflected to the eye. The two beams reaching the eye interfere. To measure a distance, one of the mirrors (M_2, say) is moved, thereby changing the interference pattern at the eye. As the mirror moves, alternately bright and dark fringes are seen. The number of fringes that appear are counted and related to the distance moved by the mirror.

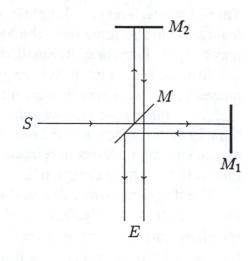

Suppose the distance traveled by the light that strikes mirror M_1 is d_1 and the distance traveled by the light that strikes mirror M_2 is d_2. If the wavelength of the light is λ, then the difference in phase of the two waves at the eye is ϕ = _____. Suppose d_1 and d_2 happen to have values so a maximum of intensity is produced at the eye. Then, mirror M_2 is moved until the intensity is again a maximum. In terms of the wavelength, the distance it moved must have been _____. Interferometers have been used to measure distances with an accuracy of about _____ the wavelength of the light used.

II. PROBLEM SOLVING

You should know the relationship between the wave speed in a material medium and the wave speed in vacuum. You should also know that when a wave enters a medium, the frequency does not change but the wave speed and wavelength do.

You should know how to calculate the phase difference of two waves. If wave 1 travels a distance r_1 in a medium with refractive index n_1 and wave 2 travels a distance r_2 in a medium with refractive index n_2, the difference in phase is $\phi = (2\pi/\lambda)(n_2 r_2 - n_1 r_2)$.

You should understand the conditions for the formation of bright and dark fringes by a double-slit arrangement. Find an expression for the phase difference between the waves from the two slits and set it equal to $2\pi m$ for a bright fringe and $2\pi(m + \frac{1}{2})$ for a dark fringe. In many cases, the phase difference is due to the difference in distance traveled by the waves. This is $d\sin\theta$, where d is the slit separation. If a medium with thickness L and refraction index n is placed in front of a slit, you must add $2\pi(n-1)L/\lambda$ to the phase of the wave through that slit.

You should also be able to calculate the intensity at points on the interference pattern, relative to the intensity at the center of the central maximum. If the waves have equal amplitudes at the screen, then $I = I_m \cos^2(\phi/2)$. Learn to use phasor diagrams to sum two or more waves. See the Mathematical Skills section below.

Some problems deal with the interference of waves reflected from the front and back surfaces of a thin film. The difference in phase comes from the difference in distances traveled and, in some cases, from the change in phase by π on reflection from a medium with a higher index of refraction. Don't forget to think about both of these sources of phase difference. Also, don't forget to use the wavelength of the wave in the film, not the wavelength in vacuum, to compute the phase difference.

Some problems deal with the Michelson interferometer. Again you calculate the phase difference for the two waves and set it equal to $2\pi m$ for a bright fringe and $2\pi(m + \frac{1}{2})$ for a dark fringe. The phase difference may be due to the different distances traveled by the waves and due to materials with different indices of refraction in the two arms of the interferometer.

III. MATHEMATICAL SKILLS

A few problems deal with the addition of two waves with different amplitudes. Suppose one wave is given by $E_1 \sin(\omega t)$ and the second by $E_2 \sin(\omega t + \phi)$. The phasor diagram is shown to the right. According to the law of cosines, $E^2 = E_1^2 + E_2^2 - 2E_1 E_2 \cos\alpha$. Since $\alpha = 180° - \phi$ and $\cos(180° - \phi) = -\cos\phi$, this can be written

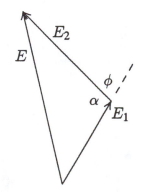

$$E^2 = E_1^2 + E_2^2 + 2E_1 E_2 \cos\phi.$$

Use this equation to calculate the amplitude of the resultant.

If the waves are in phase, then $\phi = 0$, $\cos \phi = 1$, and $E^2 = E_1^2 + E_2^2 + 2E_1 E_2 = (E_1 + E_2)^2$. If the waves are 180° out of phase, then $\cos \phi = -1$ and $E^2 = E_1^2 + E_2^2 - 2E_1 E_2 = (E_1 - E_2)^2$. These results should agree with what you expect.

If $E_1 = E_2$, then the general expression reduces to $E^2 = 2E_1^2(1 + \cos \phi)$. Now $\cos \phi = \cos(\phi/2 + \phi/2) = \cos^2(\phi/2) - \sin^2(\phi/2) = 2\cos^2(\phi/2) - 1$, where $\cos^2(\phi/2) + \sin^2(\phi/2) = 1$ was used. Thus, $E^2 = 4E_1^2 \cos^2(\phi/2)$ and $E = 2E_1 \cos(\phi/2)$, in agreement with the result given in the text.

IV. NOTES

Chapter 37
DIFFRACTION

I. BASIC CONCEPTS

Diffraction is the flaring out of light as it passes by the edge of an object or through an opening in a barrier. Pay attention to the description of this phenomenon in terms of Huygens wavelets. You should also understand how interference of the wavelets produces bright and dark fringes in the diffraction pattern of an object or opening. When you study diffraction by a double slit, pay close attention to the relationship between the single-slit diffraction pattern and the double-slit interference pattern. In particular, know what characteristics of the slits determine the maxima and minima in each pattern.

Two important applications are discussed. Diffraction patterns produced by diffraction gratings are used to study spectra and patterns produced by x rays incident on crystals are used to study crystalline structure. Your goals should be to learn what the patterns look like and to understand the details of wave interference that leads to these patterns. In addition, learn how to calculate the angles for maximum and minimum intensity.

Be sure you understand the following concepts, covered in earlier chapters: interference in Chapters 17 and 36, wavelength in Chapters 17 and 34, phasor in Chapters 33 and 36, electromagnetic waves, the intensity of an electromagnetic wave, index of refraction, law of reflection, and law of refraction in Chapter 34, and diffraction in Chapter 36.

Diffraction. The phenomenon is described in terms of Huygens wavelets. If there is no barrier, the wavelets combine to produce a wave that continues moving in its original direction. If a barrier blocks some of the wavelets, then those that are not blocked combine to produce a wave that moves into the geometric shadow. In addition, if the light is coherent, the Huygens wavelets interfere to form a series of bright and dark bands, called the diffraction pattern of the object. Figs. 37–1, 37–2, and 37–3 of the text show some diffraction patterns.

Single-slit diffraction. Consider plane waves of monochromatic light incident normally on a barrier with a single slit of width a, as shown in Fig. 37–4 of the text. To find the intensity at a point P on a screen, add the Huygens wavelets emanating from the slit and take the limit as the number of wavelets becomes infinite. We suppose the viewing screen is far away from the slit and consider parallel rays as shown in Fig. 37–5b.

Each wavelet has an infinitesimal amplitude and a phase that differs from that of a neighboring wavelet by an infinitesimal amount. The phasors form an arc of a circle, as shown to the right. The angle ϕ is the difference in phase of wavelets from the edges of the slit. It is also the angle subtended by the arc at its center. Look at Fig. 37–8 of the text. If the arc has

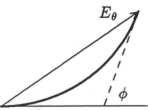

radius R, then its length is given by $E_m = R\phi$, for ϕ in radians. E_m is the sum of the amplitudes of all the wavelets and, thus, is the amplitude if they all have the same phase.

E_θ, the amplitude at P, is the chord of the arc. A little geometry shows that $E_\theta = 2R\sin(\phi/2)$. Eliminating R between these two expressions yields an expression for E_θ in terms of E_m and ϕ:

$$E_\theta =$$

Differences in the phases come about because the wavelets travel different distances to P. If the screen is far away and the slit width is a, then the lower wavelet travels $a\sin\theta$ farther than the upper wavelet. Label the slit width and mark the distance $a\sin\theta$ on the diagram to the right. If the wavelength of the light is λ, then the difference in the phases of these two wavelets is given by

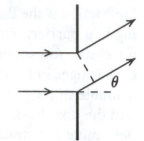

$$\phi =$$

The expression for the amplitude is often written in terms of $\alpha = \phi/2$, rather than in terms of ϕ. Then, the amplitude at P is given by

$$E_\theta =$$

where, in terms of a and θ,

$$\alpha =$$

The intensity at P is proportional to the square of the amplitude and is given by

$$I_\theta =$$

where I_m is the intensity when all wavelets are in phase.

For future reference use the axes below to make a rough sketch of the intensity as a function of the diffraction angle θ. Show two or three secondary maxima on each side of the central maximum. Label the central maximum and the secondary maxima.

0   θ

Carefully study Fig. 37–7 of the text. It shows the intensity as a function of θ for several values of the slit width a. The most prominent feature is a broad central maximum, centered at $\theta = 0$. Note particularly that $\theta = 0$ corresponds to an intensity maximum, not a minimum. In the limit as $\alpha \to 0$, $(\sin \alpha)/\alpha \to 1$. If the slit width is small, the central maximum spreads to cover the entire screen and no zeros of intensity occur. For a wide slit, the central maximum is narrow and is followed on both sides by secondary maxima. These are narrower and of considerably less intensity than the central maximum. They are roughly midway between zeros of intensity. The number that appear depends on the slit width.

In the space below, draw phasor diagrams corresponding to the peak of the central maximum, the first zero, and the second zero to one side of the central maximum. Remember that the total arc length is the same in all cases.

CENTRAL MAXIMUM   FIRST ZERO   SECOND ZERO

The first zero on either side of the central maximum corresponds to $\alpha = \pm\pi$. The angles θ for which these occur are given by

$$\sin \theta =$$

For these angles, every wavelet from the upper half of the slit can be paired with a wavelet from the bottom half, emanating from a point $a/2$ away. The phase difference of two wavelets in a pair is _____ and these wavelets sum to _____.

Notice that as a decreases, the angle θ for the first minimum increases. The central maximum is broader for narrow slits than for wide slits and diffraction is more pronounced. If a is greater than λ, no zeros of intensity occur ($\sin \theta$ cannot be greater than one) and the entire screen is within the central maximum.

In the space to the right. sketch the phasor diagram corresponding to the first secondary maximum to one side of the central maximum. In terms of the diagram explain why the intensity at a secondary maximum is less than that at the central peak: _____
_____
_____
_____

Double-slit diffraction.   Consider two identical slits, each of width a, with a center-to-center separation d. Any point P on a screen is reached by a wave from each slit, the resultant of the Huygens wavelets from that slit, and we may think of the pattern as being formed by the interference of these two waves. If light is incident normally on the slits and the observation point is far away, the wave that reaches it from each slit has amplitude $E_m(\sin\alpha)/\alpha$, where $\alpha = (\pi a/\lambda)\sin\theta$, and the two waves differ in phase by $(2\pi d/\lambda)\sin\theta$. When they are combined, the resultant amplitude is

$$E_\theta =$$

where $\beta = (\pi d/\lambda)\sin\theta$.

As you learned in the last chapter, double-slit interference minima occur for $\beta = (2m + 1)\pi/2$ or $\sin\theta = (2m + 1)\lambda/2d$. Single-slit diffraction minima occur for $\alpha = m\pi$ or $\sin\theta = m\lambda/a$. In each case, m is an integer but it may have different values in the two expressions. Since d must be larger than a, the interference minima must be closer together than the diffraction minima. The single-slit diffraction pattern forms an envelope, with the interference pattern inside. Study Fig. 37–13 of the text.

For any double-slit situation, there are three parameters you must be aware of: the wavelength λ, the slit width a, and the center-to-center slit separation d. The ratio a/λ controls the width of the central diffraction maximum, which extends from $-\sin^{-1}(\lambda/a)$ to $+\sin^{-1}(\lambda/a)$. The ratio also controls the positions of the secondary diffraction maxima, if any. The ratio d/λ controls the angular positions of the interference maxima and minima. If d/a is an integer a maximum of the two-slit interference pattern occurs at the same angle as a _____ of the single-slit diffraction pattern. The overall pattern is then said to have a missing maximum.

Finally, the ratio d/a controls how many interference maxima fit within the central diffraction maxima or any of the secondary diffraction maxima. This number is independent of the wavelength. As the wavelength increases, the central diffraction maximum widens but the interference pattern also spreads, with the result that just as many interference fringes are within the central diffraction maximum.

Diffraction from a circular aperture.   When plane waves pass through a circular aperture and onto a screen, a diffraction pattern is formed there. The pattern consists of a bright central disk, followed by a series of alternating dark and bright rings. The first minimum occurs at an angle θ, measured from the normal to the aperture and given by

$$\sin\theta =$$

where d is the diameter of the aperture. This angle can be used as a measure of the angular size of the central disk.

Stars are effectively point sources of light and lenses act like circular apertures. The image of a star formed by a lens is not a point but is broadened by diffraction to a disk and rings. Two stars do not form distinct images if the central disks of their diffraction patterns overlap too much. Describe the Rayleigh criterion for the resolution of two far-away point sources:

_____

_____

_____

Diffraction gratings. A diffraction grating consists of many thousands of closely spaced rulings on either a transparent or highly reflecting surface. When light is incident on a grating, a diffraction pattern is formed, just like the multiple-slit pattern you studied above. Because light with different wavelengths produces lines at different angles, diffraction gratings are often used to analyze the spectra of light sources.

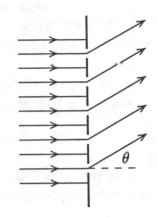

Consider a barrier in which N parallel slits have been cut, with distance d between adjacent slits. Monochromatic plane waves are incident normally on the barrier and the intensity pattern formed by waves passing through the slits is viewed on a screen far away. The slits are so narrow that single-slit diffraction can be ignored. That is, the interference pattern is well within the central maximum of the single-slit diffraction pattern.

The pattern on the screen consists of a series of intense, narrow bands, called _____. The pattern is usually described in terms of the angle θ made with the normal by a ray from the slit system to a point on the screen. A diffraction line occurs when the phases at the screen of waves from any two adjacent slits are either _____ or differ by a multiple of _____ rad. Since waves from two adjacent slits travel distances that differ by $d \sin \theta$, where d is the slit separation, the condition for a line is

$$d \sin \theta =$$

where λ is the wavelength. The integer m in this equation is called _____. High order lines occur at _____ angles than low order lines.

Notice that the locations of the lines are determined by the ratio d/λ and are independent of the number of slits. Also notice that the lines occur at different angles for different wavelengths of light. For the same order line, the angle for red light is _____ than that for violet light. If white light is incident on the barrier, the color of an observed band continuously changes from red at one end to violet at the other.

The width of a line is indicated by the angular position $\Delta\theta$ of an adjacent minimum. For the line that occurs at angle θ,

$$\Delta\theta =$$

Notice that it depends on the number of slits. In fact, as the number of slits increases without change in their separation, the width of every line _____. Also notice that maxima near the normal (small θ) are _____ than maxima farther away from the normal (larger θ).

Two parameters are used to measure the quality of a diffraction grating. The <u>dispersion</u> of a grating measures the angular separation of lines of the same order for wavelengths differing by $\Delta\lambda$. It is defined by

$$D =$$

and for order m, occurring at angle θ, it is given in terms of the slit separation d by

$$D =$$

Notice that dispersion does not depend on the number of rulings. Large dispersion means
_____ angular separation.

The second parameter is the resolving power. It measures the difference in wavelength for waves with the angular separation of their lines equal to half the angular width of a line; that is, for two lines that obey the Raleigh criterion for resolution. Mathematically, it is defined by $R = \lambda/\Delta\lambda$, where $\Delta\lambda$ is the difference in wavelength. For a system of N slits, the resolving power at the line of order m is given by

$$R =$$

Be sure you understand the difference between dispersion and resolving power. Describe in words the pattern produced by a grating with large dispersion and small resolving power:

_____

_____

_____

Consider a principal maximum near $\theta = 0$ and tell which quantities (of m, d, and N) should be large and which should be small to produce such a pattern: _____

_____

Describe the intensity pattern produced by a grating with small dispersion and large resolving power: _____

_____

_____

Tell which quantities should be large and which should be small to produce such a pattern:

_____

X-ray diffraction. Atoms in a crystal form a periodic array in three dimensions and x-ray radiation scattered by their electrons produces a diffraction pattern. Why are x rays with wavelengths on the order of 0.1 nm, rather than visible light with wavelengths on the order of 500 nm, used to form the pattern? _____

_____

_____

Diffraction occurs only when the x rays are incident at certain angles to crystal planes. A crystal plane is a plane that _____

_____.

Suppose monochromatic plane-wave x rays with wavelength λ are incident at an angle θ to a set of crystal planes with separation d. An intense spot is produced if the angle of incidence satisfies

$$2d\sin\theta =$$

where m is an integer. When this condition is met, a high intensity beam is radiated at the angle θ to the planes. Carefully note that θ is *not* the angle between the incident rays and the normal to the planes. Rather, it is the angle between the incident rays and the planes themselves.

The diagram on the right shows the edges of two crystal planes as horizontal dotted lines and two rays reflected from them. If d is the separation of the planes, the lower ray travels a distance $2d \sin \theta$ further than the upper ray. On the diagram, draw a normal to the rays through the point of reflection of the upper ray and point out the distance $d \sin \theta$. For constructive interference to occur, the difference in the distance traveled must be a multiple of _____, so $2d \sin \theta = m\lambda$. There are many more crystal planes parallel to the ones shown. If x rays from two adjacent planes interfere constructively, then x rays from all of them interfere constructively and an intense beam is formed.

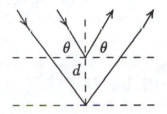

X rays are used to experimentally determine the atomic arrangements in crystals. The idea is to find the orientations and separations of a great many sets of crystal planes and use these to reconstruct the crystal. The symmetry of the diffraction spots is indicative of the symmetry of the crystal and can often be used to great advantage.

Crystals with known atomic arrangements are used as filters to separate x rays of a given wavelength from an incident beam containing a mixture of wavelengths. The crystal is oriented so that only the waves with the desired wavelength form an intense scattered beam.

II. PROBLEM SOLVING

The single-slit diffraction pattern is fundamental to this chapter. You should know how to find minima: $a \sin \theta = m\lambda$, and how to calculate the intensity: $I = I_m (\sin \alpha)^2 / \alpha^2$, where $\alpha = (\pi a / \lambda) \sin \theta$.

You should know that the diffraction angle for the first minimum of a circular aperture is given by $\theta_R = \sin^{-1}(1.22\lambda/d)$, where d is the diameter of the aperture. You should also know that this expression also gives the Rayleigh criterion for the resolution of two far-away objects. If $d \gg \lambda$ then θ_R is given in radians by $\theta_R = 1.22\lambda/d$.

A double-slit pattern combines double-slit interference and single-slit diffraction. The interference minima are given by $d \sin \theta = (2m + 1)\lambda/2$ and the maxima are given by $d \sin \theta = m\lambda$. The diffraction minima are given by $a \sin \theta = m\lambda$. When an interference maxima coincides with a diffraction minima, the order is said to be missing. You should also be able to calculate the intensity: $I = I_m (\cos \beta)^2 (\sin \alpha)^2 / \alpha^2$, where $\beta = (\pi d / \lambda) \sin \theta$ and $\alpha = (\pi a / \lambda) \sin \theta$. Don't confuse the slit width a and slit separation d.

Diffraction gratings produce intense lines. The condition is $d \sin \theta = m\lambda$, where d is the slit separation. Use this expression, for example, to find the angular positions of the lines. You should also know how to compute the angular width of a line: $\Delta\theta = \lambda/Nd \cos \theta$, where N is the number of lines. You should know the meaning of the dispersion and know how to calculate it in terms of the angular separation of lines of different wavelength ($D = \Delta\theta/\Delta\lambda$) and in terms of the grating properties ($D = m/d \cos \theta$). You should know the meaning of the resolving power and how to calculate it in terms of the difference in wavelength ($R = \lambda/\Delta\lambda$) and in terms of grating properties ($R = Nm$).

You should be able to use the Bragg condition for diffraction from a crystal: $2d \sin \theta = m\lambda$. Remember that θ is measured from the reflecting planes, not their normal. Some problems ask you to use the Bragg condition to find the atomic separation in a crystal.

III. NOTES

OVERVIEW IV
MODERN PHYSICS

This section opens with a chapter on special relativity, which tells us how measurements of the same phenomenon made by observers who are moving with respect to each other are related. The theory has forced us to redefine our concepts of time, energy, and momentum. The results are important and interesting in their own right. They are absolutely essential for the discussions of nuclear and particle physics later in the text.

You have learned that electric and magnetic fields propagate as waves. Waves are predicted by Maxwell's equations and phenomena such as reflection, refraction, interference, diffraction, and polarization are all explained in terms of electromagnetic waves. However, experiments that test the nature of electromagnetic radiation at the atomic level show that in some aspects, it behaves like a collection of particles that exchange energy and momentum in localized collisions.

You have also learned that electrons are particles. They are highly localized in space and exchange energy and momentum in collisions with other particles. However, experiments show that in some aspects, matter behaves like waves, exhibiting interference and diffraction effects.

A highly successful theory, the quantum theory, has been developed to explain in a comprehensive manner the behavior of nature at the atomic level. The fundamentals are discussed in Chapters 39 and 40. Many of the results will mystify you, but the ideas of the quantum theory will not be difficult for you to understand if you keep an open mind as you study these chapters.

Some of the experiments that force us to treat electromagnetic radiation as particles are described in Chapter 39. Here you will be introduced to relationships between particle properties such as energy and momentum and wave properties such as frequency and wavelength. The experiments that force us to associate waves with matter are also discussed in Chapter 39. Here you will learn that the same relationships hold for particles and the waves associated with them.

In Chapter 39, you will also get a glimmer of how the wave and particle natures of both light and particles are made compatible. In each case, the wave is not a manifestation of the motion of the medium in which it propagates. Instead, it is closely associated with the probability that the particle is in a given region of space. It is the probability that shows interference effects, for example.

Interesting and important consequences arise because the probability of finding the particle in a certain given region propagates as a wave. The more closely the position of the particle is defined, for example, the less closely its momentum is defined. By necessity, a measurement of the position destroys information about the momentum. Another important consequence is that a particle initially on one side of a barrier may appear on the other side. It is said to have tunneled through. Still another consequence is that the energy of a bound particle is limited to certain discrete values. Chapter 40 contains discussions of all these consequences.

Remaining chapters of this section deal with applications of the quantum principles. In Chapter 40, quantum mechanics is applied to the electron in a hydrogen atom. This is the simplest atom to understand. Here you will see in detail how the quantization of energy comes about because restrictions must be placed on the wave function. You will also see graphs of the probability density for an electron in a hydrogen atom. In Chapter 41, the ideas are extended to atoms with more than one electron.

Not only is the energy of an electron in an atom quantized but so is its angular momentum. You will learn to categorize the quantum mechanical state of an electron by giving its energy, the magnitude of its angular momentum, and one component of its angular momentum. You will be introduced to the idea of spin angular momentum, the intrinsic angular momentum of a particle.

Quantum numbers that specify the energy, the orbital angular momentum, and the spin angular momentum are used to label the states of electrons in atoms. You will learn to use these labels to distinguish one state from another. The Pauli exclusion principle plays an important role in the building of atoms. It tells us that no two electrons are ever in the same state. It is discussed in Chapter 41. This knowledge leads to an understanding of the periodic table of the chemical elements. You will also learn how x rays can be used to place elements in the table.

Quantum mechanics has enormously increased our understanding of the properties of materials. You learned in Chapter 27 that the resistivity of a material is determined by the density of conduction electrons and by the mean free time between their collisions with atoms. Quantum mechanics is used in Chapter 42 to provide a deeper understanding. You will learn how metals, insulators, and semiconductors differ in their conducting properties and about the role played by quantum mechanics in determining the differences. You will also learn about the basis for the temperature dependence of the resistivity. Chapter 42 also includes an introduction to semiconductor devices, so pervasive in modern technology.

The next two chapters deal with the important topic of nuclear physics, the fundamentals in Chapter 43 and applications to the harnessing of nuclear energy in Chapter 44. In the first of these chapters, you will learn what particles are emitted by radioactive nuclei and about the rate of emission for a collection of such nuclei. You will learn to use conservation of energy to predict the energies of the emission products.

Energy is released when a heavy nucleus fragments (fissions). The basics of this phenomenon and its utilization in modern power plants are discussed in Chapter 44. Energy is also released when two light nuclei fuse together. This phenomenon powers stars (including the Sun) and is a candidate for power plants of the future. It is also discussed in this chapter.

Chapter 45 is devoted to the very small (fundamental particles, the building blocks of matter) and the very large (the universe itself). You will learn that some particles, such as protons and neutrons, are actually made of more fundamental entities called quarks. Other particles, such as electrons, seem to be indivisible. You will learn a little about the basic interactions of the fundamental particles with each other. You will see that the basic forces and the properties of the fundamental particles played pivotal roles in the evolution of the universe from the big bang to its present state. Evidence for the big bang is discussed and a chronological description of events in the early universe is given.

Chapter 38
RELATIVITY

I. BASIC CONCEPTS

When two observers who are moving relative to each other measure the same physical quantity, they may obtain different values. The special theory of relativity tells how the values are related to each other when both observers are at rest in different inertial frames. Although the complete theory deals with all physical quantities, the ones you consider here are the coordinates and time of an event and the velocity, momentum, and energy of a particle.

All the equations you will use are linear algebraic equations, so the mathematics is quite simple. The concepts, however, are difficult for some students because they run counter to experience. You must get used to the idea, for example, that the time interval and spatial distance between two events depend on the velocity of the observer carrying out the measurements. Relativity has significantly altered our concepts of length and time.

The basis of special relativity. The theory is based on two postulates. The first deals with the laws of physics. It is: _____

_____

_____

Keep in mind that the laws of physics are relationships between physical quantities, not the quantities themselves. Newton's second law and the conservation principles are examples of laws. The momentum of a system has a different value for different reference frames but if it is conserved for one inertial frame it is conserved for all inertial frames, according to the postulate.

The second postulate deals with the speed of light. It is: _____

_____

Suppose a light source sends a pulse of light toward you. If you are stationary with respect to the source, the speed of the pulse relative to you is _____ . If you are moving at $c/2$ toward the source, the speed of the pulse relative to you is _____ . If you are moving away from the source, the speed of the pulse relative to you is _____ . This postulate is consistent with the notion that electromagnetic radiation does not require a medium for its propagation.

Special relativity deals with measurements made in <u>inertial</u> <u>reference</u> <u>frames</u>. Tell what an inertial frame is: _____

_____

As you study this chapter you will be concerned with <u>events</u>. An event has four numbers associated with it: three of them are _____ that designate the location of the event; the fourth designates the _____ of the event. To measure the coordinates of an event, meter sticks must be laid out in an inertial reference frame, at rest with respect to the frame. To measure the time of an event, a clock must be present at the location of the event and it must be at rest with respect to the reference frame. In addition, it must be synchronized with other clocks at rest in the frame. Synchronization is accomplished by _____

_____

_____ .

The coordinates and time of any event may be measured by meter sticks and clocks at rest in any inertial frame. In general, different results are obtained for different frames. Relativity tells how the measurements made in one frame are related to those made in another.

Simultaneity. Two events, separated in space and simultaneous for one observer, are NOT simultaneous to another observer, moving with respect to the first along the line joining the positions of the events. Carefully study Fig. 38–4 of the text. It shows two events, labeled Red and Blue. The events are simultaneous according to Sam. He knows they are simultaneous because the events occurred at the ends of his spaceship and electromagnetic waves from the events met at _____ . Since the waves travel at the same speed and go the same distance, they must have started at the same time.

The events also occur at the ends of Sally's space ship but she is moving away from the position of the Blue event and toward to the position of the Red event. Waves from the _____ event reach the midpoint of her spaceship before waves from the _____ event. In her reference frame, the waves move with the speed of light, just as they do in Sam's reference frame. Therefore, she knows that the _____ event occurred before the _____ event.

Time dilation. The time interval between two events is different when measured with clocks at rest in two inertial frames that are moving relative to each other. The diagram to the right shows a clock. The flash unit F emits a light pulse that travels to mirror M and is reflected back to the flash unit. It is detected there and immediately triggers the next flash. If the flash unit and mirror are separated by a distance D, then the time interval between flashes is given by $\Delta t_0 = $ _____ .

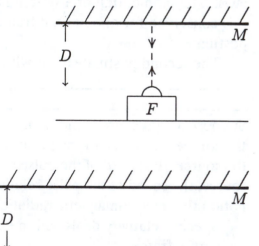

Sally carries a clock like this at speed v past Sam. Complete the diagram on the right by drawing the path of one pulse as seen by Sam. If Δt is the time interval between emission and detection of the pulse, as measured by Sam, then during this interval the flash unit moves a distance $\ell = $ _____ and the light pulse moves a distance $2L = $ _____ . This follows because L

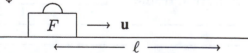

is the hypotenuse of a right triangle with sides of length D and $v\Delta t/2$. Substitute $2L = c\Delta t$ and solve for Δt:

You should obtain $\Delta t = \gamma \Delta t_0$, where the Lorentz factor γ is $1/\sqrt{1-(v/c)^2}$. The speed is often given as a fraction of the speed of light. In terms of β ($- v/c$), $\gamma =$ _____. Notice that the derivation of the result for Δt depends strongly on the second postulate of relativity. An observer comparing a moving clock with his clocks concludes that the moving clock ticks at a slower rate. It is not important that the clock utilize light as does the clock you used to derive the relationship above. Any timing device will do.

You should realize that there is perfect symmetry between the two reference frames. If Sally watches a clock at rest with respect to Sam, she sees it tick at a slower rate.

The concept of <u>proper time</u> is important for understanding the relativity of time. Suppose two events, such as the emission and detection of a light pulse, occur at the same coordinate in one frame. The time interval between them, as measured in that frame, is the proper time between the events. The time interval between the same two events, as measured in a frame that is moving relative to the first, is longer by the factor γ. Since the speed of a reference frame is always less than the speed of light, the factor γ is always greater than one.

Length contraction. If you measure the length of a rod that is moving past you at high speed, the result is less than if you measure it when it is at rest with respect to you. Which of these measurements gives the <u>proper length</u> of the rod? _____

_____

Suppose you know the speed of the rod to be v. You place a marker on your coordinate system and measure the interval from the time the front end of the rod is at the marker to the time the back end of the rod is at the marker. If that interval is Δt_0, then the length of the moving rod is $L =$ _____. Notice that Δt_0 is the proper time interval between the two events. A single clock at the position of the marker can be used. L is NOT the proper length because the rod is moving relative to the frame used to measure L.

Now consider the same events from the point of view of someone moving with the rod. He sees the marker move from the front to the back of the rod in time Δt and gives the length of rod as $L_0 = v \Delta t$. This is the proper length because the rod is at rest relative to the observer. Since Δt_0 is the proper time interval between the events, Δt and Δt_0 are related by $\Delta t =$ _____, where $\gamma = 1/\sqrt{1-(v/c)^2}$. Thus, in terms of its proper length, the length of the moving rod is $L =$ _____.

The Lorentz transformation. You should understand the phenomena of relativity not only qualitatively but also from the mathematical viewpoint of the <u>Lorentz transformation</u>. Suppose an event is viewed by two observers, one at rest in inertial frame S and another at rest in inertial frame S'. From the viewpoint of S, S' is moving in the positive x direction with velocity v. The coordinate systems and clocks are arranged so that the origins coincide at time $t = 0$ and the clocks of S' are synchronized to read 0 when the origins coincide. The coordinates of the event are x, y, z and time of the event is t, all as measured in S. The same event has coordinates x', y', z' and occurs at time t', as measured in S'. Then, according to the Lorentz transformation, the coordinates and times are related by

$$x' = $$

$$y' = $$

$$z' = $$

$$t' = $$

where $\gamma = $ _____ .

You will often deal with the time and spatial intervals between two events. Suppose the spatial interval has components Δx, Δy, and Δz in S and has components $\Delta x'$, $\Delta y'$, and $\Delta z'$ in S'. The time interval is Δt in S and $\Delta t'$ in S'. Then, the intervals are related by

$$\Delta x' = $$

$$\Delta y' = $$

$$\Delta z' = $$

$$\Delta t' = $$

You should recognize that the same equations can be used if S' is moving in the negative x direction relative to S. The velocity v is then negative.

Whether you must take special relativity into account is controlled by the value of γ. If γ is close to 1, you can safely ignore relativity. But if γ is much greater than 1, relativity is important. If $v = 0.1c$, $\gamma = $ _____ ; if $v = 0.5c$, $\gamma = $ _____ ; if $v = 0.9c$, $\gamma = $ _____ ; if $v = 0.95c$, $\gamma = $ _____ ; and if $v = 0.99c$, $\gamma = $ _____ .

You can use the Lorentz transformation equations to investigate simultaneity. Suppose two events are simultaneous in S and are spatially separated by Δx. Set $\Delta t = 0$. The time interval between them, as measured in S', is given by $\Delta t' = $ _____ if S' is moving with speed v in the positive x direction relative to S.

Now suppose two events are simultaneous in S' and are spatially separated by $\Delta x'$. Give an expression for the time interval between them, as measured in S: $\Delta t = $ _____ .

Use the Lorentz transformation equations to obtain the time dilation equation. Take $\Delta x' = 0$ (the events occur at the same coordinate in S') and solve for Δt:

You should obtain $\Delta t = \gamma \Delta t'$. The proper time interval between the events is measured in the _____ frame.

Now suppose the two events occur at the same coordinate in the S frame. Take $\Delta x = 0$ and solve for $\Delta t'$:

You should obtain $\Delta t' = \gamma \Delta t$. The proper time interval between the events is now measured in the _____ frame.

Notice that two events might occur at different coordinates in both frames. Then, the clocks in neither frame measure the proper time interval between the events.

The length contraction equation can also be derived by means of the Lorentz transformation. Suppose an object of proper length L_0 is at rest along the x axis of frame S and its length is measured in S', a frame that is moving with speed v in the positive x direction relative to S. One way of measuring the length is to place two marks on the x' axis of S', one at the front of the object and one at the back, then measure the distance between the marks. The marks, of course, must be made simultaneously. Set $\Delta x' = L$, $\Delta t' = 0$, and $\Delta x = L_0$. Use the transformation equations to solve for $\Delta x'$ in terms of Δx:

You should obtain $\Delta x' = \Delta x / \gamma$.

Now suppose the object is at rest on the x' axis of S' and its length is measured in S. Set $\Delta x' = L_0$, $\Delta x = L$, and $\Delta t = 0$. Solve for Δx in terms of $\Delta x'$:

You should obtain $\Delta x = \Delta x' / \gamma$.

Relativistic velocities. Here you deal with the relationship between the velocity of an object as measured in one frame and its velocity as measured in another, moving with speed v in the positive x direction. Carefully note that the symbol v is used for the velocity of one reference frame relative to another, and the symbol v is used for the velocity of an object.

Suppose a particle is moving with velocity u' along the x' axis of S'. Then, its velocity as measured in S is given by

$$u = $$

This expression can be derived easily from the Lorentz transformation equations. Divide $\Delta x = \gamma(\Delta x' + v \Delta t')$ by $\Delta t = \gamma(\Delta t' + v \Delta x' / c^2)$ to obtain $\Delta x / \Delta t = $ _____.
Now divide both the numerator and the denominator by $\Delta t'$, replace $\Delta x' / \Delta t'$ with u', and replace $\Delta x / \Delta t$ with u:

When $v \ll c$, the quantity $1 + u'v/c^2$ in the denominators can be approximated by 1 and the Galilean velocity transformation equation is obtained. It is $u = $ _____.

One of the most important consequences of the relativistic velocity transformation equations is: If the speed of an object, as measured in one frame, equals the speed of light, then its speed, as measured in any other frame, also equals the speed of light. No matter how fast you travel away from an approaching light pulse, its speed relative to you will be c. The same is true no matter how fast you travel toward it. Prove this for an object moving in the positive x direction: replace u' with c in the expression for u and show that the result is $u = c$.

A corollary is: If the speed of an object, as measured in one frame, is less than the speed of light, then its speed, as measured in any frame, is less than the speed of light.

The Doppler Effect. When a source of light is moving toward an observer or an observer is moving toward a source, the observed frequency is greater than the frequency f_0 measured in the rest frame of the source (the proper frequency). When the relative motion of the source and observer is one of increasing separation, the observed frequency is less than the proper frequency. Relativity predicts that the observed frequency is given by

$$f =$$

where $\beta = v/c$ and v is the relative velocity of the source and observer. If the distance between the source and observer is increasing, β is _____ ; if the distance is decreasing, β is _____ . Remember that the equation is valid only if the motion is along the line joining the source and observer.

A Doppler shift in frequency also occurs if the motion is transverse to the line joining the source and observer. Then, the observed frequency is given by

$$f =$$

This expression is a direct result of applying the relativistic time-dilation equation to the period of oscillation.

Relativistic momentum and energy. Relativity theory requires that the definitions of momentum and kinetic energy be revised if these quantities are to obey the familiar conservation laws. More precisely, if measurements taken in one inertial frame show that momentum is conserved, then measurements taken in any other inertial frame should also show that momentum is conserved. Similarly, if energy is conserved in one inertial frame then it should be conserved in all inertial frames.

If a particle has mass m and travels with velocity $\mathbf{v}$, then its momentum $\mathbf{p}$ is given by

$$\mathbf{p} =$$

If $v \ll c$, this expression reduces to $\mathbf{p} =$ _____ , the non-relativistic definition.

If a system consists of several particles, the total momentum is the vector sum of the individual momenta. If the net external force on particles of a system vanishes, then the total momentum of the system is conserved.

The kinetic energy of a particle with mass m moving with speed v is given by

$$K =$$

This expression is valid no matter what the speed v. If $v \ll c$, it reduces to the familiar non-relativistic expression $K = \frac{1}{2}mv^2$. The total kinetic energy of a system of particles is the scalar sum of the individual kinetic energies. This quantity is conserved in collisions, provided the masses of the particles do not change. Work must be done to change the kinetic energy of a particle and the change in kinetic energy equals the net work done.Ł In many nuclear scattering and decay processes, the masses do change. You must then take into the account the rest energies of the particles. The rest energy of a particle of mass m is given by $E_0 =$ _____. The total energy, the sum of the rest and kinetic energies, of a particle is given in terms of the mass and speed of the particle by

$$E =$$

The total energy and the magnitude of the momentum of any particle are related by

$$E^2 =$$

This expression replaces $E = p^2/2m = mv^2/2$, the non-relativistic relationships between E and p and between E and v. Note that the energy E in the relativistic relationship includes both rest and kinetic energies.

In several problems you are asked to find the time interval between two events in one reference frame, given the interval in another. If the two events occur at the same coordinate in one of the frames, so the time interval is measured by a single clock, then that clock measures the proper time interval Δt_0 and the interval in the other frame is given by $\Delta t = \gamma \Delta t_0$, where $\gamma = 1/\sqrt{1 - (v/c)^2}$ and v is the velocity of the second frame relative to the first. You must be able to identify which interval is the proper time interval.

You must also know that the proper length of an object is measured in the frame in which the object is at rest. Its length in a frame in which it is moving is given by $L = L_0/\gamma$.

In some cases, two events do not occur at the same place in either frame. Then, the proper time interval between the events is not measured in either frame and you cannot use the time-dilation equation. Similarly, the two events may not occur at the same time in either frame. Then, the proper distance between the events is not measured in either frame and you cannot use the length-contraction equation. You must instead use the full Lorentz transformation equations: $\Delta x' = \gamma(\Delta x - c\Delta t)$ and $\Delta t' = \gamma(\Delta t - v\,\Delta x/c^2)$.

Some examples deal with the transformation of velocities. Typically you are given the velocity of a particle as measured in one reference frame and asked for its velocity as measured in another. Use $u = (u' + v)/(1 + u'v/c^2)$. Be careful to distinguish between the particle velocity u (and u') and the velocity v of the primed frame relative to the unprimed frame.

Some problems deal with the Doppler effect. Use $f = f_0\sqrt{(1 - \beta)/(1 + \beta)}$ when the motion is along the line joining the source and observer and $f = f_0/\gamma$ if the motion is transverse to that line. In the first case, pay careful attention to the sign of β.

You should know the relativistic definition of kinetic energy: $K = mc^2(\gamma - 1)$; rest energy: mc^2; total energy: $E = mc^2\gamma$; and momentum: $mv\gamma$, as well as the relationship between total energy and momentum: $E^2 = (mc^2)^2 + (pc)^2$.

III. MATHEMATICAL SKILLS

Binomial series expansion. To understand the how some classical equations are derived from relativistic equations you need to know how an expression can be written as a binomial series. In the text this type expansion is used in discussions of addition of velocities, the Doppler shift, the definition of momentum, and the definition of kinetic energy.

The binomial series expansion deals with expressions of the form $(x + y)^n$, where y is much smaller than x in magnitude. Then

$$(x + y)^n = x^n + nx^{(n-1)}y + \frac{n(n-1)}{2!}x^{n-2}y^2 + \frac{n(n-1)(n-2)}{3!}x^{n-3}y^3 + \dots .$$

Here n can be any number whatsoever, including both positive and negative numbers. The symbol $p!$, where p is the positive integer, represents the factorial of p and is given by the product $p! = 2 \cdot 3 \cdot 4 \dots p$. Thus, for example, $2! = 2$ and $3! = 2 \cdot 3 = 6$. If n is a positive integer the series terminates after a finite number of terms, with the last term containing $x^0 y^n$. Otherwise it has an infinite number of terms.

Notice that each term in the series contains y/x to a higher power than the previous term. Thus each term is smaller in magnitude than the previous term. Often only a few terms are needed to closely approximate the original expression. In this chapter the small quantity y is sometimes the ratio v/c of the speed v of a reference frame or particle to the speed of light. In many cases it is the square of this ratio. Thus the first term of a binomial expansion, which does not contain the speed of light, gives the classical result and the second term gives the low speed relativistic correction. Thus the low speed approximation to the Lorenz factor is

$$\gamma = \frac{1}{\sqrt{1-\beta^2}} = \left(1-\beta^2\right)^{-1/2} = 1^{-1/2}(-\beta^2) + \dots = 1 + \beta^2 + \dots .$$

Here $\beta = v/c$ and the expansion is valid for $\beta \ll 1$ or $v \ll c$.

IV. NOTES

Chapter 39
PHOTONS AND MATTER WAVES

I. BASIC CONCEPTS

With this chapter, you begin your study of quantum physics. You will learn that electromagnetic radiation, which is described classically as a wave-like propagation of electric and magnetic fields, also has particle-like properties. Learn to use conservation of energy and momentum to explain the experiments that provide evidence for the particle nature of electromagnetic radiation. You will learn that particles, such as electrons, also have waves associated with them. Matter waves give rise to interference and diffraction effects and to tunneling phenomena. Pay particular attention to the relationship between the energy of a particle or photon and the frequency of the wave and to the relationship between the momentum of the particle or photon and the wavelength of the wave. For both light and matter, pay attention to the interpretation of the wave in terms of probability. Also understand the uncertainty principle relates position and momentum measurements.

Some topics from earlier chapters that you should consider reviewing are: kinetic energy from Chapter 7, momentum from Chapter 9, electrical potential energy from Chapter 25, and electromagnetic wave, wavelength of light, and frequency of light from Chapter 34, and interference from Chapter 36.

Photons. The quantum hypothesis, applied to electromagnetic radiation, says that the energy in a beam is concentrated in discrete bundles, called _____ . The energy of a single photon associated with a wave of frequency f is given by

$$E =$$

where h is the Planck constant. Its value is $h =$ _____ J·s = _____ eV·s. In terms of photons, the rate with which energy is carried by a beam is given by Rhf, where R is the number of photons per unit time that cross a plane perpendicular to their direction of travel. The intensity of a uniform beam is given by $I = Rhf/A$, where A is the cross-sectional area of the beam.

Photons carry momentum in addition to energy, the momentum carried by a single photon associated with a wave of wavelength λ being given by

$$p =$$

If we compare photons associated with waves of different frequency, the one with the higher frequency has the _____ energy and the _____ momentum. The energy and momentum are related by

$$E =$$

where c is the speed of light. For example, the energy of a single photon of visible light is about _____ eV (or _____ J) and its momentum is about _____ kg·m/s.

In some respects, electromagnetic radiation behaves as a wave while in other respects, it behaves as a collection of particles. Wavelength and frequency are characteristic of a wave; energy and momentum quanta are characteristic of particles. Carefully note that each of the equations displayed above relate a wave characteristic to a particle characteristic. In this chapter, you study some of the experiments that justify the quantum hypothesis.

You should understand the quantum mechanical interpretation of an electromagnetic wave: the square of the electric field vector at a point is proportional to the _____ that a photon will be in a small volume at that point. Carefully note that physics cannot predict where any photon will be at any time; it can only give the probability it will be in some specified region. Probability waves can interfere destructively and thereby reduce the chance that a photon will get to a given place; they can interfere constructively and thereby increase the chance that a photon will get to a given place.

The photoelectric effect. When monochromatic light shines on a sample, energy is transferred to electrons of the material. As a result, some electrons overcome the potential energy barrier that normally keeps them inside the sample and they leave the sample with various values of kinetic energy. Data taken in a photoelectric effect experiment is used to calculate the kinetic energy of the most energetic electrons emitted.

If the intensity of the incident light is increased, with the frequency remaining the same, the number of electrons ejected with maximum kinetic energy _____ but the value of the maximum kinetic energy _____. If the frequency of the incident light is increased, no matter if the intensity changes or not, the value of the maximum kinetic energy _____. Clearly, each of the electrons with maximum kinetic energy receives the same energy from the light, regardless of the intensity of the light, and the energy received is greater for higher frequency light than for lower frequency light.

The classical wave hypothesis cannot explain this phenomenon. Averaged over a cycle, the energy carried by a classical plane wave is spread uniformly over the region in which the wave exists. When the intensity is increased without changing the frequency, each electron receives more energy. If this hypothesis were valid, you would expect the kinetic energy of the most energetic electron to _____ when the intensity increases. In addition, when the light intensity is low, you would expect significant time to pass before electrons are ejected because electrons have such small _____. Experiments show that even at low intensity, electrons are ejected immediately after the light is turned on.

The photon hypothesis explains the experimental results as follows. The light is assumed to consist of a large number of photons, each with energy hf. Some fraction of the photons interact with electrons and each of those that do transfer energy _____ to an electron and disappear. If the intensity is increased without increasing the frequency, then _____ increases and the number of electrons ejected _____. Since each electron that absorbs a photon gets the same energy, the energy of the most energetic electrons does not change. Now suppose the frequency increases. Then, the energy of each photon is greater and each ejected electron gets _____ energy than before. The energy of the most energetic electron is _____ than before.

Carefully note that the fundamental event here consists of a single quantum of light interacting with a single particle of matter. Suppose a single photon, associated with frequency f, interacts with one of the most energetic electrons in the sample. If Φ is the energy needed to remove one of these electrons from the sample, then conservation of energy leads to

$$hf =$$

where K_m is the kinetic energy of the electron after leaving the sample. Φ is called the _____ of the sample. It is different for different materials. It has the SI unit _____.

You should know something about the experimental arrangement for measuring the maximum kinetic energy K_m. The sample is enclosed in a metal collector cup, which is held at a negative electric potential relative to the sample. Thus, the cup tends to repel electrons. If the potential difference is not too great, some electrons reach the cup because _____

_____

_____.

Those electrons cause an ammeter in the circuit to deflect. The potential on the cup is adjusted until _____

_____.

If this potential, called the _____ potential, is denoted by V_0, then

$$K_m =$$

gives the maximum kinetic energy.

The Compton effect. In this experiment, short-wavelength electromagnetic radiation is scattered by electrons and the wavelength of the scattered radiation is measured. It is found to be longer than the wavelength of the incident radiation and the difference depends on the scattering angle. Experiments also show that energy and momentum are transferred from the radiation to the electron.

If the classical wave hypothesis were valid, the wavelength of the scattered radiation would be the same as that of the incident radiation. Explain how this would come about: _____

_____

_____

_____

According to the photon hypothesis, the fundamental interaction, between a single photon and a single electron, can be treated as a collision in which energy and momentum are conserved. An electron essentially at rest is knocked away from its initial position by a photon. Since the energy of the electron increases in the interaction, the energy of the photon must _____. Since the photon energy is related to the frequency by $E = hf$, the frequency of the electromagnetic wave must _____ and since $\lambda = c/f$, the wavelength must _____.

The dependence on scattering angle of the energy transferred and the change in wavelength are such that momentum is conserved in the photon-electron interaction. The diagrams to the right show the situation before and after the collision. Before the collision, the electron has a momentum of _____ and a kinetic energy of _____ . If the electromagnetic radiation has wavelength λ, then the momentum of the photon is given by _____ and its energy is given by _____ .

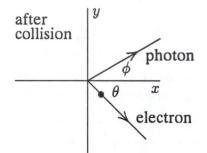

Suppose that after the collision, the scattered radiation has wavelength λ' and travels at the angle ϕ, as shown. Then, the x component of the photon momentum is _____ , the y component is _____ , and the energy of the photon is _____ . If the electron has speed v and moves off at the angle θ, then the x component of its momentum is given by _____ , the y component is given by _____ , and its kinetic energy is given by _____ . Note that relativistic expressions must be used.

Write the conservation laws below.

x component of momentum:

y component of momentum:

energy:

These equations can be solved for the change in wavelength ($\Delta\lambda = \lambda' - \lambda$) in terms of the mass m of the electron and the photon scattering angle ϕ. The result is

$$\Delta\lambda =$$

In the usual experiment, many photons (with the same frequency) are incident on many electrons. Photons leave at all angles. A detector is moved around a circle centered at the sample and the wavelength of the radiation is measured for each angular position of the detector.

The maximum change in wavelength occurs for a scattering angle of $\phi =$ _____ . For this scattering angle, the change in wavelength is given algebraically by

$$\Delta\lambda =$$

and has the numerical value $\Delta\lambda =$ _____ m (_____ pm).

Notice that the change in wavelength is independent of the wavelength itself. It is the same for visible light, gamma rays, and all other electromagnetic radiation. The reason why the Compton effect is important for gamma rays and short-wavelength x rays but not for visible light or microwaves is _____

This experiment is important because it provides evidence that photons carry momentum and that the momentum of a photon is related to the wavelength of the wave by $p = h/\lambda$.

You can find the prediction of classical physics by setting $h = 0$ in the equation for $\Delta\lambda$. The result is $\Delta\lambda = $ _____ . Classically, an electromagnetic wave does not change frequency on scattering.

For scattering from bound electrons and nuclei, the change in wavelength is much smaller than for scattering from electrons because _____

_____.

Matter waves. Particles exhibit interference phenomena. Fig. 39–9 of the text shows a uniform beam of electrons incident on a crystal. Describe in your own words the characteristics of the reflected beam that convince you that wave interference has taken place: _____

_____

Clearly a wave is associated with a particle.

Suppose the wave associated with a particle is sinusoidal, with a definite wavelength λ and a definite frequency f. The wavelength of the wave is related to the _____ of the particle and the frequency of the wave is related to the _____ of the particle. The relationships are

$$\lambda =$$

and

$$f =$$

where h is the Planck constant. These relationships for particles are exactly the same as the relationships between the energy and momentum of a photon on the one hand and the frequency and wavelength of the associated electromagnetic wave on the other.

You should be aware of the magnitudes involved. For macroscopic objects, the momentum is usually so large and the wavelength so small that interference and diffraction effects cannot be detected. For atomic particles, on the other hand, the momentum is sufficiently small and the wavelength correspondingly large that interference and diffraction effects are evident. The wavelength of the matter wave associated with a 150-g ball moving at 30 m/s is _____ m, while the wavelength of the matter wave associated with an electron (with a mass of about 9×10^{-31} kg) moving at 10^4 m/s is about _____ m. If the ball is to be diffracted by a grating, the slits should be about _____ m apart. This is clearly impossible. If the electron is to be diffracted, the slits should be about _____ m apart.

Wave functions. The wave associated with a particle has something to do with the position of the particle. Section 39–6 gives some of the details. Quantum physics deals with probabilities — the probability a particle is in a certain region at a certain time and the probability the momentum of the particle is in a certain range, for example. If the wave function of a particle moving on the x axis is represented by ψ, a function of position, then _____ gives the probability that the particle is in the small region of width dx. Here the value of the wave function at the position of the region must be used.

ψ gives only the coordinate-dependent portion of the wave function. It should be multiplied by a function of time. However, this function has magnitude 1 and $|\psi|^2$ correctly given the probability density.

Here's how to measure $|\psi|^2\,dx$, in principle. The same experiment is performed many times, always under the same conditions, and the location of the particle is detected. Now pick a small region of the x axis and calculate the fraction of experiments for which the particle is found in that region. In the limit of a large number of experiments, this fraction gives $|\psi|^2\,dx$. The quantity $|\psi|^2$ is often called a probability density. This means it is a probability per unit _____ .

The wave function ψ satisfies a differential equation, called Schrödinger's equation. Write it here:

Here $U(x)$ is _____ and m is the mass of the particle. For any given situation, if the function $U(x)$ is known, this equation is to be solved for the wave function. If the particle is free, then $U(x) = 0$ and Schrödinger's equation becomes

The wave function for a free particle traveling in the positive x direction is

$\qquad \psi(x) =$

and the wave function for a free particle traveling in the negative x direction is

$\qquad \psi(x) =$

In terms of the energy E of the particle, $k =$ _____ and in terms of the momentum p of the particle, $k =$ _____ . Notice that $E = p^2/2m$.

Uncertainty. If the wave function of a particle does not have the form Ae^{ikx}, then the particle does not have a definite momentum. If we measure the momentum a large number of times, with the particle always having the same wave function, we obtain a distribution of values. Quantum physics can be used to predict the probability that the particle momentum will be in any given range but it cannot predict the result of any of the momentum measurements. The situation is quite similar to the measurement of position: quantum physics can predict the probability that a particle is in a given region of space but it cannot predict its position at any given time.

The uncertainties in position and momentum are related. If Δx is the uncertainty in the position of a particle and Δp is the uncertainty in its momentum, then

$\qquad \Delta x \cdot \Delta p \geq$

where $\hbar = h/2\pi$. Similar relationships hold for each coordinate and the corresponding momentum component when the motion is in three dimensions. In words, the principle tells us

_____

_____

_____

Carefully note the appearance of the Planck constant h in the uncertainty relations. If the classical limit is obtained by setting $h = 0$, then the relationship for the x component becomes $\Delta p \cdot \Delta x = 0$. If the same experiment, under the same conditions, is run many times, then classically for every trial it is possible for the particle to have the same position and the same momentum at the same time after the start of the experiment. Quantum mechanically, it is not possible.

Barrier tunneling. A matter wave extends into the region beyond any potential energy barriers if the barriers have finite height and width. This means a particle may escape from such a region. Consider the one-dimensional case. The upper diagram to the right shows a potential energy barrier of height U_0 and width L. On the axes below it, sketch the probability density for a particle approaching the barrier from the left with energy less than the barrier height. Classically, this particle cannot travel to the other side of the barrier. Quantum mechanically, it has a non-zero probability of being on either side. If the barrier is made higher or wider, the wave function outside the barrier becomes _____ in magnitude.

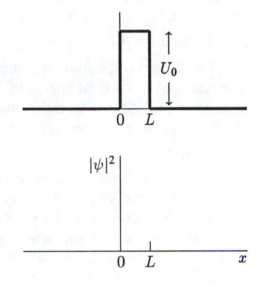

Tunneling through a barrier is described quantitatively by a transmission coefficient T and a reflection coefficient R. The transmission coefficient is defined so that it gives the probability the particle will _____ the barrier, while the reflection coefficient gives the probability the particle does not. The sum of the two is _____ .

If the transmission coefficient is small, it is given by

$$T =$$

where

$$k =$$

and E is the energy of the particle. This expression tells us that if the barrier is made wider (L is increased), then T _____ and if the barrier is made higher (U_0 is increased), then k _____ and T _____ . Notice also that the parameter k depends on $U_0 - E$. If the energy of the particle is increased (but is still not greater than U_0), then k _____ and T _____ .

The text gives some examples in which barrier tunneling plays an important role. List two of them here:

1. _____

2. _____

II. PROBLEM SOLVING

You should know the relationship between the energy E of a photon and the frequency f of the wave associated with it: $E = hf$. You should also know the relationship between the magnitude of the photon momentum and the wavelength of the wave: $p = h/\lambda$. In both classical and quantum physics the momentum and energy are related by $E = pc$. This follows immediately from $f\lambda = c$ and the relationships given above.

You should understand that a collection of N monochromatic photons (in a light beam, for example) has a total energy of Nhf. If a uniform beam of monochromatic photons has cross-sectional area A then the rate at which energy is transmitted through the area is $P = Rhf$, where R is the number of photons that pass through per unit time. The intensity is $I = Rhf/A$.

The photoelectric effect is described by $hf = K_{max} + \Phi$, where f is the frequency of the incident radiation, K_{max} is the kinetic energy of the most energetic photoelectron ejected, and Φ is the work function of the target material. You should also know that the value of K_{max} is found by measuring the stopping potential V_0 and that $K_{max} = eV_0$. The work function can be measured experimentally by finding the frequency of the light for which the stopping potential is zero. Then $hf = \Phi$.

Upon scattering from an electron initially at rest, the wavelength associated with a photon changes from λ to $\lambda' = \lambda + \Delta\lambda$, where $\Delta\lambda = (h/mc)(1 - \cos\phi)$. Here m is the mass of an electron, h is the Plank constant, c is the speed of light, and ϕ is the scattering angle (measured from the direction of incidence). The energy of the photon is reduced from $E = hf = hc/\lambda$ to $E' = hf' = hc/\lambda'$. The energy lost by the photon appears as an increase in the kinetic energy of the electron.

You should know the relationship between the energy of a particle and the frequency of its wave ($E = hf$), the relationship between the momentum of a particle and the wavelength of its wave ($p = h/\lambda$), and the relationship between the energy and momentum of a particle ($E = p^2/2m$). The latter expression is valid for a *nonrelativistic* particle of mass m. The analogous expressions for a photon are $E = hf$, $p = h/\lambda$, and $E = pc$. Only the last is different for photons and electrons.

Some problems deal with solutions to Schrödinger's equation and with the nature of the probability density $|\Psi|^2$ when Ψ is complex. Remember that the wave function for a free particle traveling in the positive x direction is $\psi = Ae^{ikx}$, where $k = 2\pi/\lambda = 2\pi p/h$, where p is the momentum of the particle and λ is the wavelength of its wave.

Some problems deal with the uncertainty principle $\Delta x \cdot \Delta p \geq h$. Usually Δx or Δp is given and you are asked to compute the other quantity.

Barrier tunneling problems usually involve a computation of the transmission coefficient $T = e^{-2kL}$, where $k = \sqrt{8\pi^2 m(U_0 - E)/h^2}$. Here L is the width of the barrier (a length), U_0 is the height of the barrier (an energy), and E is the kinetic energy of the particle incident on the barrier. T gives the probability that the particle will tunnel through the barrier. If N particles, all with the same mass and energy, are incident on a barrier, then on average, TN tunnel through. The others are reflected.

III. MATHEMATICAL SKILLS

Complex numbers. Matter wave functions are often complex quantities. This means they can be written as the sum of a real and an imaginary part: $\psi(x) = \psi_R(x) + i\psi_I(x)$, where ψ_R is the real part and ψ_I is the imaginary part. The symbol i stands for $\sqrt{-1}$.

For example, the wave function for a free particle traveling in the positive x direction is $\psi(x) = Ae^{ikx}$, where k is related to the momentum p of the particle by $p = hk/2\pi$. Since $e^{i\alpha} = \cos\alpha + i\sin\alpha$, the free-particle wave function can be written

$$\psi = Ae^{ikx} = A\cos(kx) + iA\sin(kx).$$

Both the real and imaginary parts are sinusoidal and both have the same wavelength λ, related to k by $k = 2\pi/\lambda$.

The complex conjugate of a complex number has the same real part as the original number but its imaginary part is the negative of the imaginary part of the original number. Thus, the complex conjugate of $\psi = \psi_R + i\psi_I$ is $\psi^* = \psi_R - \psi_I$ and the complex conjugate of e^{ikx} is e^{-ikx}.

The square of the magnitude of a complex number is found by multiplying the number by its complex conjugate: $|\psi|^2 = \psi\psi^*$. Thus, the square of the magnitude of ψ is

$$|\psi|^2 = (\psi_R + i\psi_I)(\psi_R - i\psi_I) = \psi_R^2 + \psi_I^2.$$

It is the sum of squares of the real and imaginary parts. The square of the magnitude of the free-particle wave function is

$$|Ae^{ikx}|^2 = Ae^{ikx} Ae^{-ikx} = A^2 e^0 = A^2,$$

where we have assumed A is real.

If a particle has energy E, its complete wave function Ψ is the product of a coordinate-dependent function and a time-dependent function. The time-dependent function has the form $e^{i\omega t}$, where ω is the angular frequency. Since $\omega = 2\pi f$ and $f = E/h$, where f is the frequency, the angular frequency is related to the energy by $\omega = 2\pi E/h$. Thus, $\Psi = \psi e^{i\omega t}$. Ψ and ψ lead to the same probability density:

$$|\Psi|^2 = |\psi e^{i\omega t}|^2 = \psi\psi^* e^{i\omega t} e^{-i\omega t} = \psi\psi^* = |\psi|^2.$$

IV. NOTES

Chapter 40
MORE ABOUT MATTER WAVES

I. BASIC CONCEPTS

In this chapter you learn some details about matter waves that will prove useful when you study atoms. Learn that the energy of a particle is quantized when the particle is confined in space, as electrons in atoms are. Also learn the mathematical forms of matter waves for an electron in a one-dimensional trap and in a hydrogen atom. Pay close attention to the quantum numbers used to designate hydrogen atom states.

Here are some topics from previous chapters that are used in this chapter: momentum of a particle in Chapter 9, wavelength in Chapter 17, and photon momentum, photon energy, particle wave function, and Schrödinger's equation in Chapter 39.

One-dimensional electron traps.   Section 40–3 gives a simple example of a trap for a particle that moves along the x axis. The diagram to the right shows the potential energy. We first assume that U_0 is infinite so the particle is trapped in a well that extends from $x = 0$ to $x = L$. No force acts on it when it is in the interior but when it reaches either side of the well, an infinite force acts on it, pushing it toward the interior. As $U_0 \to \infty$, the diagram becomes like Fig. 40–2 of the text.

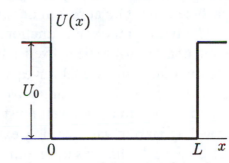

Inside the well (from $x = 0$ to $x = L$) the coordinate-dependent part of the wave function ψ satisfies the Schrödinger equation

$$\frac{d^2\psi}{dx^2} + \frac{8\pi^2 m}{h^2}E\psi = 0.$$

Outside the well, $\psi = 0$. ψ obeys the boundary conditions $\psi(0) =$ ___ and $\psi(L) =$ ___.

There are many possible wave functions for the particle. Equation 40–10 gives an expression for them. Write it here:

$$\psi_n(x) =$$

with $n = 1, 2, 3, \ldots$. The integer n is an example of what is known as a *quantum number*. The constant A in the expression you wrote above is determined by the normalization condition. The equation for that condition is

$$\int_{-\infty}^{+\infty} |\psi|^2 \, dx =$$

This condition is derived from the statement that $|\psi|^2 \, dx$ is the probability that the particle can be found between ___ and _____. Since the particle must be somewhere on the x axis, the sum of probabilities for all segments of the axis must be ___.

On the axes below, sketch the probability density $|\psi|^2$ for $n = 1, 2,$ and 3:

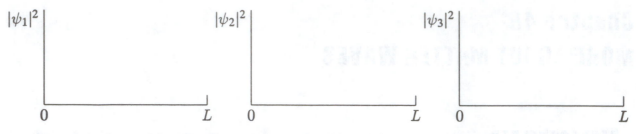

$$|\psi_1|^2 \qquad\qquad |\psi_2|^2 \qquad\qquad |\psi_3|^2$$

0   L 0 L 0 L

An energy is associated with each wave function. If the particle is in the state with wave function ψ_n, then its energy is (see Eq. 40–4):

$$E_n =$$

The energy is said to be quantized. It can have any of the values given above but it can have no others. This is quite generally true of any particle that is confined to a limited region of space. Write the confinement principle here: _____

_____

Notice that the particle has non-zero energy when it is in its lowest (or ground) state. This energy, which is given by _____, is called its _____ energy. The particle cannot be at rest under any conditions.

Notice that the wave functions are quite similar to the functions that describe the displacement of a vibrating string fixed at both ends. Both the matter wave and the string displacement are the sum of sinusoidal waves with the same frequency and wavelength, traveling in opposite directions. For the standing wave to have zero amplitude at $x = 0$ and $x = L$, the wavelength λ and the width of the well must be related by $\lambda =$ _____, where n is an integer. The traveling matter waves, if they extend throughout the entire x axis, would be associated with a free particle that has momentum with magnitude (in terms of n and L)

$$p_n = \frac{h}{\lambda} =$$

Use this result to find an expression for the allowed energy values:

$$E_n = \frac{p_n^2}{2m} =$$

The trapped particle may absorb a photon and, thus, increase its energy, but this can occur only if the photon energy equals the difference in energy between the initial state of the particle and some higher state. Suppose the electron is initially in the state with quantum number n_i when it absorbs a photon and goes to the state with quantum number n_f. Write an expression for the photon energy in terms of n_i, n_f, L, and other constants:

$$E_{\text{photon}} =$$

Similarly, a particle in an excited state can make the transition to a state with lower energy by emitting a photon. The energy of the photon equals the difference in energy between the initial and final states.

Now suppose the barriers that define the ends of the well have finite heights. On the axes below, make a sketch of the probability densities for the $n = 1, 2,$ and 3 states:

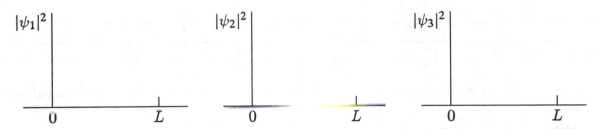

Notice that the wave function is not zero at points just outside the well, in the region forbidden to the particle by classical mechanics. Energy is still quantized, but only for those energies that are less than the barrier height. If the particle has greater energy, it is not confined by the well and its energy can have any value (greater than U_0).

Two- and three-dimensional electron traps. Suppose the electron is trapped in the plane rectangular region with sides of length L_x (along the x axis) and L_y (along the y axis. Its potential energy is zero inside the region and is infinite at the region boundary (and outside). The allowed values of the energy are then given by

$$E_{n_x, n_y} =$$

where n_x and n_y are positive integers. It is impossible for either n_x or n_y to be zero since both of these conditions lead to a wave function that is zero everywhere. That is, there is no particle in the trap.

If the region of the trap is a rectangular solid with sides of length L_x, L_y, and L_z, then the allowed values of the energy are given by

$$E_{n_x, n_y, n_z} =$$

where n_x, n_y, and n_z are positive integers. Again none of them may be zero since the wave function would then be zero.

The hydrogen atom. The wave functions appropriate to an electron in a hydrogen atom differ from those appropriate to an electron in another situation (trapped in a well, for example) because the _____ acting on the electron is different for the two cases. The important quantity for a quantum physics calculation of a particle wave function is the _____ function and this is different if different forces act. For hydrogen, which consists of an electron and proton, this function is

$$U =$$

where r is the _____ of the two particles.

The electron is trapped by the proton and quantum physics predicts a discrete set of energy levels. They are given by Eq. 40–24:

$$E_n =$$

where n is a positive integer. Notice that even in the lowest energy state ($n = 1$), the electron has energy and is not at rest. The energy is negative because the potential energy was chosen to approach zero as r approaches ___.

Wave functions for electrons in atoms are identified by a set of three quantum numbers: the principal quantum number n, the orbital quantum number ℓ, and the orbital magnetic quantum number m_ℓ. Tell what physical quantity each is most closely associated with:

n: _____

ℓ: _____

m_ℓ: _____

For the ground state of hydrogen, $n =$ ___, $\ell =$ ___, and $m_\ell =$ ___. The allowed values of the energy for a hydrogen atom depend only on ___, not the other quantum numbers. For other atoms, they also depend on ℓ.

The values of ℓ are restricted by the value of n. For a given value of n, ℓ can have any integer value from ___ to ___. The values of m_ℓ are restricted by the value of ℓ. For a given value of ℓ, m_ℓ can have any integer value from ___ to ___, including zero.

All states with the same value of n (but different values of ℓ and m_ℓ) belong to the same _____. All states with the same value of n and the same value of ℓ (but not the same value of m_ℓ) belong to the same _____.

The ground state ($n = 1$) probability density for the electron in a hydrogen atom is

$$|\psi|^2(r) =$$

where a is called the _____. In terms of fundamental constants, it is given by _____ and its value is _____ pm.

The electron now moves in three-dimensional space, rather than along the x axis and the square of the wave function gives the probability per unit _____ that it can be found in a small volume. If dV is an infinitesimal volume a distance r from the proton, then the probability the electron can be found in that volume is given by $|\psi^2|\,dV$. The <u>radial probability density</u> $P(r)$ is defined so that $P(r)\,dr$ gives the probability that the electron can be found in a spherical shell of width dr a distance r from the proton. Since the volume of the shell is $dV = 4\pi r^2\,dr$, the radial probability density is given in terms of the wave function by

$$P(r) =$$

The radial probability density for the ground state wave function of hydrogen is

$$P(r) =$$

Use the axes to the right to sketch a graph of P as a function of the distance r from the proton. Label the r axis to indicate the distance scale. In terms of the Bohr radius, the maximum of the function occurs at $r =$ _____ a. The fraction of the time the electron can be found inside a sphere of radius a is about _____ and the fraction of the time it can be found outside the sphere is about _____.

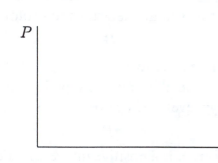

The $n = 1$ state is spherically symmetric; the wave function depends only on the distance r from the proton and not on any angular variables. The same is true of the $n = 2$, $\ell = 0$ state and, in fact, of any state for which $\ell = 0$. Wave functions for the three $n = 2$, $\ell = 1$ states, however, depend on the angular coordinates θ and ϕ as well as on r. But the sum of the probability densities for these three states is _____. Thus, the probability distribution for any completely filled subshell is spherically symmetric.

A hydrogen atom in an excited state can emit a photon and make a transition to a lower energy state. If n_i is the principal quantum number associated with the initial state and n_f is the principal quantum number associated with the final state, then the energy of the photon is given by

$$E_{\text{photon}} = \underline{\hspace{3cm}}$$

The frequency of the electromagnetic wave associated with the transition is given in terms of the photon energy by $f = \underline{\hspace{2cm}}$. Possible transitions are grouped into series, with all members of a series having the same value of ____. If $n_f = 1$, the transition is a member of the _____ series; if $n_f = 2$, the transition is a member of the _____ series; and if $n_f = 3$, the transition is a member of the _____ series. The series limit of any series is obtained by setting $n_i = $ ____. For any series, this radiation has the _____ frequency and the _____ wavelength of any radiation in the series.

II. PROBLEM SOLVING

Some problems deal with the allowed values of the energy of a particle trapped in a well with infinite barriers at the ends: $E_n = n^2 h^2 / 8mL^2$, where L is the width of the well and n is an integer that designates the quantum mechanical state of the particle. Other problems use a particle trapped in a well to demonstrate some general properties of wave functions. If the particle is in the state n, then its wave function is $\psi = \sqrt{2/L}\sin(\pi x/L)$. Remember that $|\psi|^2(x)\,dx$ gives the probability that the particle can be found between x and $x + dx$ and that $\int_0^L |\psi|^2\,dx = 1$ is a statement that the particle is guaranteed to be found somewhere in the well.

Some problems deal with the allowed values of the energy of a hydrogen atom, with the possible wavelengths and frequencies of the radiation emitted when a hydrogen atom makes a transition from one state to a lower energy state, or with the frequency and wavelength of radiation that can be absorbed by a hydrogen atom. The allowed values of the energy are given by

$$E_n = -\frac{me^4}{8\epsilon_0^2 h^2}\frac{1}{n^2} = -\frac{13.6\,\text{eV}}{n^2},$$

where n is a positive integer. Remember $E = 0$ means the electron is just free of the proton and has no kinetic energy. The energy of the photon is the magnitude of the difference in energy of the two states involved in the transition: $hf = |E_f - E_i|$. $E_f - E_i$ is positive for an absorption event and negative for an emission event.

Some problems involve the wave functions for the electron in a hydrogen atom. The electron moves in three-dimensional space and $|\psi|^2\,dV$ gives the probability it can be found in the infinitesimal volume dV. The radial probability density is given by $P(r) = 4\pi r^2|\psi|^2$ and $P(r)\,dr$ gives the probability that the particle can be found in the spherical shell with inner radius r and outer radius $r + dr$.

III. NOTES

Chapter 41
ALL ABOUT ATOMS

I. BASIC CONCEPTS

Here you learn about the structure of atoms and, in particular, about the states of multi-electron atoms. Atomic structure is reflected in the periodic table of chemistry and is fundamental to our understanding of the properties of materials. An important new idea is the Pauli exclusion principle, which greatly influences the structures of atoms and their chemical interactions and determines many important details of the periodic table. The chapter closes with an explanation of how a laser works.

Some topics from previous chapters that are used in this chapter are: angular momentum (Chapter 12), electrical potential energy (Chapter 25), magnetic dipole (Chapters 29 and 30), atomic magnetism (Chapter 32), and photon momentum, photon energy, particle wave function, and probability density (Chapter 39).

Properties of atoms. The text mentions five important properties of atoms that require quantum mechanics to understand them. They are:

1. _____
2. _____
3. _____
4. _____
5. _____

Angular momentum and magnetic dipole moment. The magnitude of the orbital angular momentum of the electron in an atom is quantized. The allowed values are given by

$$L =$$

where ℓ is zero or a positive integer, called the _____ quantum number. It can be no greater than ___, the _____ quantum number. The constant $\hbar$ is the Planck constant divided by _____. Its value is _____ J·s.

Each component of the angular momentum is also quantized. The z component, for example, may have only the values

$$L_z =$$

where m_ℓ is an integer (positive, negative, or zero) and is called the _____ quantum number. If the orbital quantum number is ℓ, the possible values of m_ℓ run from _____ to _____ and include all integer values between. Identical expressions are valid for the x and y components. Note carefully that if the z component of the angular momentum has a definite value, the other components do not. The quantum number m_ℓ refers to a single component; you cannot specify all three. The quantization of L_z is often called space quantization.

Because the z component of the angular momentum is quantized, the angle θ between the angular momentum vector and the z axis can have only certain values, given by

$$\cos^{-1}\theta = \frac{L_z}{L} = \frac{m_\ell}{\sqrt{\ell(\ell + 1)}}.$$

For a given value of ℓ, the smallest angle occurs when $m_\ell =$ _____.

A magnetic dipole moment is associated with the orbital motion of the electron. At the atomic level, magnetic dipole moments are conveniently measured in units of the Bohr magneton μ_B. In terms of fundamental constants of nature,

$$\mu_B =$$

Its value is _____ J/T or _____ eV/T. If an electron is in a state with magnetic quantum number m_ℓ, then, in terms of m_ℓ and μ_B, the z component of its orbital magnetic dipole moment is

$$\mu_{orb,z} =$$

Notice that the direction of the dipole moment is opposite to the direction of the angular momentum. This is because the electron is _____ charged.

Spin angular momentum. An electron also has an <u>intrinsic</u> angular momentum, called its spin angular momentum. Its magnitude is given by

$$S =$$

where s is the spin quantum number of the electron and has the value $s =$ ____. The z component is quantized in the same manner as orbital angular momentum:

$$S_z =$$

where m_s has either of the values ____ or ____.

A magnetic dipole moment is associated with spin angular momentum. If an electron has spin quantum number m_s, then, in terms of m_s and μ_B, the z component of its spin dipole moment is

$$\mu_{S,z} =$$

Except for a factor of two, this is the same as the relationship between the z component of the orbital dipole moment and the z component of the orbital angular momentum.

Spin angular momentum and its quantum number are not predicted by Schrödinger's equation but they are predicted by _____ quantum physics.

The total angular momentum of an atom is denoted by **J** and is the vector sum of the orbital angular momenta and spin angular momenta of all its electrons. Similarly, the magnetic dipole moment of an atom is the vector sum of the orbital and spin dipole moments of all the electrons. The dipole moment may not be parallel to **J**.

The Stern-Gerlach experiment. This experiment provides an important experimental verification of space quantization. A beam of atoms passes through a non-uniform magnetic field to a screen. If the field is in the z direction and varies with z, then it exerts a force on the atoms in either the positive or negative z direction, depending on the direction of the magnetic dipole moment of the atom.

If μ_z is the z component of the magnetic moment of an atom in a magnetic field **B** that is in the positive z direction, then the potential energy of the dipole-field interaction is

$$U = $$

If dB_z/dz is the rate at which the field varies with z, then the force is given by

$$F_z = -\frac{dU}{dz} = $$

Suppose the magnetic force is vertically upward for atoms with a positive z component of their magnetic moment and vertically downward for particles with a negative z component. Consider a beam of atoms such that the orbital and spin magnetic moments of all the electrons in each atom cancel except for the moment of a single electron in an $\ell = 0$ state and suppose that half the atoms in the beam are in $m_s = +\frac{1}{2}$ states and half are in $m_s = -\frac{1}{2}$ states. Describe the pattern on the screen: _____

_____

For contrast, describe the pattern you would expect to see if angular momentum were not quantized: _____

_____

Magnetic resonance. Each proton also has an intrinsic angular momentum and an intrinsic spin magnetic dipole moment, in the same direction as its spin angular momentum. In a magnetic field, the dipole moment can either be parallel or antiparallel to the field. Suppose the magnetic field is in the positive z direction and has magnitude B. In terms of the z component μ_z of the dipole moment, the energies of the two orientations differ by $\Delta E = $ _____.

If electromagnetic radiation of the appropriate frequency is incident on a proton in the lower energy state, it can absorb a photon and jump to the higher energy state. This process is called magnetic resonance. The presence of protons can be detected by monitoring the electromagnetic beam for absorption. In terms of μ_z and B, the resonance frequency f is given by $f = $ _____.

The magnetic field that appears in this equation is actually the sum of the externally applied field and the local field produced by the immediate environment of the protons being irradiated. Magnetic resonance experiments are used to measure the local magnetic fields of atoms in materials. Magnetic resonance is also used to map the proton density in human bodies and, as a result, find widespread medical use.

The Pauli exclusion principle. Electrons obey the Pauli exclusion principle, which states that _____

For practice think about electrons in a square trap with sides of length L. The single-particle energy levels are given by

$$\frac{h^2}{8mL^2}\left(n_x^2 + n_y^2\right),$$

where the quantum numbers n_x and n_y are positive integers. No two electrons in the trap can have the same set of values for the quantum numbers n_x, n_y, and m_s. That is, the value of at least one of these four quantum numbers must be different for every electron.

The lowest single particle energy is $2(h^2/8mL^2)$, corresponding to $n_x = 1$ and $n_y = 1$. The number of electrons that can have this energy is ____ (don't forget the spin quantum number). The next highest single-particle energy is $5(h^2/8mL^2)$, corresponding to $n_x =$ ____, $n_y =$ ____ and to $n_x =$ ____, $n_y =$ ____ . . The number of electrons that can have this energy is ____ .

Suppose there are five electrons in the trap and the system is in its ground state (the state with the lowest total energy). Then two of the electrons are in states with energy $2(h^2/8mL^2)$. Both have $n_x = 1$ and $n_y = 1$. One has $m_s = +1/2$ and the other has $m_s = -1/2$. Three electrons have energy $5(h^2/8mL^2)$. One might have $n_x = 1$, $n_y = 2$, and $m_s = -1/2$, a second might have $n_x = 1$, $n_y = 2$, and $m_s = +1/2$, and the third might have $n_x = 2$, $n_y = 1$, and $m_s = -1/2$. There are other possibilities. In any event, the energy of the five-electron ground state is $(2)(2)(h^2/8mL^2) + (3)(5)(h^2/8mL^2) = 19(h^2/8mL^2)$. The electrostatic potential energy associated with the interactions of the electrons with each other are neglected here.

To find the energy of the first excited state of the system start with the ground state configuration and jump one of the electrons to a higher energy single-particle state. In the case you are considering the some of the states with energy $5(h^2/8mL^2)$ are empty, so one or two electrons might jump from a state with energy $2(h^2/8mL^2)$ to a state with energy $5(h^2/8mL^2)$. On the other hand, an electron might jump from a state with energy $5(h^2/8mL^2)$ to a state with energy $8(h^2/8mL^2)$, the next highest single-particle energy. Examine all possibilities and see which produces the lowest energy above the ground state energy. Fill in the following table for five electrons in the square well. For each of three lowest values of the single-particle energy tell how many states (including spin) are associated with that energy and how many electrons have that energy.

Energy	Number of States	Number of electrons
$2h^2/8mL^2)$	_____	_____
$5h^2/8mL^2)$	_____	_____
$8h^2/8mL^2)$	_____	_____

Verify that the total energy is $22(h^2/8mL^2)$.

The periodic table of chemical elements. The quantum mechanical state of an electron in an atom is specified by giving the values of four quantum numbers. They are:

1. the principal quantum number n, which specifies _____ .

2. the orbital quantum number ℓ, which specifies _____
_____ .

3. the magnetic quantum number m_ℓ, which specifies _____
_____ .

4. the spin quantum number m_s, which specifies _____
_____ .

All electrons with the same values of ____ and ____ are said to belong to the same subshell. When labeling a subshell, the various values of ℓ are designated by lower case letters: $\ell = 0$ is designated ____, $\ell = 1$ is designated ____, $\ell = 2$ is designated ____, $\ell = 3$ is designated ____, $\ell = 4$ is designated ____, and $\ell = 5$ is designated ____. A particular subshell is designated by a symbol such as 3d, which indicates that $n =$ ____ and $\ell =$ ____. The number of electrons in the subshell is written as a superscript. Thus $3d^2$ indicates that there are two electrons in the 3d subshell.

Because electrons obey the Pauli exclusion principle, no two electrons have all four quantum numbers with the same values. For example, electrons with the same values of n and ℓ have different values of m_ℓ or m_s (or both).

Furthermore, when an atom is in its ground state, the electrons fill the various states in such a way that the total _____ is the least possible. You may think of starting with an atom with atomic number $Z - 1$ and constructing an atom with atomic number Z by adding a single proton to the nucleus and a single electron outside. If the atom is in its ground state, the electron goes into the state that gives the atom the lowest possible energy without violating the Pauli exclusion principle. It goes into a previously unoccupied state.

The text considers a neon atom, with ten electrons. _____ of these occupy the $n = 1$, $\ell = 0$ subshell, _____ occupy the $n = 2$, $\ell = 0$ subshell, and _____ occupy the $n = 2$, $\ell = 1$ subshell. The 1s, 2s, and 2p subshells each have as many electrons as possible and are said to be closed.

For every electron in a closed subshell with any given value of the orbital magnetic quantum number, there will be another with the negative of that value; for every electron with any given value of the magnetic spin quantum number, there will be another with the negative of that value. This means that the electrons in a closed subshell have _____ total angular momentum and _____ total magnetic dipole moment.

A sodium atom has eleven electrons. In the ground state of the atom, ten of them are in the same subshells as the ten electrons of a neon atom. The other is in the $n =$ ____, $\ell =$ ____ subshell. In terms of $\hbar$, the magnitude of the z component of the total angular momentum of this atom is _____ and in terms of μ_B, the magnitude of the z component of the magnetic dipole moment is _____ .

Neon is an inert gas. A neon atom interacts extremely weakly with other atoms and does not enter into chemical reactions. Sodium, on the other hand, is very active chemically. Explain why the addition of a single electron changes the chemical properties so dramatically:

A chlorine atom has seventeen electrons. In the ground state, ten have the same configuration as neon, _____ fill the 3s subshell, and _____ are in the 3p shell. Chlorine is very active chemically. It combines easily with atoms that have an additional electron outside a closed shell. This electron can occupy the sixth state in the 3p subshell.

An iron atom has twenty-six electrons. Eighteen of them occupy closed subshells. These are the ___, ___, ___, ___, and ___ subshells. Six are in the ___ subshell and two are in the ___ subshell. Note that the 4s subshell starts filling before the 3d subshell is full. The $3d^6 4s^2$ configuration has a lower energy than the $3d^8$ configuration.

The periodic table of the chemical elements groups atoms with similar chemical properties in the same column, with atoms in neighboring columns of the same row differing by _____ in the number of electrons. Application of quantum mechanics and the Pauli exclusion principle explains the similarities of the properties of atoms in the same column and the variation of properties from column to column.

The number of columns is different for different rows (periods) of the table. Explain why:

_____

_____

X-ray spectra.   Look at Fig. 41–14, which shows the spectrum of x rays that are emitted from a molybdenum target when 35-keV electrons strike it. There are two main parts: the relatively broad, low-level, continuous spectrum and the two peaks, called the characteristic spectrum. In addition, an important feature of the continuous spectrum is the cut-off wavelength; no x rays with wavelengths less than λ_{min} are ever emitted when electrons with energy of 35 keV are incident. You should be able to explain the origin of both the continuous and characteristic spectra and explain the reason for a cut-off wavelength.

The basic fact you must know to understand the existence of a continuous spectrum is that an accelerating charge emits electromagnetic radiation. Describe what happens to an incident electron that causes it to decelerate after it enters the target: _____

_____

After an electron is accelerated through a potential difference V, its kinetic energy is $K = $ _____ and after it enters the target, it may lose any amount from 0 to K in a decelerating event. Radiation with the greatest possible frequency (and so the shortest possible wavelength) is emitted by those electrons that lose _____ their kinetic energy in a single event. The maximum frequency is, therefore, given by $h f_{max} = eV$, so the shortest possible wavelength is given by $\lambda_{min} = $ _____. Note that the value of λ_{min} is independent of the target material.

Radiation of the characteristic spectrum is emitted when an incoming electron knocks another electron from a low-energy state, an $n = 1$ state for example. An electron from a state with higher energy makes the transition to the empty state and emits a photon. For most atoms the difference in energy of the two states is sufficient to produce a photon in the x-ray region of the electromagnetic spectrum. The K_α x-ray line is produced when electrons fall from an $n =$ _____ state (labeled an L state) to an $n =$ _____ state (labeled the K shell). The K_β line is produced when electrons fall from an $n =$ _____ state (labeled an M state) to a K state. Radiation with other frequencies is produced when electrons with still higher energies fall into the vacated states.

X-ray energy level diagrams are usually drawn as in Fig. 41–16 of the text. Carefully note that $E - 0$ represents an atom with all electrons in the states with the lowest possible energies consistent with the Pauli exclusion principle. The energy labeled K represents an atom with an empty K state, the energy labeled L represents an atom with an empty L state, etc. A downward arrow represents a transition in which the energy of the atom decreases and a photon is emitted.

X-ray emissions can be used to determine the position of a chemical element in the periodic table. In his pioneering work, Moseley showed that the frequency of the K_α line, for example, changes in a regular way from element to element in the table. If f is the x-ray frequency associated with this line, then a plot of _____ as a function of position in the table produces a nearly straight line. Moseley concluded that the order of atoms in the table depends on the number of _____ in the nucleus.

You should be able to use the results of quantum physics to understand this result. If there are Z protons in the nucleus, then the effective charge that acts on an electron in a K state is somewhat less than Ze. Explain why: _____

_____

Write $(Z - 1)e$ for the effective charge. According to quantum physics, the energy associated with a deep lying state with principal quantum number n is

$$E_n =$$

This is the same as the expression for hydrogen atom energies except that e^4 in the numerator has been replaced by $(Z - 1)^2 e^4$. The frequency associated with K_α-photons is given by

$$f =$$

This shows that $\sqrt{f}$ is proportional to $Z - 1$. A Moseley plot is essentially a graph of $\sqrt{f}$ vs. Z and is nearly a straight line. Z is called the atomic number of the chemical element. The expression you wrote for E_n gives a very poor approximation to the energies of outer electrons in many-electron atoms but it is quite good for the innermost electrons responsible for x-ray emission.

Explain why Moseley plots are of great historical interest: _____

_____

_____

Lasers. List the four properties of laser light that distinguish it from light emitted by an ordinary tungsten filament:

1. _____

2. _____

3. _____

4. _____

Laser light is produced in the following process. An electron in a state with high energy is stimulated by an incoming photon to drop to a state with lower energy and emit a photon. The energy of the incoming photon must match the difference in energy of the two states ($hf = E_2 - E_1$) and the energy of the stimulated photon is exactly the same. After the emission, there are two photons, identical in every way. If there are sufficient electrons in the higher energy state, each of these photons can stimulate the emission of another photon and both the original photon and the new photon can in turn stimulate the emission of more. The number of identical photons quickly multiplies.

Laser beams are highly monochromatic because all the stimulated photons have the same _____ and all the waves associated with them have the same _____ . Laser beams are highly coherent because all the waves associated with the stimulated photons have the same _____ . Laser beams do not spread significantly because all the stimulated photons travel in the same _____ . Actually some spreading does occur because the waves are diffracted as they pass through the window of the laser.

To produce laser light, the following conditions must hold:

1. The higher in energy of the two states must be metastable. Explain what this term means.

_____

_____

Explain why it is important that the higher energy state be metastable. _____

_____

2. The populations of the two states must be inverted. Explain what population inversion is.

_____

_____

Note that an incoming photon with the correct energy to stimulate emission also has the correct energy to be absorbed. Use this to explain why population inversion is important.

_____

_____

Explain how population inversion is accomplished in a helium-neon laser: _____

_____

_____

_____

3. Lasers usually have a mirror at each end. Explain why. _____

_____

4. One of the mirrors is an excellent reflector while the other is a partial reflector and lets some of the laser light out.

II. PROBLEM SOLVING

You should know that the magnitude of the orbital angular momentum of an electron in an atom is given by $L = \sqrt{\ell(\ell + 1)}\hbar$ and that for a given value of the principal quantum number n the allowed values of the orbital quantum number ℓ are the integers 1, 2, ..., $n - 1$. You should know that the z component of the angular momentum is $L_z = m_\ell\hbar$ and that for a given value of ℓ the allowed values of m_ℓ are the integers from $-\ell$ to $+\ell$, inclusive. You should also know that the allowed values of the z component of the spin angular momentum are $m_s = -1/2$ and $m_s = +1/2$.

Some problems ask you to compute the magnitude of the orbital or spin contributions to the magnetic dipole moment of an electron in an atom. Use $\mu_{orb} = \sqrt{\ell(\ell + 1)}\mu_B$, $\mu_{orb,\,z} = -m_\ell\mu_B$, and $\mu_{s,z} = -m_s\mu_B$, where μ_B is the Bohr magneton.

Some problems deal with the Stern-Gerlach experiment. You should know that states with different values of m_ℓ have different energies in a magnetic field and know how to compute the energy difference. Use $U = -\mu_z B$ for a magnetic field in the z direction. You should know how to calculate the force of a non-uniform magnetic field on a magnetic dipole. Use $F_z = -dU/dz$. You should be able to interpret a given pattern on the observing screen in terms of the allowed values of the z component of the magnetic dipole moment of the atoms passing through the apparatus. There will be one spot for each different value of the z component of the magnetic dipole moment. In many cases the dipole moment of the atom is the same as the spin dipole moment of a single electron.

To find the deflection of a given atom by the Stern-Gerlach apparatus use Newton's second law the equations for motion with constant acceleration.

Other problems deal with magnetic resonance. You should know that the magnitude of the change in energy when the spin flips is given by $\Delta E = 2\mu_z B$, where the magnetic field was taken to be in the z direction. At resonance, ΔE equals the energy of the absorbed photon and the frequency f of the associated electromagnetic wave is given by $hf = \Delta E$.

Some problems deal with consequences of the Pauli exclusion principle. To find which single-particle states are occupied when the system is in its ground state, list the energies of the single-particle states in order of increasing energy and note any degeneracies, including those due to spin. Then assign particles to the states in order of increasing energy until all the particles have been assigned. Be sure that no two particles have the same set of values for the quantum numbers that describe the single-particle states. The total energy of the system is the sum of the energies of the particles. A similar procedure can be used for excited states. In this case you search for total energies of the system that are greater than the ground state energy. The same technique can be used for electrons in one- and two-dimensional traps and for electrons in atoms.

You should be able to distinguish between continuous and characteristic x-ray spectra, be able to compute the wavelength cutoff for a given energy of the incident electrons, and be able to compute the characteristic wavelengths for a given set of energy levels. You should also know how to use quantum physics to predict a characteristic spectrum for a given target material and be able to interpret a Moseley plot.

Some problems require you to understand the operation of a three-level laser. You may be asked to calculate the wavelength or frequency of the laser light, given the energy levels or you might be asked to calculate the populations of the levels, in thermodynamic equilibrium or after pumping has occurred. If atoms in the laser jump from energy E_i to lower energy E_f, then the frequency f of the light is given by $hf = E_i - E_f$. In thermodynamic equilibrium the number of atoms with an energy E is proportional to $e^{-E/kT}$, where T is the temperature on the Kelvin scale and k is the Boltzmann constant.

III. NOTES

Chapter 42
CONDUCTION OF ELECTRICITY IN SOLIDS

I. BASIC CONCEPTS

Here the ideas of modern physics are used to understand one of the important properties of solids, their electrical conductivity. As you study the chapter, be sure to pay attention to distinctions between metals, insulators, and semiconductors. These materials differ greatly in the fraction of their electrons that participate in electrical conduction and in the changes that occur in their resistivities when the temperature changes. You should understand how the differences come about. Later sections of the chapter are devoted to solid-state devices, so pervasive in modern technology.

Topics that have previously been covered and are used in this chapter include temperature from Chapter 19, conductor, insulator, and semiconductor from Chapter 22, resistivity, temperature coefficient of resistivity, electric current, and electron relaxation time from Chapter 27, photon energy from Chapter 39, and quantized energy levels from Chapters 40 and 41.

Resistivity.   Much of this chapter is based on Eq. 27–20 for the resistivity. It is

$$\rho = \underline{\hspace{3cm}}$$

where m is the _____ of an electron and e is its _____. n is number of _____ electrons per unit volume and τ is the average time between _____.

The quantity n enters the expression for ρ because the more conduction electrons there are, the more will pass through a unit area per unit time when an electric field exists in the sample, all else being the same. That is, if the same electric field exists in two samples for which the mean free time is the same, the one with the greater number of conduction electrons per unit volume will have the larger current. τ enters the expression because it gives the average time that the field accelerates an electron before the electron is stopped by a collision. The greater the mean free time, the greater the drift speed. A greater drift speed means more electrons pass through a unit area per unit time, all else being equal.

In the remainder of the chapter, pay close attention to how quantum physics is used to determine n.

Energy bands and gaps.   The allowed values of the electron energy for a crystalline solid form bands of energy levels, separated by gaps. A crystalline solid is one in which the equilibrium positions of the atoms form a pattern that is repeated with uniform spacing and the same orientation throughout the solid. Look at Fig. 42–1 to see two such patterns. The entire crystal can be generated by placing cubes side-by-side and on top of each other with their edges aligned. A band of levels is _____

In one view, energy bands arise as atoms are brought close together because the wave functions for electrons originally associated with different atoms _____. In another view, an electron originally associated with one atom experiences electrical forces that arise from charges on neighboring atoms.

The diagram on the right shows several groups of closely spaced energy levels, separated by gaps in which there are no levels. Label the bands B and gaps G. In reality, the number of states in a band is huge, about the same as the number of atoms in the crystal. The true separation between levels cannot be shown on the diagram.

When you have finished reading about the distinguishing properties of metals, insulators, and semiconductors, mark on the diagram the highest occupied state for electrons in a metal at the absolute zero of temperature, assuming it to be in the third band from the bottom. Do the same for an insulator or semiconductor. Label these levels so they are meaningful to you when you review.

Metals. For a metal, the states in one band, called the conduction band, are partially occupied. States in lower energy bands are completely occupied and states in higher energy bands are all unoccupied. Electrons in the conduction band are responsible for the current when an electric field is turned on.

The energies of these electrons are chiefly kinetic energies, the variations in the potential energy function being small over most of the volume of the crystal. To a first approximation, we may assume that the conduction electrons of a metal are simply trapped in a box the size of the sample and take the potential energy to be zero on the inside and to be infinitely high at the surface. The energy levels are given by an expression similar to Eq. 40–21. The levels are extremely close together because the box has macroscopic dimensions (several centimeters, for example). You should realize that this model is applicable only to electrons that are not bound to atoms and that are, therefore, free to move throughout the solid. For most metals, this amounts to roughly one electron per atom.

When a system such as a solid has a great many closely spaced energy levels, a quantity called the <u>density</u> <u>of</u> <u>states</u> and denoted by $N(E)$ is used to describe them. The quantity $N(E)\,\mathrm{d}E$ tells us how many quantum mechanical _____ per unit _____ of sample have energies between _____ and _____. For a free electron metal, it is given by

$$N(E) =$$

You should know how to use this function. The number of states with energies between E_1 and E_2, for example, is given by the integral

$$N =$$

where V is the volume of the sample. If the energy interval $\Delta E\ (=\ E_2 - E_1)$ is extremely small, the integral can be approximated by $N = N(E)\,\Delta E\,V$, as it is in Sample Problem 42–3 of the text.

Not all states are filled. The probability that, at absolute temperature T, a state with energy E contains an electron is given by the Fermi-Dirac occupancy probability:

$$P(E) =$$

where E_F is a parameter called the Fermi energy. This function takes the Pauli exclusion principle into account: no state may be occupied by more than one electron. The density of *occupied* states N_o (which is the same as the number of electrons per unit volume of sample per unit energy interval) is given by the product of $N(E)$ and $P(E)$:

$$N_o(E) =$$

At the absolute zero of temperature, the probability that a state with energy less than _____ is occupied is one; these states are guaranteed to be filled. The probability that a state with energy greater than _____ is occupied is zero; these states are guaranteed to be empty. The Fermi energy at $T = 0\,\mathrm{K}$ is determined by the condition that the number of occupied states per unit volume between $E = 0$ and $E = E_F$ is the same as the number of conduction electrons per unit volume in the metal. The determining condition is written mathematically as Eq. 42–8 of the text. Copy it here:

$$n =$$

It leads to the relationship between the Fermi energy and the electron concentration n (see Eq. 42–9 of the text):

$$E_F =$$

When the temperature is raised, an extremely small fraction of the electrons, with energies quite near _____, receive energy. This means that some states with energies slightly above the Fermi energy become occupied while some states with energies slightly below the Fermi energy become unoccupied. The energies of the vast majority of electrons do not change.

When there is no electric field in the metal, the average electron velocity is zero because for every electron traveling in any direction another is traveling with the same speed in the opposite direction. When a field exists, slightly more electrons occupy states with momentum in the direction opposite that of the field than occupy states with momentum in the direction of the field. There is then a current. If we examine the electron distribution when a field exists, we see that some states, near the Fermi energy and with momentum opposite the field, are occupied while other states, also near the Fermi energy but with momentum in the direction of the field, are unoccupied.

If the electrons did not suffer collisions with atoms of the solid, the momenta of the electrons would continue to increase in the direction opposite the field. Electrons with energies near the Fermi energy, however, do suffer collisions. Explain why electrons with less energy do not: _____

You should know that interactions with ions in a perfectly periodic arrangement, as in a crystal, cannot cause an electron to change its state. Electrons have collisions only when the periodicity is destroyed. This might be because there are impurities or other defects present or simply because the atoms are vibrating. Collisions with vibrating atoms are responsible for the temperature dependence of the resistivity of a metal. As the temperature is raised, the amplitude of these vibrations _____ and the mean time between collisions _____ . A smaller value for τ means a larger value for ρ. The other quantities in the expression for ρ, including the concentration of free electrons n, are comparatively insensitive to the temperature of a metal. Thus, the resistivity _____ as the temperature increases.

Insulators and semiconductors. For both insulators, such as carbon in the form of diamond, and semiconductors, such as silicon and germanium, there are precisely enough electrons in the solid to completely fill all the states in an integer number of bands, with none left over. At $T = 0\,\mathrm{K}$, the most energetic electron is in the highest energy state of one of the bands and is separated in energy from the empty state above it by a gap. The highest filled band is called the _____ band and the lowest empty band is called the _____ band.

At higher temperatures, electrons are thermally promoted across the gap from the valence to the conduction band. You use the Fermi-Dirac occupancy probability to find the probability that a state in the conduction band is occupied and the probability that a state in the valence band is unoccupied. The difference between an insulator and a semiconductor is that the gap for _____ is much larger than the gap for _____ , so at any given temperature, there are far fewer electrons in the conduction band of _____ than in the conduction band of _____ .

You should understand that electrons in a completely filled band do not contribute to an electrical current. This is because _____
_____ .

Electrons in semiconductors and insulators must be promoted across the gap to the conduction band before an electric field will generate a current. For an insulator, exceedingly few electrons cross the gap, even at high temperatures. For semiconductors, many more electrons cross the gap, although the number is still far less than the number of free electrons in a typical metal. At room temperature, the resistivity of a semiconductor is much greater than the resistivity of a metal because _____

_____

_____ .

When electrons have been excited to the conduction band of a semiconductor, both the conduction and valence bands are partially filled and both contribute to the current when an electric field is turned on. There are two contributions to the conductivity, one from the conduction band and one from the valence band.

Rather than deal with the contributions of the vast number of electrons in the valence band of a semiconductor, the band may be thought to consist of a much smaller number of fictitious particles, called _____ , one for each empty state. These particles have positive charge and are accelerated in the direction of an applied electric field. When semiconductors are discussed, the word "electron" usually means an electron in the conduction band and the word "hole" means a hole in the valence band.

As the temperature increases, the mean free time decreases for both electrons in the conduction band and holes in the valence band but, unlike for a metal, this does not mean the resistivity increases. Both the number of electrons in the conduction band and the number of holes in the valence band increase dramatically with temperature. Because of this, the resistivity of a semiconductor _____ as the temperature increases.

The number of electrons in the conduction band or the number of holes in the valence band can be greatly increased by doping a semiconductor. Explain what a doped semiconductor is: _____

_____

_____

To increase the number of electrons in its conduction band, a semiconductor is doped with donor replacement atoms. Each such atom normally has _____ electrons in its outer shell. When a donor atom is substituted for a host atom (silicon or germanium, for example), _____ of these electrons form bonds with neighboring atoms. The other electron is in a hydrogen atom-like state around the impurity. Its energy is in the gap between the valence and conduction bands and it is easily promoted to the conduction band. Semiconductors that are doped with donors are said to be _____-type semiconductors.

To increase the number of holes in the valence band, a semiconductor is doped with acceptor atoms. Each of these normally has _____ electrons in its outer shell and can easily accept an electron from the valence band, thereby creating a hole in that band. Semiconductors that are doped with acceptors are said to be _____-type semiconductors.

Semiconductor devices. The basic building block of nearly all semiconductor devices is the p-n junction, consisting of p-type and n-type semiconducting materials in contact. Electrons from the _____-type material diffuse to the _____-type material, where they fall into holes. Holes from the _____-type material diffuse to the _____-type material, where electrons combine with them. The electron and hole currents, called _____ currents, are in the same direction, from the _____ side toward the _____.

Electron and hole diffusion leaves a small region near the boundary on the n side with _____ charged ions, the donors that have lost their electrons, and a small region near the boundary on the p side with _____ charged ions, the acceptors that have gained electrons (or lost holes). As a result, an electric field exists near the boundary. It points from the _____ side toward the _____ side. The electric potential on the _____ side is higher than the electric potential on the _____ side. This electric field pushes electrons toward the _____ side and pushes holes toward the _____ side. This is the _____ current. If the circuit is not complete, the drift and diffusion currents cancel and the net current is _____. The difference in the electric potential of the two sides is called the _____ potential difference. The region in which the electric field exists is called the _____ region because it contains few charge carriers.

p-n junctions are often used as rectifiers. Explain what rectification is: _____

_____

An *ideal* rectifier has _____ resistance for current in one direction and _____ resistance for current in the other direction.

When a source of emf is connected to a *p-n* junction, with the positive terminal at the *p* side, the junction is said to be _____ biased. The electrical resistance of the junction is small and the current is large. If the positive terminal of the emf is connected to the *n* side, the junction is said to be _____ biased. The electrical resistance is large and the current is small. On the axes to the right, sketch the current as a function of the potential difference V across a junction. Take V to be positive for forward bias and negative for back bias.

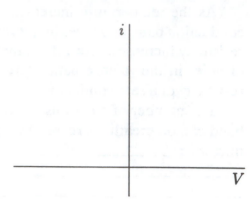

A forward bias on a *p-n* junction lowers the electron energy levels on the _____ side relative to those on the other side. This has two effects: the population of electrons in the conduction band on the *n* side and the population of holes in the valence band on the *p* side both increase dramatically and the barrier for forward current is lowered. A back bias on the junction raises the electron levels on the _____ side relative to those on the other side. The populations decrease and the barrier is raised.

For most semiconductors the energy released when an electron in the conduction band falls into an empty state in the valence band ends as thermal energy. For some, light is emitted. The most used light-emitting semiconducting material is _____. For most light-emitting semiconducting materials, the photon energy equals the energy gap between the two bands. If the gap is E_g, then the frequency of the light is $f =$ _____ and its wavelength is $\lambda =$ _____ . By using appropriate semiconducting materials the gap width and hence the color of the emitted light can be controlled.

To obtain reasonable intense light there should be a large number of electrons in the conduction band and a large number of empty states in the valence band. Doping with donors can produces a large electron population but it reduces the number of empty states in the valence band. Instead forward-biased *p-n* junctions are used. Explain how this solves the problem: _____

_____

_____

_____

The same idea is used to construct solid-state lasers but now the device is designed so that the photon produced when one electron falls to the valence band stimulates a downward jump by another electron.

Field-effect transistors make use of the dependence of the depletion region width on an applied bias. Two *n*-type regions, for example, are imbedded in a *p*-type semiconductor and are connected by a channel of *n* type material. A potential difference applied to the regions generates a current in the channel. The width of the channel are controlled by a back bias applied to the junction formed by the channel and the substrate. The depletion region has high electrical resistivity because few charge carriers are there and small changes in the bias potential can produce large changes in the resistance and thus large changes in the current in the channel. As a result, the device can be used for amplification.

II. PROBLEM SOLVING

You should know how to compute the density of states, the occupation probability, and the density of occupied states for the conduction electrons of a metal. You should also know how to compute the Fermi energy. Use

$$N(E) = \frac{8\sqrt{2}\pi m^{2/3}}{h^3} E^{1/2}$$

for the density of states,

$$P(E) = \frac{1}{e^{(E-E_F)/kT} + 1}$$

for the occupation probability,

$$N_o(E) = N(E)P(E)$$

for the density of occupied states, and

$$E_F = \left(\frac{3}{16\sqrt{2}\pi}\right)^{2/3} \frac{h^2}{m} n^{2/3}$$

for the Fermi energy. Some problems may ask you for the Fermi momentum p_F or Fermi speed v_F, related to the Fermi energy by $E_F = \frac{1}{2}p_F^2/m = \frac{1}{2}mv_F^2$. Here are some examples.

Some problems deal with semiconductors. The same probability function is valid and it is important for describing the thermal promotion of electrons across the gap between the valence and conduction bands. The Fermi energy for a pure semiconductor is determined by the condition that the density of occupied states in the conduction band equals the density of unoccupied states in the valence band. Doping with donor or acceptor replacement atoms changes the density of occupied states for the two bands and also changes the Fermi energy but the same general principles apply.

III. NOTES

Chapter 43
NUCLEAR PHYSICS

I. BASIC CONCEPTS

This chapter is about the nuclei of atoms. A nucleus takes up only an extremely small fraction of the atomic volume but accounts for most of the mass of an atom. You will learn about the constituents. You will also learn that one nucleus might change into another by the emission of alpha or beta particles. Pay attention to the energy considerations that allow you to predict whether a nucleus is stable and, if it is not, to find the energy of the decay products. Because nuclear decay is a random process, the rate of decay for a large collection of nuclei can be described in statistical terms. Learn the mathematics of random decay.

Here are some topics from previous chapters that you might review in preparation for studying this chapter: radioactive decay in Chapter 10, Coulomb's law in Chapter 22, energy associated with rest mass in Chapter 38, and quantum mechanical energy states and energy levels in Chapters 40 and 41.

The atomic nucleus. Rutherford's experiment determined that the nucleus of an atom contains positively charged particles compressed into an extremely small fraction of the atom's volume. The experiment consisted of firing a beam of _____ particles at a gold foil. The distinguishing feature of the results that led him to his conclusion was _____ _____ .

You should know some facts about atomic nuclei. They are made up of nucleons, a term that encompasses both _____ and _____ . The number of protons is called the _____ number and is denoted by _____, the number of neutrons is called the _____ number and is denoted by _____, and the total number of nucleons is called the _____ number and is denoted by _____. The word <u>nuclide</u> is used to designate a nuclear species. Two nuclides are different if they differ in either their atomic numbers or their neutron numbers. If two nuclides have the same atomic number but differ in their neutron numbers, they are said to be _____ .

All nucleons attract each other via the _____ force. This force is extremely strong but it also has an extremely short range. Two nucleons must be separated by less than about _____ m to experience the force. As a result, a nucleon interacts with only close neighbors. In addition, protons repel each other via long range electrical forces. As a result, stable nuclei with small mass numbers have the same number of neutrons as protons but stable nuclei with large mass numbers have more _____ than _____ . The extra neutrons participate in the _____ interactions, which hold the nucleus together, but not in the _____ interactions, which tend to tear it apart.

The nucleus of an atom contains most of the mass of the atom but takes up only a small fraction of the volume. The surface of a nucleus is somewhat ill defined but an average radius

R can be measured. In terms of the mass number A, it is given by

$$R =$$

where R_0 is about _____ fm. The radius of a mid-size nucleus is less than the radius of an atom by a factor of about _____. A femtometer is _____ m.

Nuclear masses are commonly measured in <u>atomic</u> <u>mass</u> <u>units</u>, abbreviated _____. 1 u is about _____ kg. The mass of a nucleus in u, rounded to the nearest integer, is its _____ number. Because the volume of a nucleus and its mass are both proportional to A, the densities of all nuclei are nearly the same, about _____ kg/m^3.

The mass of a stable nucleus is less than the sum of the masses of the constituent nucleons. The difference accounts for the binding together of the nucleons to form the nucleus. If Δm is the mass difference, then the binding energy is given by $\Delta E_{\text{be}} =$ _____, where c is the speed of light. This is the energy that must be given a nucleus to separate it into its constituent particles, well-separated and at rest. In detail, if m is the mass of a nucleus with Z protons and N neutrons, then

$$\Delta E_{\text{be}} =$$

where m_p is the mass of a proton and m_n is the mass of a neutron. You should be aware that atomic, not nuclear, masses are usually tabulated. The mass of an appropriate number of electrons must be subtracted from the atomic mass to obtain the nuclear mass or else the atomic mass of a nucleus with Z protons and N neutrons must be compared to the sum of the masses of Z hydrogen atoms and N neutrons.

Fig. 43–6 of the text shows the binding energy per nucleon as a function of mass number. The shape of this curve is important. Nuclides in the vicinity of iron have the greatest binding energy per nucleon and the curve drops as A becomes either larger and smaller. Energy is released when two nuclei with low mass numbers combine to form a single nucleus. This process is called nuclear _____. Energy is also released if a nucleus with high mass number breaks into smaller fragments. This process is called _____.

An internal energy is associated with a nucleus and this energy has a discrete set of allowed values, a different set for each type nucleus. The difference in energy of adjacent low-lying states is on the order of _____ eV and when a nucleus changes state from a higher to a lower energy, the photon emitted is in the _____ portion of the electromagnetic spectrum.

Nuclei also have intrinsic angular momenta and magnetic dipole moments. Observed values of the angular momentum are roughly the same as for electrons in atoms but the magnetic dipole moments are smaller by a factor of about _____ because _____ _____.

Radioactive decay. Most conceivable nuclei are not stable. An unstable nucleus may spontaneously turn into another nucleus with the emission of one or more particles. Alpha decay, for example, involves the emission of a _____ nucleus and beta decay involves the emission of either an _____ or a _____. Spontaneous decay occurs only if the mass of the nucleus is _____ than the sum of the masses of the decay products.

Although any nucleus in a collection of identical unstable nuclei might decay in any given time interval, it is impossible to predict which will actually decay. This means that the number that decay in any small time interval Δt is proportional to the number N of undecayed nuclei present at that time and to the interval itself: $\Delta N = -\lambda N \Delta t$, where λ is a constant of proportionality, called the _____ constant. The negative sign appears because ΔN is negative (N decreases). In the limit as $\Delta t \rightarrow 0$, this equation becomes

$$\frac{dN}{dt} =$$

It has the solution

$$N(t) =$$

where N_0 is the number of undecayed nuclei at time $t = 0$.

The decay rate or activity R is defined as $R = $ _____. It also follows an exponential law:

$$R =$$

where R_0 is the decay rate at $t = 0$. Notice that the exponent is the same as the exponent in the law for N. A common unit of the decay rate is the becquerel (abbreviated ___). $1\,\text{Bq} = $ _____ disintegrations/s.

A radioactive decay is often characterized by its <u>half-life</u>. Define this term: _____
_____

_____

The half-life $T_{1/2}$ is related to the disintegration constant λ by

$$T_{1/2} =$$

The decay rate R and the number N of undecayed nuclei both decrease by a factor of _____ in every half-life interval. Study Sample Problems 43–4 and 43–5 of the text to see how the half-life is calculated from experimental data when it is long and when it is short compared to the observation time.

The reduction in mass that occurs with a decay, multiplied by c^2, gives the _____ energy and is denoted by _____. This energy appears as the kinetic energy of the decay products and as the excitation energy of the daughter nucleus, if it is left in an excited state. A nucleus in an excited state decays to its ground state with the emission of one or more photons. The excitation energies involved usually lead to photons in the _____ portion of the electromagnetic spectrum.

When a nucleus decays by emission of an alpha particle, the daughter nucleus has an atomic number that is _____ less than that of the parent nucleus, a neutron number that is _____ less, and a mass number that is _____ less.

The half-lives of alpha emitters range from less than a second to times that are longer than the age of the universe. The combination of the strong nuclear force of attraction and the electrical repulsion of the daughter nucleus for the alpha particle leads to a high potential energy barrier that tends to hold the alpha inside. Classically, it can never escape because its energy is less than the barrier height, but alphas are emitted because quantum mechanical _____ is possible. The half-life depends sensitively on the _____ and _____ of the barrier. The disintegration energy is shared by the alpha and the recoiling daughter nucleus.

In β^- decay, a _____ is converted to a _____ with the emission of an _____ and a _____. The daughter nucleus has an atomic number that is _____ greater than that of the parent nucleus, a neutron number that is _____ less, and a mass number that is the same. In β^+ decay, a _____ is converted to a _____ with the emission of a _____ and a _____. The daughter nucleus has an atomic number that is _____ less than that of the parent nucleus, a neutron number that is _____ greater, and a mass number that is the same. Neutron-rich nuclides tend to decay by _____ emission while proton-rich nuclides tend to decay by _____ emission.

The disintegration energy is shared by the β particle, neutrino, and recoiling daughter nucleus, with nearly all of it going to the first two. For a collection of identical parent nuclei, the energies of the emitted β particles span a wide range but the sum of the β and neutrino energies is the same for every decay event. The β particle and neutrino are created in the decay process. They do not exist inside the nucleon before decay.

Neutrinos are rather ephemeral particles. The mass of a neutrino is too small to be measured and may be zero. It interacts only extremely weakly with matter. Enormous numbers, most from the Sun but many from other stars, pass through Earth and our bodies every second. They are extremely difficult to detect.

Two different units are used in the study of radiation damage. For each of them, tell what quantity it is used to measure:

gray   (abbreviated ___)   1 gray = _____ mJ/kg
Used to measure: _____

_____

sievert   (abbreviated ___)
Used to measure: _____

_____

Nuclear models.   Three models are described in the text. Briefly explain how the nucleus is viewed in each.

The collective model: _____

_____

_____

The independent particle model: _____

_____

_____

The combined model: _____

_____

_____

II. PROBLEM SOLVING

Some problems examine details of the Rutherford experiment. Remember that momentum and energy are conserved in all scattering, reaction, and decay events. For most (but not all) of these events you should use the relativistic expressions for momentum ($\vec{p} = \gamma m \vec{v}$ and energy $E = \gamma m c^2$). Here $\gamma = 1/\sqrt{1 - v^2/v^2}$. You should also know the relationship between kinetic energy and momentum ($(pc)^2 = K^2 + 2Kmc^2$) and the relationship between total energy and momentum ($E^2 = (pc)^2 + (mc^2)^2$).

Some problems test your understanding of nuclear properties. You should know the meaning of the terms mass number, atomic number, and neutron number and the relationship between them. You should also know the relationship between the radius of a nucleus and its mass number and how to calculate the binding energy of a nucleus, given its mass and the masses of its constituent nucleons.

Some problems deal with the number and rate of decays. You should know the relationship between the disintegration constant λ and the half-life $T_{1/2}$. The half-life is often given but you need the disintegration constant to carry out the calculation. In an original collection of N_0 nuclei, the number that remain undecayed at the end of time t is given by $N = N_0 e^{-\lambda t}$. The rate of decay is given by $R = \lambda N$ and by $R = R_0 e^{-\lambda t}$, where R_0 is the decay rate at $t = 0$. You should also know that $R = -\lambda N$. $R = R_0 e^{-\lambda t}$ is used to find the disintegration constant from experiment data when the half-life is short compared to the observation time and $R = -\lambda N$ is used when the half-life is long.

Some problems deal with the energetics of α and β decays. You should know that a helium nucleus (two protons and two neutrons) is emitted in an α decay and that either an electron or a positron is emitted in a β decay. In each case, you should be able to find the resulting heavy nucleus if the original nucleus is given. You should also be able to calculate the energy Q released in a decay if the various masses are given. Recall that Q is the energy equivalent of the reduction in mass that occurs in a decay. If masses are given in atomic mass units, the conversion factor $931.5\,\text{MeV/u}$ can be used. Remember that the masses given are often the masses of atoms, not nuclei, and you must, in principle, subtract an appropriate number of electron masses. These may cancel in the equation you are evaluating but you should check in each case. You should also know how to compute the height of the Coulomb barrier to α decay.

Some problems deal with radiation dosages. You should be able to distinguish between a becquerel, a gray, and a sievert.

Some nuclear reactions occur, not spontaneously like decays, but only after an incoming nucleus brings energy to a target nucleus. The total mass of the products is greater than the total original mass, the initial kinetic energy of the incoming nucleus or the excitation energy of the compound nucleus providing the additional energy. Another set of problems deals with the nuclear models discussed in the text.

III. NOTES

Chapter 44
ENERGY FROM THE NUCLEUS

I. BASIC CONCEPTS

Two important nuclear phenomena, the fission of a heavy nucleus into two lighter nuclei and the fusion of two light nuclei into a heavier nucleus, both convert internal energy associated with nucleon-nucleon interactions into kinetic energy of the products. In each case, your chief goal should be to understand the basic process. Learn what the products are, what energy is released, and what inhibits the process. In addition, you will learn about occurrences of the phenomena in nature and about human attempts to produce and control nuclear energy in a sustained fashion.

Some topics that have been discussed in earlier chapters and are used in this chapter are: temperature (Chapter 19), distribution of molecular speeds (Chapter 20), energy associated with rest mass (Chapters 38 and 43), nuclear binding energy, binding energy per nucleon, alpha decay, and beta decay (Chapter 43).

Nuclear fission. In the basic fission process, a thermal neutron is absorbed by a heavy nucleus to form a compound nucleus and the compound nucleus breaks into two medium-mass nuclei (fission fragments), with the emission of one or more neutrons. Look at Fig. 43–6 of the text and notice that the binding energy per nucleon is greater for nuclei with mass numbers closer to $A = 56$ than for heavier nuclei. The total internal energy of the fragments is less than the internal energy of the original compound nucleus. Internal energy is converted chiefly to kinetic energy of the fragments.

Describe what is meant by a thermal neutron: _____

_____

The kinetic energy of a thermal neutron is about _____ eV.

Starting with the same nucleus, the mass numbers of the fragments may be different for different fission events and experiments give the relative probabilities for the various possible outcomes. Look at Fig. 44–1 of the text to see the distribution of fragments for the fission of ^{236}U. Notice that the fragments are not usually identical. In fact, the probability for identical fragments is near the minimum of the curve. The most likely fragments have mass numbers of _____ and _____ .

Fission takes place in several steps. First, the compound nucleus breaks into what are called the primary fragments and these fragments immediately emit one or more neutrons, called the prompt neutrons. The fragments are usually still neutron-rich and they decay by _____ emission. This decay often leaves a fragment nucleus in an excited state and if the excitation energy is sufficient it may decay via the emission of neutrons, although most times the decay is via gamma emission. Neutrons, if they are emitted, are called <u>delayed</u> neutrons because _____

_____ .

The disintegration energy can be computed from the change in total mass. Consider the fission of the heavy compound nucleus F into the fragments X and Y, with the production of b neutrons:

$$F \rightarrow X + Y + b\text{n} .$$

Let m_F be the mass of F, m_X be the mass of X, m_Y be the mass of Y, and m_n be the mass of a neutron. Then, the disintegration energy Q is given by

$$Q =$$

Q is positive, indicating the release of energy. Most of this energy appears as the kinetic energy of _____ but some appears as the kinetic energy of the neutrons, and if X and Y are the final stable nuclei, some appears as the kinetic energies of electrons and neutrinos produced in the β decays.

Not all neutron-rich heavy nuclides fission. As the two fragments pull apart, they must overcome the strong mutual attraction of nucleons for each other. The potential energy as a function of the separation of the fragments is shown in Fig. 44–3 of the text. Note the barrier height E_b. The neutron absorbed by the target nucleus must supply about this much energy for fission to take place. Once the fragment separation is beyond the peak in the potential energy curve, the potential energy decreases with separation because the fragments are both _____ charged and repel each other.

The most energy that can be supplied by a thermal neutron is equal to its _____ energy E_n. This energy can be calculated from the difference in mass of the target nucleus and the so-called _____ nucleus formed by adding a neutron to the target. If this energy is not about the same as the barrier height or greater, fission does not occur and the compound nucleus loses the excitation energy by means of _____ emission. On the other hand, fission is favored if $E_n \geq E_b$. A more energetic neutron can sometimes be used to produce a fissionable nucleus for which the barrier is high and the excitation energy is low. Its kinetic energy is transferred to the compound nucleus.

Fission reactors.   To produce a large number of fission events in a reactor, the fissionable material is made to undergo a <u>chain reaction</u>. Explain what this is: _____

_____

_____

If fission takes place in a solid or liquid, most of the released energy appears as internal energy and raises the temperature of the material. In a reactor power generator, the energy is used to raise the temperature of _____ and the resulting steam is used to drive a turbine.

The text discusses three problems that arise in the design of a fission reactor and their solutions. Carefully read Section 44–4, then describe each of the problems in your own words and briefly tell how it is solved.

1. Neutron leakage: _____

_____

_____

Solution: _____

_____

_____

2. Neutron energy: _____

_____

Solution: _____

_____

_____

3. Neutron capture: _____

_____

_____

Solution: _____

_____

_____

_____

If the number of neutrons present in a reactor remains constant with time, the reactor is said to be _____; if the number decreases, it is said to be _____; and if the number increases, it is said to be _____. The value of the <u>multiplication</u> factor k indicates the tendency. If N neutrons that will participate in fission events are present at some time, then after the fission events occur, there will be _____ neutrons present to participate in the next round of fission events. The value of k is _____ if the reactor is critical, _____ if it is subcritical, and _____ if it is supercritical. For steady power generation, $k =$ _____.

The multiplication factor is varied by inserting and withdrawing <u>control</u> <u>rods</u>, which readily absorb _____. To increase the number of fission events per unit time, some rods are _____ for a short time, during which k is _____ than one. The rods are then _____, so k becomes one again. Because some neutron emissions are delayed, the rods are effective in controlling power generation even though there is a response time for inserting and withdrawing them.

Thermonuclear fusion. Nuclear fusion occurs when two light nuclei fuse together to form a single heavier nucleus. Recall from your study of Chapter 43 that for low mass numbers, the binding energy per nucleon is much greater for heavier nuclei than for lighter. The heavy nucleus has much less internal energy and the excess appears as kinetic energy of the products. Products include the final nucleus and perhaps neutrons, β particles, neutrinos, and photons.

The nuclei, of course, are positively charged and repel each other electrically. They must overcome a potential energy barrier before they get close enough for the strong force of attraction to be effective. The barrier height is about _____ keV for two deuterons.

In thermonuclear fusion, thermal energy is the chief means by which the light nuclei obtain enough energy to overcome the barrier. Notice that the mean thermal energy need not be nearly as great as the barrier height. There are two reasons: the nuclei can tunnel through the barrier rather than go over the top and, in a gas at temperature T, there are many nuclei with energies greater than the mean ($\frac{3}{2}kT$). Look at Fig. 44–10 of the text to see the distribution of kinetic energies for molecules in a gas.

In Fig. 44–10 of the text the curve labelled $n(K)$ gives the number of nuclei per unit energy range as a function of their kinetic energy. The curve labelled $p(K)$ gives the _____ of barrier penetration for protons with kinetic energy K. The kinetic energy that maximizes the product $n(K)p(K)$ and so maximizes the number of fusion events is about _____ keV. At higher energies, the probability of barrier penetration is greater but there are fewer protons with sufficient energy. There are more protons with lower energies but the probability of barrier penetration is less.

Fusion processes are responsible for the internal energy of the Sun. The proton-proton cycle is diagramed in Fig. 44–11 of the text. Complete the symbolic statements of the process below and tell what happens at each stage.

1. $^1H + {}^1H \rightarrow$
   Two protons fuse to form a _____ nucleus. A _____ (e^+) and a _____ (ν) are also formed. The e^+ annihilates with a free _____ with the result that two _____ are produced.

2. $^2H + {}^1H \rightarrow$
   Here a deuterium nucleus and a proton fuse to form a _____ nucleus with the emission of another _____.

3. $^3He + {}^3He \rightarrow$
   Two light helium nuclei fuse to form an _____ particle and two _____.

The overall process can be written

$$4^1H + 2e^- \rightarrow$$

The disintegration energy for the entire process is about _____ MeV. Some of this energy, about _____ MeV, is carried away by the neutrinos but most is available to maintain the internal energy and keep the Sun shining. It is estimated that the proton-proton cycle can continue for about _____ y before the Sun's hydrogen is gone.

When all the hydrogen has been fused into helium, the Sun's core will collapse and as gravitational potential energy is converted to kinetic energy, it will get hotter. In the hot dense core helium can fuse into heavier elements. Explain why a higher temperature is required for these fusion events than for the proton-proton cycle: _____

_____

_____

Chemical elements with mass numbers beyond $A =$ _____ cannot be formed in fusion events. Explain why: _____

_____

_____

Controlled fusion. Currently, attempts are being made to produce sustained controlled fusion for purposes of generating electric power. The three fusion processes that hold the most promise are

$$^2H + {}^2H \rightarrow$$

$$^2H + {}^2H \rightarrow$$

$$^2H + {}^3H \rightarrow$$

The goal of fusion research is to maintain a high-temperature, high-density gas (a plasma) of particles for sufficient time that a significant number of fusion events take place. If n is the particle concentration and τ is the confinement time, then Lawson's criterion for a successful fusion reactor is

$$n\tau \geq$$

In addition, the temperature must be high enough to allow particles to overcome the electrical barrier to their fusion.

Two types of confinement technologies are presently being developed. Magnetic confinement makes use of a _____ field to maintain high particle concentrations and to keep the hot plasma away from walls. In the inertial confinement method, a high-power laser beam is used to ionize and compress small pellets of deuterium and tritium. Fusion takes place in the core of the pellet. Both methods attempt to reach the critical value of the Lawson number $n\tau$.

II. PROBLEM SOLVING

In a primary fusion event, before any beta decay of the primary fragments, the number of protons and the number of neutrons are separately conserved. You can use the conservation of neutrons and protons to predict one of the fission products if the others are known. Don't forget to count any neutrons that are emitted in the fission process.

The primary fission fragments may decay, via the emission of beta particles and neutrons, to the final stable products. Since a neutron turns into a proton in the beta decays of interest here you can tell how many such decays occur by comparing the number of protons in the original nucleus or in the primary fission fragments with the number in the final stable products.

The energy released in a fission event can be calculated from the change in the total mass of the system. Don't forget that atomic masses are usually given and you must convert to nuclear masses if the electron masses do not cancel.

The fissioning fragments attract each other via the strong nuclear force and repel each other via the electrostatic force. The maximum in the barrier occurs approximately when their surfaces are touching. You can use the expression for the electrostatic potential energy of two spherical charge distributions to estimate the height of the barrier. Use $r = r_0 A^{1/3}$ to compute the nuclear radius.

Some problems deal with the operation of fission reactors. You should know that the multiplication factor k allows you to calculate the change in the number of neutrons from one generation to the next. You should also know how to calculate the power output from the number of neutrons present and the generation time.

You should know how the total energy released by a collection of fissionable nuclei or by a collection of fusing protons depends on the value of Q for a single fission or fusion event.

III. NOTES

Chapter 45
QUARKS, LEPTONS,
AND THE BIG BANG

I. BASIC CONCEPTS

This chapter deals with the ultimate constituents of matter, as far as known, and with the evolution of the universe from its start at the Big Bang. As you will learn, these topics are closely related. Pay attention to the types of particles that exist and to the mechanisms by which they interact. This information can be used to unravel the history of the early universe. Also learn of the evidence for the Big Bang and for dark matter, on which the future of the universe may depend.

Here are some topics that were discussed in earlier chapters and are used in this one: conservation of energy in Chapter 8, conservation of momentum in Chapter 9, particle decay in Chapter 10, angular momentum in Chapter 12, spin angular momentum in Chapters 32 and 41, relativistic energy and momentum in Chapter 38, and beta decay in Chapter 43.

Classification of particles.   Particles may be classified according to their intrinsic angular momentum (or spin). The maximum value of any cartesian component of the intrinsic angular momentum is given by

$$S_z =$$

where m_s is either an integer or half integer and $\hbar$ is the Planck constant divided by 2π.

Particles that have m_s equal to 0 or an integer are called _____ while particles that have m_s equal to half an odd integer are called _____. _____ obey the Pauli exclusion principle; _____ do not. Particles that obey the exclusion principle must be in different _____. Suppose a system has four possible states, with energies E_1, E_2, E_3, and E_4 (in increasing order). If there are three bosons in the system, the least total energy is obtained if _____ particles are in the state with energy E_1, _____ are in the state with energy E_2, _____ are in the state with energy E_3, and _____ are in the state with energy E_4. If the three particles are fermions, the least total energy is obtained if _____ particles are in the state with energy E_1, _____ are in the state with energy E_2, _____ are in the state with energy E_3, and _____ are in the state with energy E_4.

Give some examples of fermions: _____

Give some examples of bosons: _____

Particles can be classified according to the types of interactions they have with other particles. There are _____ different basic interactions (or forces).

1. Every particle attracts every other particle with a <u>gravitational</u> force, the weakest of all the basic forces. It is important for large macroscopic bodies but it is insignificant for interactions on the subatomic scale. Give some phenomena for which gravity is important:

   _____

   _____

2. The <u>weak</u> interaction is the next strongest interaction. Give at least one phenomenon for which it is important: _____

   _____

3. The next strongest interaction is the <u>electromagnetic</u> interaction. Give some phenomena for which it is important: _____

   _____

   Particles that carry _____ interact electromagnetically.

4. The <u>strong</u> interaction is the strongest force of all. Give at least one phenomenon for which this interaction is important: _____

   _____

   Not all particles interact via the strong force. Those that do are called _____ . These particles are further categorized according to their spin. Particles that interact strongly *and* have integer (or zero) spin are called _____ . Give some examples: _____

   _____

   Particles that interact strongly *and* have half-integer spin are called _____ . Give some examples: _____

   _____

   Particles that do not interact via the strong force but do interact via the weak are called _____ . Give some examples: _____

   _____

   An <u>antiparticle</u> is associated with each particle. Tell what is meant by this term: _____

   _____

Except for the positron (e^+), an antiparticle is denoted by placing a _____ over the symbol for the particle.

Leptons. Counting the neutrinos but not counting the antiparticles, there are six members of the lepton family. They are: _____ (symbol: _____), _____ (symbol: _____), _____ (symbol: _____), _____ (symbol: _____), _____ (symbol: _____), and _____ (symbol: _____). The _____, _____, and _____ are charged _____ ; the neutrinos have _____ charge. All have intrinsic angular momentum _____ $\hbar$.

None of the leptons participate in the _____ interaction; all participate in the _____ interaction. Charged leptons participate in the _____ interaction, while uncharged leptons do not. No evidence of any internal structure has been observed for any lepton. They are thought to be truly fundamental.

Notice that there is a neutrino associated with each of the other types of leptons (an electron neutrino, a muon neutrino, and a tauon neutrino). Although they have the same charge (0) and spin ($\frac{1}{2}$) and perhaps have the same mass, we know they are different particles because _____

_____

_____ .

Mesons.

List some charged mesons: _____

List some neutral mesons: _____

All have antiparticles associated with them, although for some, the particle and antiparticle are identical. Not all mesons have an intrinsic angular momentum but for those that do, the angular momentum is an integer multiple of _____. The distinguishing characteristic of a meson is that it is an integer spin particle that participates in the strong interaction.

Baryons.

List some charged baryons: _____

List some neutral baryons: _____

Baryons have spin angular momentum that is an odd multiple of _____. The distinguishing characteristic of a baryon is that it is a half-integer spin particle that participates in the strong interaction.

Conservation laws. There are several conservation laws that hold in every particle inter-action or decay. You have already learned about the conservation of energy, momentum, angular momentum, and charge. Carefully read Section 45–3 to see how these conservation laws are used to determine the properties of particles. In this section, two new conservation laws are described.

A <u>baryon number</u> $B = +1$ is assigned to each _____, a baryon number $B = -1$ is assigned to each _____, and a baryon number $B = 0$ is assigned to every other particle. The sum of the baryon numbers before an event is always equal to the sum of the baryon numbers after the event. This means that if a baryon decays, a baryon must be among the products. If an antibaryon decays, an antibaryon must be among the products. If two baryons interact, two baryons, perhaps different from the original baryons, must result.

Carefully note that there is no similar conservation law for mesons. Mesons can be created or destroyed in any number without violating any conservation law.

<u>Strangeness</u> is a particle property that is conserved in strong and electromagnetic inter-actions but not in weak interactions. Particles are assigned strangeness S by observing the results of decays and interactions. K^+ and K^0, for example, have $S =$ _____; K^- and $\overline{K}^0$ have $S =$ _____.

When a decay or reaction takes place via either the strong or electromagnetic interaction, the total strangeness of the products must be the same as the total strangeness of the original particles. If the decay or reaction takes place via the weak interaction, the total strangeness may change.

The quark model. A model has been developed to explain the properties of baryons and mesons, the particles that interact via the strong interaction. These particles are thought to be composite — made of smaller, more fundamental particles called _____. Every meson is a combination of a _____ and an _____ whereas every baryon is a combination of three _____ and every antibaryon is a combination of three _____. Quarks have baryon number _____ and antiquarks have baryon number _____, so the model predicts that the baryon number of a meson is _____, the baryon number of a baryon is _____, and the baryon number of an antibaryon is _____, in agreement with observation.

So far six quarks (and the associated antiquarks) have been discovered, in the sense that they are required for the construction of observed mesons or baryons. Many physicists believe that there are no more. The lowest mass mesons and baryons are combinations of the u (up), d (down), and s (strange) quarks, along with the associated antiquarks. The quark content of the π^- meson, for example, is _____; the quark content of the π^+ meson is _____. Kaons, which are strange mesons, have an s or $\bar{s}$ quark. The quark content of a proton is _____ and the quark content of a neutron is _____. Σ baryons are strange and have s or $\bar{s}$ quarks. Carefully compare Figs. 45–4 and 45–5 to see the quark content of other mesons and baryons.

Quarks are charged and so participate in electromagnetic interactions, but the charge on a quark is a fraction of the fundamental charge e, not a multiple of it. A u quark has a charge of _____ e, a d quark has a charge of _____ e, and an s quark has a charge of _____ e. The charge on an antiquark is the negative of the charge on the associated quark. The charges on quarks that make up a proton sum to _____ e, the charges on quarks that make up a neutron sum to _____ e, the charges on quarks that make up a π^+ meson sum to _____ e, and the charges on quarks that make up a π^- meson sum to _____ e.

Quark number is conserved in strong interactions. That is, the strong interaction can create and destroy quark-antiquark pairs (such as d$\bar{d}$ or u$\bar{u}$) but it cannot change one type quark into another. The weak interaction, however, can change one type quark into anther. In a beta decay of a proton ($p \rightarrow n + e^+ + \nu$), a _____ quark is changed into a _____ quark. In a beta decay of a neutron ($n \rightarrow p + e^- + \bar{\nu}$), a _____ quark is changed into a _____ quark.

Messenger particles. In modern quantum physics, the basic interactions are pictured as exchanges of <u>messenger</u> particles. The messenger particles associated with the weak interaction are called _____, the messenger particle associated with the electromagnetic interaction is called _____, and the messenger particles associated with the strong interaction are called _____.

When a particle emits a messenger particle, its rest energy is reduced. Nevertheless, it retains its character as long as it absorbs a messenger particle a short time later. Energy is not conserved during the interval between emission and absorption. For a given loss in energy ΔE, the time interval between emission and absorption must be less than the minimum predicted by the Heisenberg uncertainty principle. Thus,

$$\Delta t <$$

This consideration limits the time for events, such as decays, and also limits the range of the interaction.

Quarks interact with each other via the strong interaction, which involves the exchange of gluons. Inside a baryon or meson, the interaction energy of two quarks is very weak but the interaction becomes extremely strong if the separation is larger than the particle size. Free quarks have never been observed. When a high-energy bombarding particle strikes a baryon or meson, other quarks are created (in quark-antiquark pairs) and these form combinations with existing quarks and are emitted as other mesons or baryons.

The property of quarks that causes them to emit and absorb gluons (participate in the strong interaction) is called _____. It is somewhat similar to charge, the property that causes particles to absorb and emit photons and thus participate in the electromagnetic interaction. The chief difference is that a gluon carries color away from the quark that emits it and thereby changes the color of the quark. On the other hand, photons do not carry charge.

The three types of color are called _____, _____, and _____. The condition that the net color of quarks in a hadron be color neutral (or colorless) explains the possible combinations of quarks observed (three quarks of different color, three antiquarks of different color, or a quark and antiquark of the same color).

The basic interactions may, in fact, all be different manifestations of a single interaction. The _____ and _____ interactions have already been successfully unified into what is known as the _____ interaction. When particles interact at high energy, these two interactions are quite similar. A theory that unifies the strong and electroweak interactions is called a _____ theory (GUT) and a theory that unifies all four interactions is called a theory of _____ (TOE).

Cosmology. You should recognize that when you look at distant cosmological objects (stars, quasars, etc.) you are seeing them as they were when the light left them, perhaps more than 10^9 y ago.

Most physicists believe that the universe started in a state of extremely high density and temperature, perhaps a highly concentrated mixture of quarks, leptons, and messenger particles. A "big bang" initiated an expansion (and cooling) that will continue for some time into the future, perhaps forever. By understanding fundamental particles and their interactions, physicists hope to understand the early universe.

Two independent experimental observations lend credence to the idea of the Big Bang. They are the expansion of the universe and the microwave background radiation.

Hubble's law relates the _____ with which a galaxy is receding from us to the _____ between us and the galaxy. Mathematically, it is

$$v =$$

where H is the Hubble parameter. Its numerical value is somewhat uncertain but it is about _____ m/(s·ly). An observer in *any* part of the universe sees galaxies at the same distance receding with the same speed.

If the rate of expansion of the universe has been constant, the reciprocal of the Hubble constant gives the _____ of the universe. It is about _____ y.

The cosmic microwave background radiation fills the entire universe with an intensity that is nearly the same in every direction. There is no distinguishable source. It is believed to have originated shortly after the Big Bang. Although the background radiation started as a hot gas of highly energetic photons in the gamma ray region of the electromagnetic spectrum, today it has a thermal spectrum corresponding to a temperature of about _____ K and is chiefly in the _____ region of the spectrum. The decrease in frequency accompanied the _____ of the universe.

Whether the universe will continue to expand indefinitely or will collapse (perhaps in preparation for another big bang) depends on the amount of mass it contains. The universe may contain much more mass than is visible and the postulated invisible matter is called <u>dark matter</u>. The chief evidence for dark matter comes from a study of _____

_____ .

The text describes seven eras in the history of the universe. Tell what you can about each of them.

Big Bang to 10^{-43} s: _____

_____

10^{-43} s to 10^{-34} s: _____

_____

10^{-34} s to 10^{-4} s: _____

_____

10^{-4} s to 1 min: _____

_____

1 min to 3×10^5 y: _____

_____

3×10^5 y to present: _____

_____

II. PROBLEM SOLVING

To solve some of the problems, you will need to know the relativistic definitions of kinetic energy and momentum. If a particle has velocity $\vec{v}$, then its momentum is given by

$$\vec{p} = \frac{m\vec{v}}{\sqrt{1 - (v/c)^2}}$$

and its kinetic energy is given by

$$K = \frac{mc^2}{\sqrt{1 - (v/c)^2}} - mc^2 .$$

Its rest energy is mc^2 and its total energy is

$$E = K + mc^2 = \frac{mc^2}{\sqrt{1 - (v/c)^2}} .$$

In addition, its energy and momentum are related by

$$E^2 = (mc^2)^2 + (pc)^2 .$$

Some problems make use of these definitions and relationships.

You should be able to apply the principles of the conservation of energy, momentum, angular momentum, charge, baryon number, lepton numbers, and strangeness to given decays and reactions. Remember that the conservation of strangeness is not absolute: strangeness need not be conserved in weak interactions.

In most cases insufficient data is given to prove that energy, momentum, and angular momentum are actually conserved in a given situation. You can however show that they *could* be conserved. If the sum of the masses of the initial particles, assumed to be nearly at rest, is greater than the sum of the masses of the final particles, then kinetic energy can account for the difference in rest energy and energy can be conserved. To check on conservation of momentum, see if the speeds and directions of the original particles and of the products can be arranged so that momentum can be conserved. The decay of a single particle at rest, for example, to a single moving particle is prohibited by the conservation law. Conservation of angular momentum requires that the sum of the spins of the products must be half an odd integer if the sum of the spins of the original particles is half an odd integer and must be an integer if the sum of the spins of the original particles is an integer.

Remember that there are three lepton numbers, one for muons and muon neutrinos, one for electrons and electron neutrinos, and one for tauons and tauon neutrinos. They are separately conserved.

Related problems ask you to compute the disintegration energy for a decay. It is, of course, the initial total rest energy minus the final total rest energy.

You should be able to use Table 45–5, which gives the properties of quarks, to construct baryons and mesons with given properties. Conversely, given the properties of a particle you should be able to predict its quark content.

You should know how to use Hubble's law to find the distance to an astronomical object, given its speed of recession. Some examples give Doppler shift data, so you can find the speed.

III. NOTES

EXAM SUMMARY

Exam number: _____   **Date:** _____   **Chapters:** _____

<u>Definitions:</u>

QUANTITY   DEFINITION

_____   _____

_____   _____

_____   _____

_____   _____

_____   _____

_____   _____

<u>Physical Laws:</u>

_____

_____

_____

<u>Other Important Relationships:</u>

_____

_____

_____

_____

_____

<u>Important Applications:</u>

_____

_____

_____

_____

_____

<u>Notes:</u>

EXAM SUMMARY

Exam number: _____   **Date:** _____   **Chapters:** _____

<u>Definitions:</u>

QUANTITY   DEFINITION

_____   _____
_____   _____
_____   _____
_____   _____
_____   _____
_____   _____

<u>Physical Laws:</u>

_____
_____
_____

<u>Other Important Relationships:</u>

_____
_____
_____
_____

<u>Important Applications:</u>

_____
_____
_____
_____
_____

<u>Notes:</u>

EXAM SUMMARY

Exam number: _____   **Date:** _____   **Chapters:** _____

Definitions:

QUANTITY   DEFINITION

_____   _____
_____   _____
_____   _____
_____   _____
_____   _____
_____   _____

Physical Laws:

_____
_____
_____

Other Important Relationships:

_____
_____
_____
_____
_____

Important Applications:

_____
_____
_____
_____
_____

Notes:

EXAM SUMMARY

Exam number: _____   **Date:** _____   **Chapters:** _____

<u>Definitions:</u>

QUANTITY   DEFINITION

_____   _____
_____   _____
_____   _____
_____   _____
_____   _____
_____   _____

<u>Physical Laws:</u>

_____
_____
_____

<u>Other Important Relationships:</u>

_____
_____
_____
_____
_____

<u>Important Applications:</u>

_____
_____
_____
_____
_____

<u>Notes:</u>

EXAM SUMMARY

Exam number: _____   **Date:** _____   **Chapters:** _____

<u>Definitions</u>:

QUANTITY   DEFINITION

_____   _____
_____   _____
_____   _____
_____   _____
_____   _____
_____   _____

<u>Physical Laws</u>:

_____
_____
_____

<u>Other Important Relationships</u>:

_____
_____
_____
_____
_____

<u>Important Applications</u>:

_____
_____
_____
_____
_____

<u>Notes</u>: